Gunnar Backstrom

Fields of Physics on the PC

by Finite Element Analysis

Studentlitteratur

Chartwell-Bratt

British Library Cataloguing in Publication Data
A catalogue record for this book is available from the British Library

Art.No. 4867
ISBN 91-44-00293-9 Studentlitteratur
ISBN 0-86238-460-5 Chartwell Bratt Ltd

Second edition

Printed in Sweden
Studentlitteratur, Lund
www-address: http://www.studli.se/

Printing/year 1 2 3 4 5 6 7 8 9 10 | 2000 1999 98 97 96

Preface

During my early years as a student at Uppsala University, Sweden, I took a special interest in partial differential equations. These equations seemed to me to be all-important keys to the predictions in space and time that physics is all about. Nevertheless undergraduates still regard them as esoteric objects that are part of the curriculum but sadly useless.

Partial differential equations are indeed cumbersome, since few analytical methods of solution exist, and numerical tools of reasonable generality have been lacking until quite recently. During a visit in the San Francisco Bay area I happened to see an advertisement about a "Solver for Partial Differential Equations", which caught my attention, since I was just planning a course in applied finite element analysis. When I started to work with this software I had the feeling that a curtain was drawn, and that I could suddenly see the fields of physics through the graphics on the PC screen. I was absolutely stunned to find that almost any problem could be solved, even if it involved several complicated, simultaneous equations and often even if they were not linear. I simply could not keep this extraordinary experience to myself.

I am deeply indebted to my old friend Dr. Russell Ross, formerly of the University of East Anglia, UK, for reading and criticizing the typescript, as well as working through most of the examples. His intelligence and vigilance, combined with his rare lack of computer experience, made him the ideal pilot user.

I also wish to thank Dr. Said K. Doss of the Lawrence Livermore Laboratory warmly for reading the chapters on hydrodynamics.

Finally I gratefully acknowledge the support of Mr. Robert Nelson, SPDE, Inc., the demon programmer behind PDEase®. His prompt attention to my innumerable queries about the operation of the software was essential to the success of this work.

Preface to the 2nd Edition

During the time that has elapsed I have applied PDEase to a few other areas of physics, i.e. to elastic and electromagnetic waves as well as to atomic wave functions.

The software also continues to develop, and the first chapters could have been expanded using the new tools available, but I preferred to leave them more or less intact, just adding notes about the additional possibilities.

There is no entirely satisfactory way of structuring a book covering so many disparate items. Elastic waves could have been made to follow static elasticity, since the equations used are closely related. On the other hand, there are several similarities between elastic waves and electromagnetic waves. Finally, the modes in wave guides are somewhat reminiscent of atomic wave functions.

Faced with these conflicting affinities, I simply put the new chapters at the end. The reader may easily skip chapters which are not of immediate concern.

Gunnar Backstrom

Contents

Introduction

The fields of physics are all governed by partial differential equations. Static problems of electricity, magnetism, thermal conduction and atomic diffusion, for instance, lead to Poisson's equation

$$\mathrm{div}(-k\,\mathrm{grad}(\varphi)) - h = 0$$

where k is a property of the material and h is a source term. Some exact solutions are available, but rarely in cases of practical interest. Even when solutions exist, they tend to be complicated and must be evaluated numerically for graphical presentation in digestible form.

Finite element analysis (FEA) is a numerical approach, in which the system under study is divided into hundreds of regions, each one giving rise to one or more equations. The main task is to solve all these hundreds or thousands of simultaneous equations, which was impracticable before the days of transistorized computers. The tools of FEA are now on any one's desk, since the average personal computer is quite adequate for solving many such problems.

Programs for FEA calculations on a PC have been available for several years, but they have generally been of a "black box" type. Now a program exists which permits you to enter the equations required by your mathematical model, solve them automatically, and present the results graphically in a variety of ways. This program, PDEase®, is available from Macsyma Inc., 20 Academy Street, Arlington, MA 02174, USA.

One of the most attractive applications of this software is in university education. An extremely large range of problems involving classical fields may easily be studied in detail, and with realistic boundary conditions. Non-linear equations and boundary conditions as well as space-dependent materials properties are no longer serious limitations. The number of cases for study is virtually unlimited, and by solving problems under various conditions the student may develop an intuitive feeling for how fields behave.

A large number of textbooks on finite element analysis exist. They all present the various numerical methods in detail and assume the reader to make calculations on simple systems by hand, then enter FORTRAN code or other, provided in an appendix, in order to handle more complex situations.

Some twenty years ago, the choice was between writing one's own programs or abstaining completely from FEA. Today, expert software saves students from worrying about programming strategy, formatting, indices and graphics.

No one would even consider taking courses in Object Oriented Programming, with applications to advanced string, windows and mouse handling, before using a modern word processor. We all leave these chores to specialists, who spend years learning the tricks of their profession, then devote more years to produce the software we use every day.

Numerical algorithms and programming have little to do with the concepts of physics, and if months or even years may be gained by skipping these items, there is no reason to hesitate. Every scientist has to make a choice about what to learn and what to leave alone. No one can pretend to master all of physics, not even all of classical physics. The real choice is between additional ignorance within the discipline of physics or within some adjacent field.

On the other hand, no scientist would be content to use a numerical toolbox, such as PDEase®, without knowing at least the principles of operation. Some details are even essential to the formulation of boundary conditions. In an attempt to include a general description of the method and at the same time de-emphasize its importance to the reader, I have included an appendix on this subject.

The main purpose of this volume is to illustrate how classical fields change when you modify the geomctry, use different materials and specify different boundary conditions. It should be regarded as a companion to textbooks in the various branches of physics, rather than a textbook on its own. Even if you have not purchased the software package, you can use this book for information on fields and about the potential value of FEA.

Graphical Facilities of PDEase®

When a solution to a partial differential equation has been obtained, one would normally want to plot the results. Even if this is the last stage of the process, it is the easiest one to learn. Hence we shall begin by studying the various modes of graphical presentation available in PDEase.

Installing the Software

PDEase may be used either under Windows® or under DOS®. Although the Windows alternative may well become viable in later versions of the program, the most convenient way so far is to operate under DOS. Once you have started your PC and see a command prompt, such as C:\>, on the screen, insert the PDEase floppy disk in one of the slots. If this is named A:, simply type A: <enter>, then INSTALL <enter>, and follow the instructions. The program asks you for the name of your floppy disk drive, and there you do not answer A: but simply A. After installation the program will be found in the directory \PDEASE and some sample files in subdirectories.

In order to run the sample problems provided by the software firm, or your own problems, you need to establish communication between the main program and the directory where you choose to work. For this, as well as for presenting problems to the program, you need a text editor, which could really be of any kind.

An excellent text editor comes with DOS 5 and 6. You first need an editor to modify the AUTOEXEC file, which controls the starting conditions of your PC system. At the DOS prompt, simply type EDIT \AUTOEXEC.BAT, and the contents of that file will appear on the screen. Help is available by means of the <F1> key, but in fact you only need to know a few commands. By the arrow keys or a mouse move your cursor to the end of the PATH statement, where you add ;\PDEASE. Push <Alt> and F at the same time, followed by X to exit from the editor, and confirm that you want to save the file by Y. Run the file immediately by typing \AUTOEXEC <enter>.

If you have never used the DOS text editor before, please refer to an appendix, which contains more details.

Plotting Functions of One Variable

You should now be ready to start calculating and plotting. Even if you could run your applications from any directory, consider creating a special one for your own files. For instance, type CD \ at the prompt to change to the root directory, and then type MD PDE. Finally change to the new directory by typing CD PDE. Use the text editor to enter the following descriptor file, which defines your task: just plotting a function of one variable and its derivatives. When starting the editor, also give the file a name, e.g. by typing EDIT FUNC1.PDE. The extension PDE is the standard one with this software.

The left-adjusted words are reserved for the main program and will hence be used again and again in all descriptors. These reserved words mark the beginning of a segment, where you enter instructions and data pertinent to the problem at hand. Although the present function depends only on one variable, the program requires that we define a region in (x,y) space, here of size 10 by 10.

```
title
  'sin(x)+ x* cos(x)'
definitions
  Lx= 10     Ly= 10              { Define extent of (x,y) space }
  f= sin(x)+ x* cos(x)           { Function to be plotted }
  fx= dx(f)                      { First derivative }
  fxx= dxx(f)                    { Second derivative }
boundaries
region 1                         { Define the main boundary }
  start(-Lx,-Ly)
  line to (Lx,-Ly) to (Lx,Ly)  to (-Lx,Ly) to finish
plots
  elevation(f, fx, fxx) from (-Lx,0) to (Lx,0)
end
```

The mathematical notation in PDEase is much like that of some standard programming languages. The sign * means "multiply" and ** means "raise to a power". Standard functions are also available, as shown in the syntax chapter at the end of the book. In addition, partial derivatives are allowed. The derivative $\partial f/\partial x$ is simply denoted dx(f). A higher derivative may be written dx(dx(f)), or simply dxx(f), etc..

The last section specifies an *elevation* plot, which just means a curve. You may plot several functions in the same figure, but that is convenient only if the function values are comparable. An *elevation* plot is always taken along a line in (x,y) space, specified by the *from-to* statement, using increasing coordinate values if possible.

Now run the program using the above descriptor. If the program is named PDEase2 on your diskette, type PDEASE2 FUNC1 <enter>. In order to avoid waiting for the logotype screen to disappear, push <enter> an extra time. (In the improbable case that the

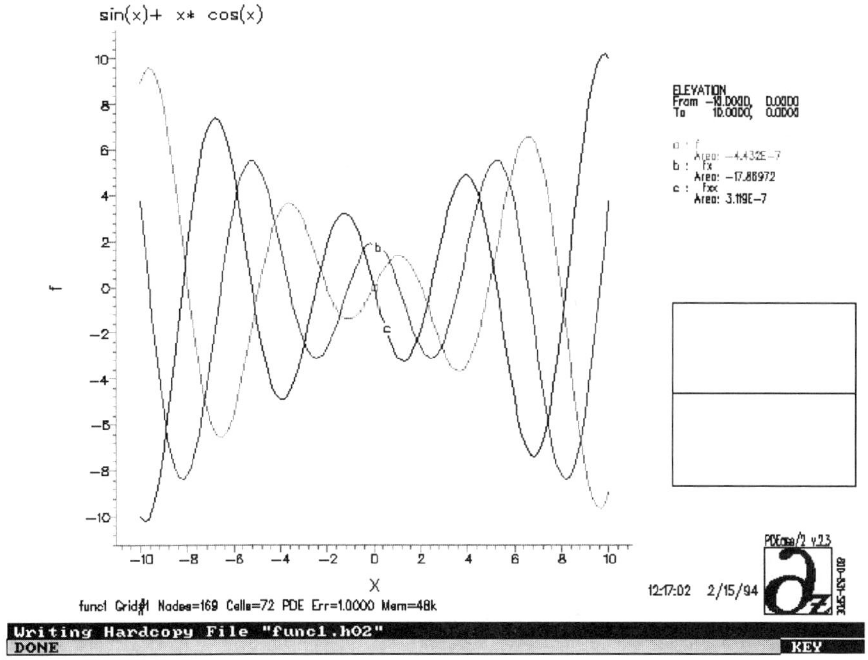

resulting figure should be garbled, exit from the program by pushing <ESC> and then try typing PDEASE2 /V FUNC1 instead. The switch /V refers to the video mode VGA, and if that should not render a clear output, replace it by /S or /E.) The result should look as shown above, except that your screen probably shows colors and hence will be more readable. The three curves *a*, *b* and *c* may by identified by means of the table on the right. The plot permits you, for instance, to relate zero crossings to minima and maxima. The *Area* value is the integral of the corresponding curve.

After this you may want to explore other functions *f*(*x*) in a similar manner. If an expression is more complicated than in the above model descriptor, divide it into groups of terms and let it run over two lines or more. The program interprets the contents of each segment in an intelligent manner, and no continuation signs are needed. If it would make an expression more readable, you may separate terms by extra spaces. As shown in the *definitions* segment, a line may contain two or more assignments, without any marks (except spaces) to separate the statements.

If you modify the descriptor again and again under the same name, or a slightly different name, it will be convenient to use a device for repeating previous DOS command lines, such as PDEASE2 /V FUNC2. Your operating system normally includes the DOSKEY program, which may be called at the start of a session, or better still included in the AUTOEXEC.BAT file. The *up* arrow key will then permit you to recall a command, and to repeat it by pushing <enter>. The PDEase manual contains additional commands that may be typed in at starting time to modify the colors, etc.. The longer the command the more interest you have in using DOSKEY.

Plotting Functions of Two Variables

Next we shall study a function of two variables, which is what this software is really designed to handle. Type COPY FUNC1.PDE FUNC2.PDE to make a copy of the previous descriptor under a different name. Then modify FUNC2.PDE as follows.

```
title
  'x**2+ 2*y**2'
select
  contours= 8
  surfacegrid= 30        paint= off
definitions
  Lx= 1     Ly= 1
  f= x**2+ 2*y**2
  laplace= dxx(f)+ dyy(f)
boundaries
region 1
  start(-Lx,-Ly)    line to (Lx,-Ly) to (Lx,Ly)  to (-Lx,Ly) to finish
plots
  contour(f)                                      { Contour plot of function }
  contour(laplace)                                { Test if f(x,y) is harmonic }
  vector(dx(f), dy(f)) as 'Gradient vector'
  surface(f) interactive                { May be rotated, then hardcopied (h) }
end
```

When you run this descriptor the program displays the first three plots as quickly as possible, but you will be able to recall the figures later by the SEEPLOT command.

FUNC2 uses three new graphics objects: *contour*, *vector* and *surface* plots. In a new segment, *select*, we request that the number of contour curves be 8. As shown below, there are obviously more than the 8 curves requested, and the discrepancy arises because the levels are to be rounded off to even values. The adjoining table includes the minimum and maximum values within the region defined under *boundaries*.

The gradient is a measure of the local steepness of a function. It is a vector quantity and is usually written

$$\nabla f \equiv \mathbf{i}\frac{\partial}{\partial x} + \mathbf{j}\frac{\partial}{\partial y}$$

where **i** and **j** are unit vectors along the x and y axes.

The Laplace function in the descriptor is derived from the gradient as follows

$$laplace \equiv \nabla^2 f = \left(\mathbf{i}\frac{\partial}{\partial x} + \mathbf{j}\frac{\partial}{\partial y}\right)\cdot\left(\mathbf{i}\frac{\partial}{\partial x} + \mathbf{j}\frac{\partial}{\partial y}\right) f = \frac{\partial^2 f}{\partial x^2} + \frac{\partial^2 f}{\partial y^2}. \quad \bullet$$

The corresponding *contour* plot, which is not shown here, exhibits no curves at all, and the maximum and minimum values are both equal to 6.0. This means that the function *laplace* is constant. The fact that the value is not zero indicates that the function f(*x*) is not harmonic[1p477].

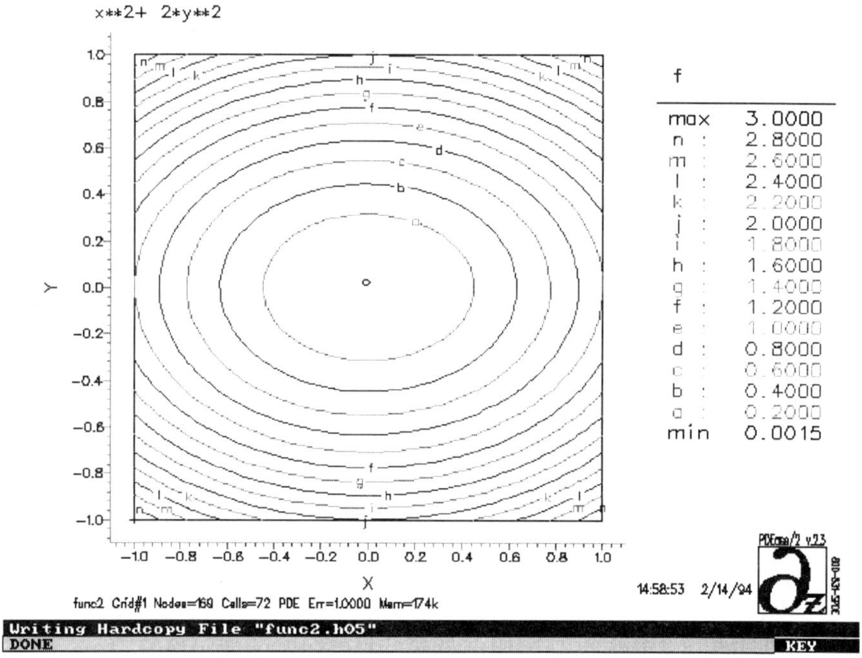

The *vector* plot consists of arrows pointing in the direction of steepest increase, and the length of an arrow gives a rough (non-linear) measure of the slope of the surface in that direction. Looking at a plot of this kind one can easily see whether a function has a minimum. The length of the gradient vector must vanish at an extreme point, and the directions of the arrows permit us to tell a minimum from a maximum, or even a saddle point. It is clear from the following figure that f(x) has a minimum at the center.

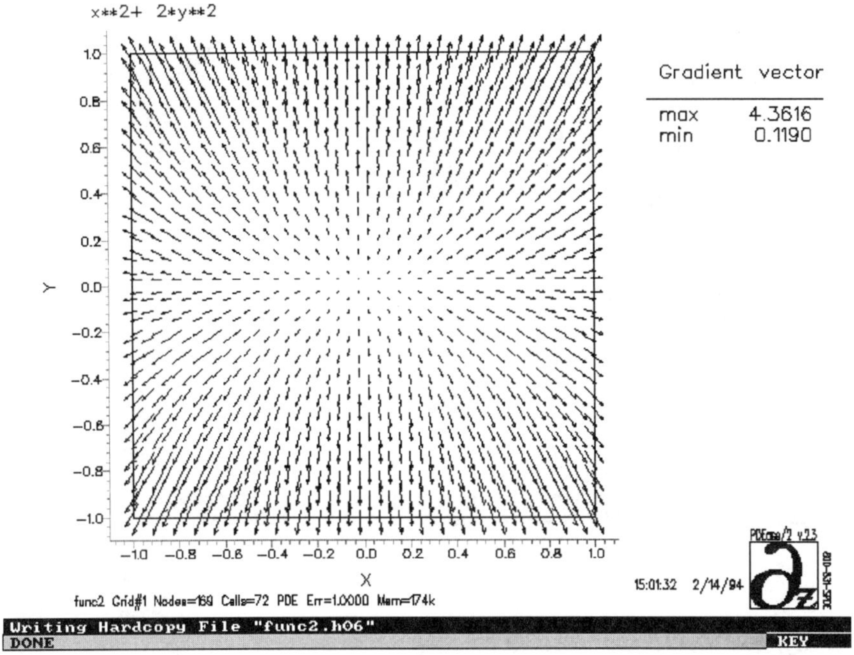

In the *select* segment we also requested a smaller surface grid density than the default value, which is 60. To decrease plotting time even further, we also refrained from painted surface plots, which are beautiful but take a long time.

The surface plot will appear as follows on the screen. The command *interactive* allows the viewer's eye to move sideways or vertically in increments of 10 degrees in order to obtain the most informative view. Try typing *R* a few times, followed by the usual <enter>. For each orientation you may store the figure by typing *h*. Some functions may need several views to show its variation clearly. The viewing distance may also be modified if necessary.

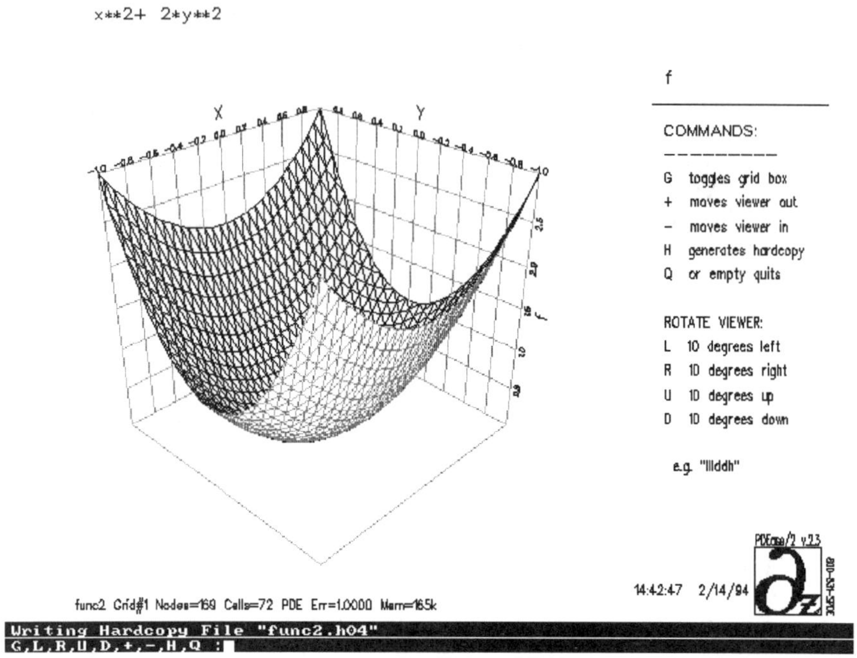

Browsing and Printing Plots

At a later stage you will probably prefer to start the program by a batch file, P.BAT, which should be placed in the directory \PDEASE and could look as follows.

```
del *.h*
del *.out
PDEASE2 /v  %1
seeplot %1
```

With this batch file installed you can start PDEase by the simple command P func2 <enter>, while deleting some previous files at the same time in order to avoid disk overflow.

The last statement in this batch file calls SEEPLOT, which permits you to scan through all the figures you have created. The important advantage about SEEPLOT is that you can scan through your figures again and again, and in any order. In response to the prompt *Next Page* at the bottom of the screen, push <enter> if you want the next figure, or type 3 for figure 3. Pushing <ESC> will end the inspection.

There are different ways of printing a figure. SEEPLOT creates ASCII files in HPGL format, which may be sent directly to a pen plotter. Such a device is slow, but it produces high-resolution plots.

As an alternative, the PDEase package includes routines for translating HPGL files so that they may be output to popular printers. With a HP LaserJet (or DeskJet) you proceed as follows. Having noted the number of your figure, 3 say, simply type

```
LJXPLOT /150 /S.7 FUNC2.H03
```

to obtain a reasonably good figure on a HP LaserJet, or even on an inexpensive HP DeskJet 500. The switch /150 refers to the resolution in number of points per inch, whereas /S.7 means that the sides of the figure are reduced by the factor 0.7. Both switches reduce the printing time. There are corresponding command lines for Epson printers, as described in the PDEase manual.

Word for Windows® also offers a filter for copying HPGL figures to a document page.

A convenient way of printing with modest resolution is to push the <PrintScreen> key, while SEEPLOT shows a graph. For this to succeed the AUTOEXEC.BAT file should contain the line GRAPHICS for a simple dot matrix printer or GRAPHICS LASERJETII for a HP LaserJet or similar printer.

Hewlett-Packard also provides a free program, HPSCREEN, for dumping screen plots onto their printers.

Exercises

Explore the following functions, possibly with modified values of *Lx* and *Ly*.

- ❑ *f*= *x***2* *y*
- ❑ *f*= *x***2* *y***2

"Curly" Velocity Fields

In the preceding chapter we demonstrated various aspects of more or less arbitrary functions. We are now about to study more realistic fields, where a velocity vector is defined at every point of the domain. Similar fields are known from the mechanics of fluids.

We shall consider a few simple cases where a solid or a liquid rotates around an axis. This means that the velocity **v** at every point is perpendicular to the radius and that its magnitude is $|\mathbf{v}| = \omega r$. These facts may be summarized by the vector expression

$$\mathbf{v} = \mathbf{w} \times \mathbf{r} = \begin{vmatrix} \mathbf{i} & \mathbf{j} & \mathbf{k} \\ \omega_x & \omega_y & \omega_z \\ x & y & z \end{vmatrix} = \begin{Bmatrix} \omega_y z - \omega_z y \\ \omega_z x - \omega_x z \\ \omega_x y - \omega_y x \end{Bmatrix} \qquad \bullet$$

The software restricts our study to the *xy* plane, i.e. the third row must generally be zero. This implies $\omega_x = \omega_y = 0$, leaving us with only the *z* component of **w**, which we shall simply call ω.

Rotating Solid Disk

In this case ω will be the same over the entire object. The following descriptor is what we need to explore the various features of this motion. In order to define a circular domain we need the new command *arc*, which permits us to draw a series of quarter-circles. In the *plot* segment we have included the special differentials div(**v**) and $(\nabla \times \mathbf{v})_z$, i.e. the *z* component of curl(**v**). The definitions of these operations in Cartesian coordinates are

$$\nabla \cdot \mathbf{v} \equiv \mathrm{div}(\mathbf{v}) = \frac{\partial v_x}{\partial x} + \frac{\partial v_y}{\partial y} + \frac{\partial v_z}{\partial z} \qquad \bullet$$

and

$$\nabla \times \mathbf{v} \equiv \mathrm{curl}(\mathbf{v}) = \begin{vmatrix} \mathbf{i} & \mathbf{j} & \mathbf{k} \\ \dfrac{\partial}{\partial x} & \dfrac{\partial}{\partial y} & \dfrac{\partial}{\partial z} \\ v_x & v_y & v_z \end{vmatrix} = \begin{Bmatrix} \dfrac{\partial v_z}{\partial y} - \dfrac{\partial v_y}{\partial z} \\ \dfrac{\partial v_x}{\partial z} - \dfrac{\partial v_z}{\partial x} \\ \dfrac{\partial v_y}{\partial x} - \dfrac{\partial v_x}{\partial y} \end{Bmatrix}$$ •

```
title
  'Rotating Disk'
select
  contours= 10            surfacegrid= 30          paint= off
definitions                                                      { SI units }
  r1= 1.0
  rad= sqrt(x**2+y**2)                 omega= 1.0
  vx= -omega*y      vy= omega*x                                  { Velocity }
  vabs= sqrt(vx**2+vy**2)                                { Magnitude of v }
boundaries
          region 1
  start(r1,0)
  ARC  to (0,r1) to (-r1,0) to (0,-r1) to finish
plots
  contour(vx)             contour(vy)              contour(vabs)
  contour( dx(vy) - dy(vx) )  as 'curl_z'
  contour( dx(vx)+ dy(vy) )  as 'div_v'
  vector( vx, vy)
end
```

The figure below shows a *vector* plot of the velocity. The *contour* plot of *curl_z* consists only of one point, since the function has the constant value 2.0000. The divergence *div_v* also turns out to be constant, but with the value zero. Of course we could easily calculate these functions exactly in this simple case, but we will find similar plots convenient with other, more complicated velocity fields.

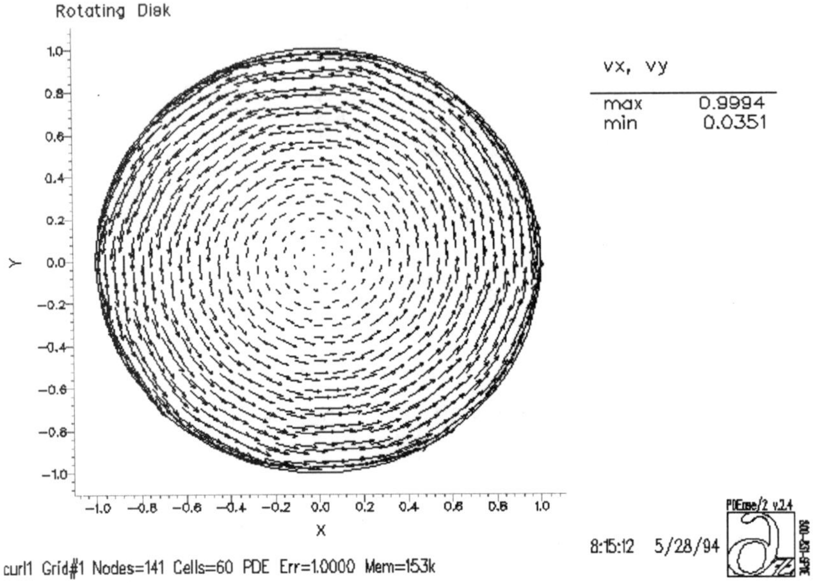

Rotating Fluid

We shall now explore a situation where the angular velocity, ω, varies with the radius. This could happen in a liquid, but we need not discuss exactly what means would be required to make the liquid move the way we specify. In the descriptor lines below we choose $\omega = 1/r^2$ and substitute this function for *omega* in the expressions for the velocity components.

```
...
rad= sqrt(x**2+y**2)
vx= -[1/rad**2]*y      vy= [1/rad**2]*x                         { Velocity }
small= 1e-6
...
contour( log10[abs(vx)+small] )          contour( log10[abs(vy)+small] )
```

The contour plot of *vy* turns out not to be the same as that for *vx*, rotated by 90^0, although that could be expected for reasons of symmetry. This rather artificial effect is caused by round-off errors when the program interpolates near the center, where the function takes very high values. The set of levels for the contours is based on the highest and lowest values found, and hence independent plots may look different. The corresponding *surface* plots, which cover the region without gaps, are in fact similar and the same is true of the logarithmic plots of *vx* and *vy*.

The following plot of *curl_z* was produced by filtering the SEEPLOT file from HPGL to Word, thus improving the resolution. Astonishingly, it consists of a number of irregularly shaped zero contours. Other contours correspond to values of the order of 10^{-12}, whereas we obtained 2.0 in the first example. In view of the round-off errors always present in numerical calculations we take it that the value of *curl_z* is zero over the whole region, which we can readily verify by exact calculus. The divergence *div_v* also seems to be zero.

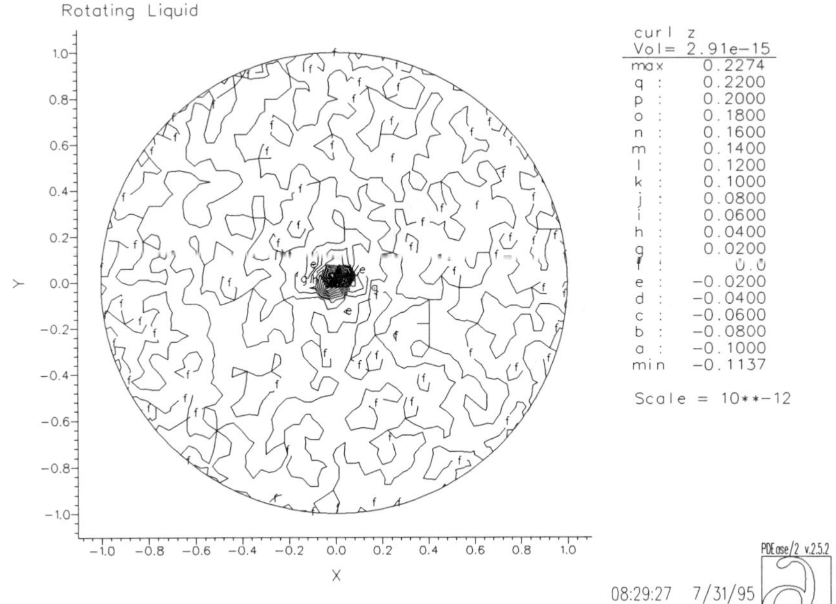

We proceed to a similar case with $\omega = \sin(r)$, which requires the following change of the descriptor.

```
vx= -[sin(rad)]*y      vy= [sin(rad)]*x                              { Velocity }
```

As illustrated by the figure below *curl_z* now is clearly different from zero, although the liquid appears to circulate much the same way as it did before. The value of the divergence *div_v*, however, still vanishes.

What kinds of rotary motion yield finite values of *curl_z*? As explained in textbooks on vector analysis[2p54]

$$(\nabla \times \mathbf{v})_z \equiv \lim_{S \to 0} \frac{1}{S} \oint_S \mathbf{v} \cdot d\mathbf{l} \qquad \bullet$$

The situation is depicted in the next figure, where we have drawn an area element which will eventually shrink to negligible size. The element is limited by two radial and two circular segments. The velocity $\mathbf{v}$ is directed along the circular path; hence the radial parts

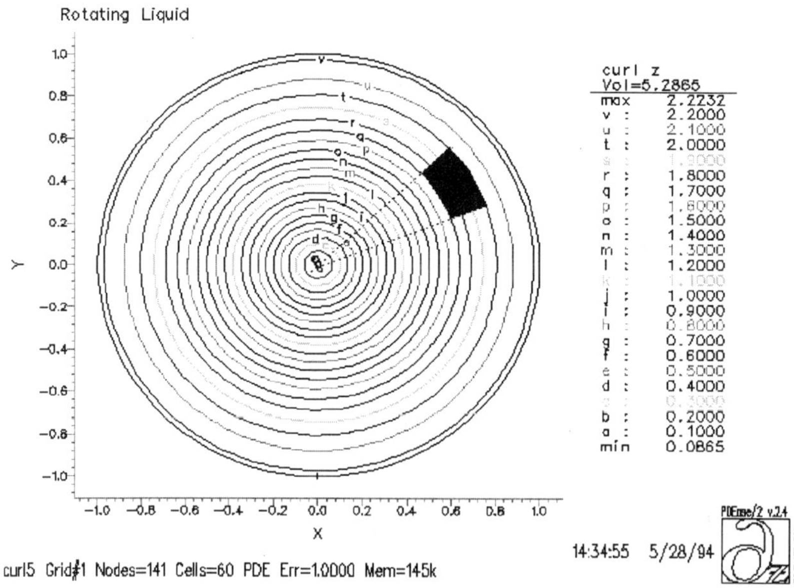

contribute nothing to the integral. The contribution from a circular segment is just the speed v multiplied by the length of the arc, and the two terms are of opposite sign and might accidentally cancel. In the case of the solid disk the outer path was longer, but in addition the speed was larger there. Each of these enhancing factors was proportional to the radius r.

In the case of the liquid with $\omega = 1/r^2$, the smaller velocity on the outer path compensates for the increase of both the radius and the path length, which makes *curl_z* equal to zero. Generally one might say that vanishing *curl* is an accidental event.

In all the past examples we noticed that the divergence of the velocity field was zero. In order to understand why this happens we consider the integral definition for the divergence

$$\nabla \cdot \mathbf{v} \equiv \lim_{V \to 0} \frac{1}{V} \oiint_V \mathbf{v} \cdot \mathbf{n} \, da,$$

where the integral is now taken over the surface of a volume element as shown in the above figure. Two of the limiting surfaces are parallel to the *xy* plane. For this integral the cylindrical surfaces of the box give no contribution, since **v** is perpendicular to the normal. The surfaces parallel to the *xy* plane also add nothing, since our velocity is always confined to the *xy* plane. As regards the radial surfaces they give equal contributions, but of opposite signs, since the velocity depends on r in the same way on both of them. Thus, whenever ω depends on r only, the divergence must be zero.

More General Velocity Fields

In previous examples we were concerned with rotation about an axis, and the angular velocity depended only on the radius. That was clearly a special case of the general motion of a fluid. We shall now look at a more general field, where the velocity components are independent functions of r. Let us introduce the following definitions.

```
vx= rad*tan(rad)      vy= exp(rad)                    { Velocity }
```

The resulting *contour* plot of *curl_z* shows finite values, and so does the plot of the divergence, as seen in the figure below. The *vector* plot of the velocity, however, exhibits no signs of any circulating motion that could lead us to suspect large values of *curl_z*.

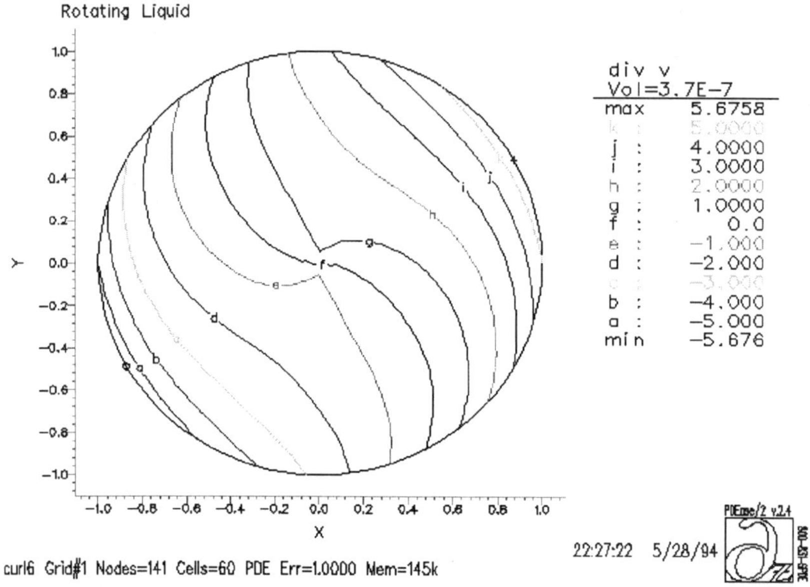

The lesson to be drawn from the examples of this chapter is that the appearance of a velocity field, as given by a *vector* plot of the motion, is not enough to assess whether *curl* or *div* is zero. We always need to apply the differential or integral definitions in order to decide.

Exercises

- ❑ Try $\omega = 1/r$ in one of the above descriptors.
- ❑ Try $\omega = r^2$ as well.
- ❑ Invent your own functions of (x,y) for *vx* and *vy*.

Harmonic Functions

The function that we presented graphically on p.8 proved to be non-harmonic. A harmonic function must satisfy the Laplace equation

$$\nabla^2 f \equiv \frac{\partial^2 f}{\partial x^2} + \frac{\partial^2 f}{\partial y^2} = 0 .$$

This is one of the simplest second-order partial differential equations, which governs several kinds of fields in physics. It is unique in having an immense variety of exact solutions.

A First Example

We know how to generate solutions, since virtually any function of a complex variable[1p477], $z \equiv x + iy$, may be separated into a real part and an imaginary part, both of which are harmonic functions of x and y. Let us try this with the analytic function $w = z\exp(z)$.

```
title
  'Re and Im of  z* exp(z)'
select
  contours= 16
  surfacegrid= 30          paint= off
definitions
  Lx= 5     Ly= 5
  fr= exp(x)* (x*cos(y)-y*sin(y))          {Real part }
  fi=  exp(x)* (y*cos(y)+x*sin(y))          { Imaginary part }
  lgradr= sqrt( dx(fr)**2+ dy(fr)**2 )   { Length of gradient vector }
  lgradi= sqrt( dx(fi)**2+ dy(fi)**2 )
  laplacer= dxx(fr)+ dyy(fr)              laplacei= dxx(fi)+ dyy(fi)
boundaries
                    region 1
  start(0,0)     line to (Lx,0) to (Lx,Ly)  to (0,Ly) to finish
plots
  contour(fr)                                   { Contour plot of function }
```

```
  contour(fi)
  vector(dx(fr)/lgradr, dy(fr)/lgradr)   as 'Direction of grad(fr)'
  vector(dx(fi)/lgradi, dy(fi)/lgradi)   as 'Direction of grad(fi)'
  contour(laplacer)                       contour(laplacei)
  surface(fr) interactive         { May be rotated, then hardcopied }
  surface(fi) interactive
end
```

The *definitions* segment of this descriptor contains the real and imaginary parts, *fr* and *fi*, as well as definitions of the lengths of the gradient vectors for these functions. The results of applying the Laplace operator to the two functions is also included. In addition we have modified the vector plot to show the gradient direction.

Let us now look at the *contour* plots of *fr* and *fi* below. If you print these figures and superimpose them, you will discover that the two sets of curves intersect at right angles. This is an example of conformal mapping[1,494]. The function $w = u + iv = z\,\exp(z)$ is an analytic

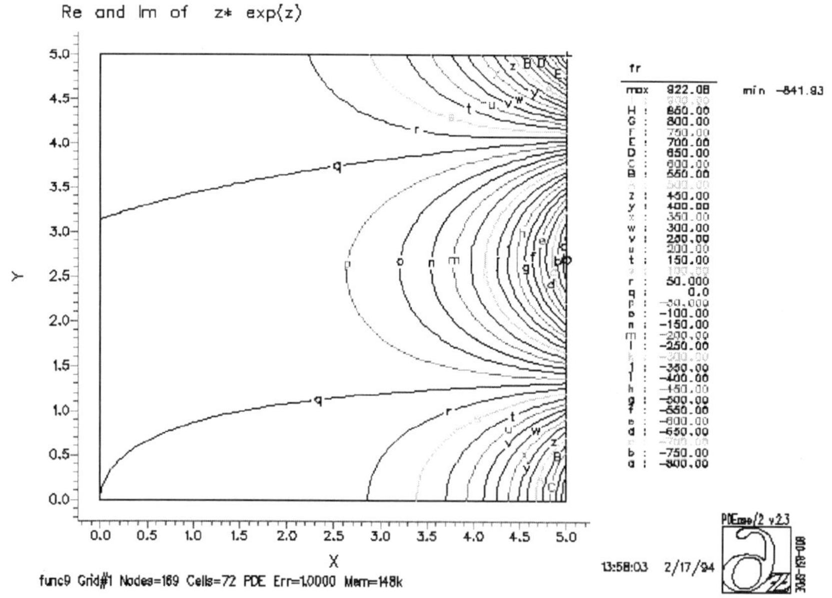

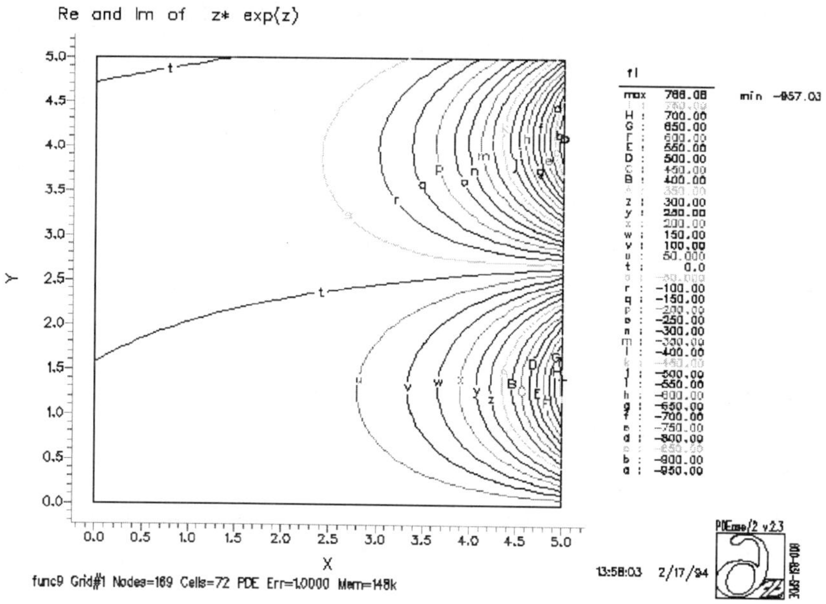
Re and Im of z* exp(z)
fi
max 788.08
min -957.03
H : 700.00
G : 650.00
E : 550.00
D : 500.00
B : 400.00
z : 300.00
y : 250.00
w : 150.00
v : 100.00
u : 50.000
t : 0.0
r : -100.00
q : -150.00
o : -250.00
n : -300.00
l : -400.00
j : -500.00
i : -550.00
g : -650.00
f : -700.00
d : -800.00
b : -900.00
a : -950.00
Y
X
func9 Grid#1 Nodes=169 Cells=72 PDE Err=1.0000 Mem=148k
13:58:03 2/17/94
PDEase/2 v2.3

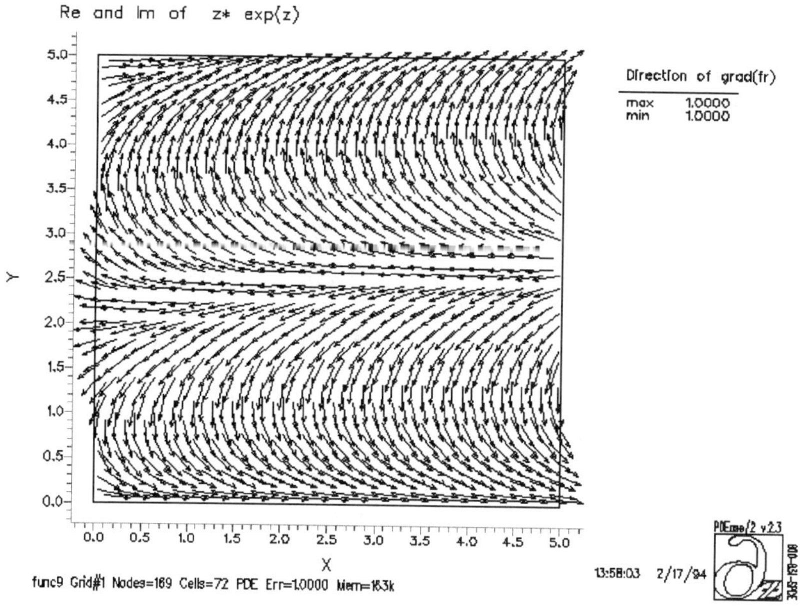
Re and Im of z* exp(z)
Direction of grad(fr)
max 1.0000
min 1.0000
Y
X
func9 Grid#1 Nodes=169 Cells=72 PDE Err=1.0000 Mem=163k
13:58:03 2/17/94
PDEase/2 v2.3

function, i.e. it has a complex derivative dw/dz. The inverse function $z(w)$ then also exists, as well as its derivative, which means that the inverse is analytic as well. The lines of constant u and constant v in the w plane are of course perpendicular, and hence the corresponding curves of $x(u,v)$ and $y(u,v)$ will also meet at right angles, according to the theory of conformal mapping.

The *vector* plot shown on the preceding page illustrates the advantage of using normalized gradient vectors. It turns out that the function *fr* is rather flat in the left half of the domain, and if we had displayed the gradient vector itself the plot would almost have degenerated into a set of points. By plotting this vector divided by its own length we clearly indicate the direction of the slope everywhere.

The following *contour* plot of $\nabla^2(fr)$ confirms the vanishing value that we could expect. The value is not exactly zero, since numerical differentiation involves some error, but certainly the value is small compared to the maximum value (1000) of *fr*. We could estimate a typical value of the gradient from the maximum variation of the *fr* and the size of the domain in the *xy* plane. The gradient

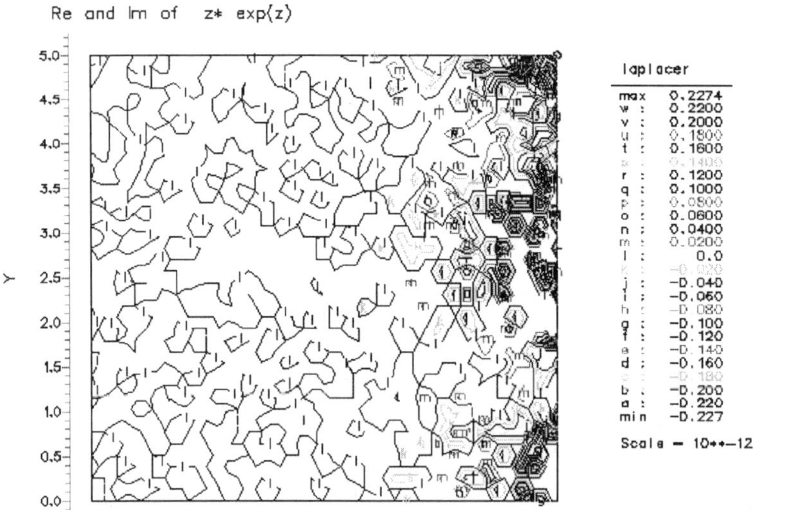

values in turn range from essentially zero to that typical value, so a typical second derivative would be 1000/5/5= 40. Hence, $\nabla^2(fr)$ is smaller than the typical second derivative by a comfortable margin of 10^{14}. The fact that the contours are irregular is further indication that the function we are plotting may be exactly equal to zero everywhere.

Of course, the contour plot of $\nabla^2(fr)$ does not prove that the function is harmonic. It would be easy to find examples to the contrary. On the other hand, if we were to find regular contours, with values different from zero, that would permit us to conclude that the function is not harmonic.

Managing Vectors

Recent versions of PDEase let us write the above descriptor more concisely by defining vectors directly, such as

```
gradr= grad(fr)
```

or

```
gradr= vector( dx(fr), dy(fr))
```

We may conveniently operate on vectors in the (x,y) plane by commands such as

```
d= div( v1)                          { Divergence of v1 }
wz= curl( v1)                             { z-component }
v1abs= magnitude( v1)
s= dot( v1, v2)          { Scalar product of vectors v1 and v2 }
vz= cross( v1, v2)          { Vector product (z) of v1 and v2 }
```

Although these short forms are useful in many circumstances, complete expressions are preferable in educational work. Hence we adhere to the latter style in the beginning of this volume.

More About Harmonic Functions

The real and imaginary parts, *u* and *v*, of an analytic function also have to obey the Cauchy-Riemann equations

$$\begin{cases} \dfrac{\partial u}{\partial x} - \dfrac{\partial v}{\partial y} = 0 \\ \dfrac{\partial u}{\partial y} + \dfrac{\partial v}{\partial x} = 0 \end{cases}$$

These relations may be verified by *contour* plots in a similar manner, and here the computational errors should be even smaller.

The algebraic work of separating the real and imaginary parts of an analytic function is rather tedious, but there are now excellent tools for such operations. One is the compact program Derive® from Soft Warehouse, Inc., Honolulu, Hawaii. The user hardly needs to know any syntax rules, except that #i is the notation for $\sqrt{-1}$. After starting the program by typing DERIVE and opening the Author window by <A> he only needs to type in the definition *z*:=*x*+#i**y*, followed by <enter>, then the function $f(z)$ in BASIC-like form and push S for *Simplify*. A large number of examples of harmonic functions can thus be generated with little effort.

Exercises

Further functions to explore in a similar manner, including tests of the Cauchy-Riemann relations (where applicable) are

- ❑ x^2+y^3
- ❑ $\mathrm{Re}(z^2)$ and $\mathrm{Im}(z^2)$
- ❑ $\sin z$

Fields of Gravitation

Massive objects create a force field, acting on any other mass. This force may be expressed in terms of a potential function[2p77]. Gravity is one of the simplest fields that we can study.

Earth and Moon as Point Masses

Let us choose the Earth-Moon system as a concrete example, using the following descriptor. Throughout this volume we shall use the fundamental SI units for all physical quantities, in the present case meter and kilogram.

```
title
  'Earth-Moon Gravitational Field'
select
  contours= 10      surfacegrid= 30        paint= off
definitions                                        { Basic SI units: m, kg }
  d= 3.84e8        Lx= d         Ly= 2*d
  m1= 5.98e24       m2= 7.35e22             { Masses of Earth and Moon }
  U= -6.67e-11* [m1/sqrt(x**2+y**2)+ m2/sqrt(x**2+(y-d)**2)]
  gx= -dx(U)      gy= -dy(U)                            { Field strength }
  gabs= sqrt(gx**2+gy**2)                             { Magnitude of g }
  small= 1e-15
boundaries
region 1
  start(-Lx,-Ly/4)
  line to (Lx,-Ly/4) to (Lx,3/4*Ly)  to (-Lx,3/4*Ly) to finish
plots
  grid(x,y)          contour( U)
  contour( log10( abs(U)+ small))         contour( log10( gabs+ small))
  contour( log10( abs(gx)+ small))        contour( log10( abs(gy)+ small))
  vector( gx/gabs, gy/gabs)  as 'Gravitational field'
  contour( arctan( gy/gx)/ pi*180)  as 'G_field direction'
  surface( log10( abs(U)+ small))
end
```

For the potential U we may simply write a sum of terms of the form $-Gm_i / r_i$, where G is the gravitational constant, m_i the masses of the objects and r_i the distances to the field point. The equation $\mathbf{g} = -\nabla U$ then yields the gravitational field strength. We place the origin at the center of the Earth and the Moon at a distance d on the y-axis.

Notice that the expression for U in the *definitions* segment contains square brackets as well as parentheses. Although brackets have the same effect, clever use of pairs of them may make an expression more readable.

The *plots* segment contains a new graphics object, *grid*(*x*,*y*). This shows the division of the xy domain into triangles which PDEase carries out automatically before it starts calculations. The program only calculates functions at corners and midpoints, also known as nodes, and it then obtains values inside the triangles by quadratic polynomial interpolation.

The *contour* plot of U yields curves which are all close to the Earth. In order to study the function over the full domain we display the logarithm of U, rather than the direct value. If log10 should encounter a zero argument during plotting, however, calculations come to a stop, and it will become necessary to restart. To avoid this painful experience we add the term *small*, chosen to be small enough for the problem at hand.

The plot of the magnitude of **g** on the next page demonstrates how the Moon perturbs the Earth's gravitational field. In fact, there is a point close to the Moon where the field even becomes zero. This is also evident from the *vector* plot, where the field direction at that point seems to be undetermined.

Since a field normally has a direction at each point in space, we may also request a *contour* plot of that direction. This we do in the second plot on the next page, just inverting the equation $\tan\theta = g_y / g_x$ and then converting from radians to degrees. This leaves an ambiguity concerning the sign of the direction, but in this case this does not cause serious confusion. We can do better with the

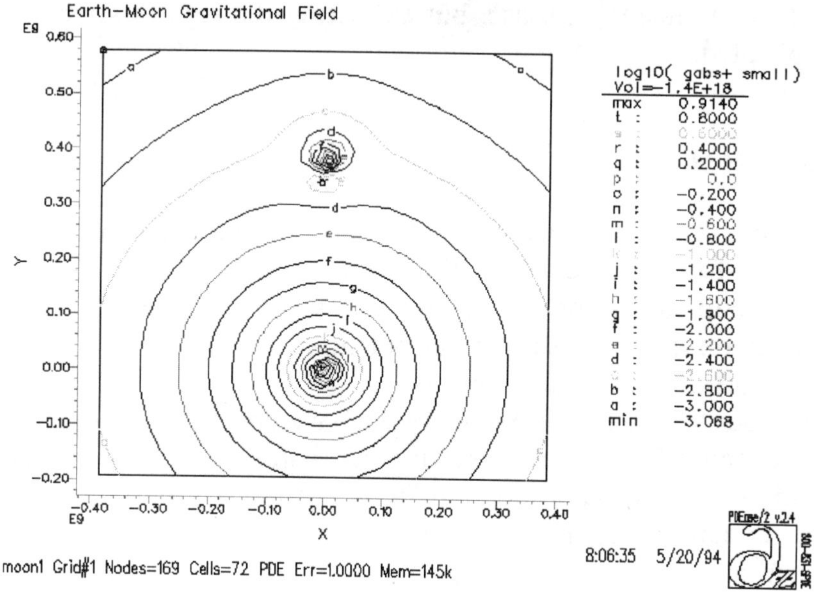

Earth–Moon Gravitational Field
log10(gabs+ small)
Vol=−1.4E+18
max 0.9140
t : 0.8000
s : 0.6000
r : 0.4000
q : 0.2000
p : 0.0
o : −0.200
n : −0.400
m : −0.600
l : −0.800
k : −1.000
j : −1.200
i : −1.400
h : −1.600
g : −1.800
f : −2.000
e : −2.200
d : −2.400
c : −2.600
b : −2.800
a : −3.000
min −3.068
X
Y
E9
8:06:35 5/20/94
PDEase/2 v2.4
moon1 Grid#1 Nodes=169 Cells=72 PDE Err=1.0000 Mem=145k

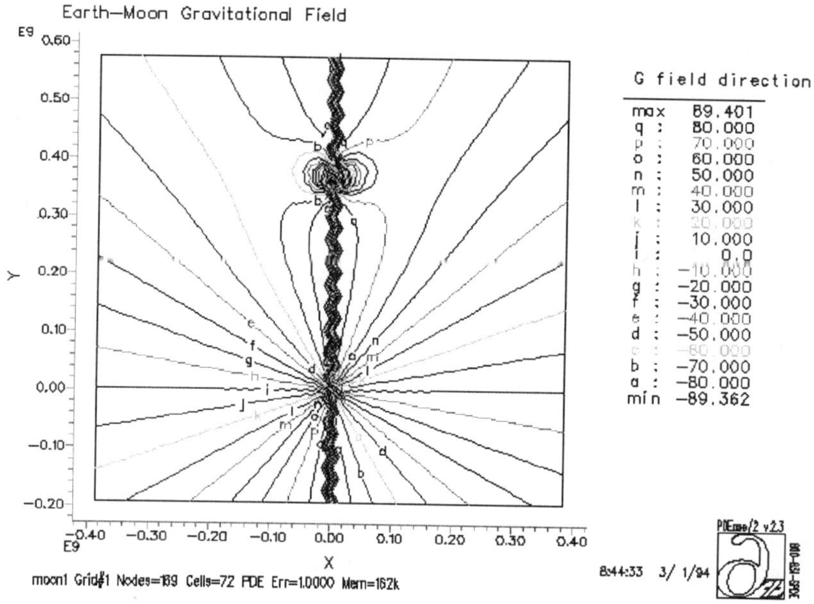

Earth–Moon Gravitational Field
G field direction
max 89.401
q : 80.000
p : 70.000
o : 60.000
n : 50.000
m : 40.000
l : 30.000
k : 20.000
j : 10.000
i : 0.0
h : −10.000
g : −20.000
f : −30.000
e : −40.000
d : −50.000
c : −60.000
b : −70.000
a : −80.000
min −89.362
X
Y
E9
8:44:33 3/ 1/94
PDEase/2 v2.3
moon1 Grid#1 Nodes=169 Cells=72 PDE Err=1.0000 Mem=162k

functions built into PDEase, but extracting the sign is slightly more complicated.

Planets of Finite Size

In the next descriptor we *zoom* the fields around each planet. Since we shall then be able to see the planets themselves we cut out the areas occupied by the Earth and the Moon. This is also a better presentation in the sense that our potential strictly applies only to the region outside the massive objects.

In order to describe a boundary we have encircled the outside in a positive sense (anti-clockwise). In other words, in walking around the boundary we have kept the domain on our left. Cutting out a portion of the domain so defined we still trace the boundary with the domain on our left, which means that we walk in a negative sense. For drawing circles we use the command *arc*, and the usage should be obvious from the file list.

```
title
  'Earth-Moon Gravitation, Zoomed'
select
  contours= 10          surfacegrid= 30         paint= off
definitions                                          { Basic SI units: m, kg }
  d= 3.84e8        Lx= d          Ly= 2*d
  r1= 6.37e6        r2= 1.74e6                                    { Radii }
  m1= 5.98e24      m2= 7.35e22                                  { Masses }
  U= -6.67e-11* [m1/sqrt(x**2+y**2)+ m2/sqrt(x**2+(y-d)**2)]
  gx= -dx(U)       gy= -dy(U)                            { Field strength }
  gabs= sqrt(gx**2+gy**2)                               { Magnitude of g }
  small= 1e-15
boundaries
region 1
  start(-Lx,-Ly/4)
      line to (Lx,-Ly/4) to (Lx, 3/4*Ly) to (-Lx, 3/4*Ly) to finish
  start(r1,0)                                        { Cut-out for the Earth }
      arc to (0,-r1) to (-r1,0) to (0,r1) to finish
  start(r2,d)                                        { Cut-out for the Moon }
      arc to (0,d-r2) to (-r2,d) to (0,d+r2) to finish
```

```
plots
  grid(x,y)
  contour( gabs) zoom(-2*r1,-2*r1,4*r1,4*r1)  as 'Magnitude of g'
  contour( gabs) zoom(-10*r2,d-10*r2,20*r2,20*r2)
  contour( arctan( gy/gx)/ pi*180)
       zoom(-50*r2,d-50*r2,100*r2,100*r2)  as 'G_field direction'
  vector( gx/gabs, gy/gabs) zoom(-20*r2,d-20*r2,40*r2,40*r2)
  elevation( arctan( gy/gx)/ pi*180) from (0,d) to (5*r2,d)  as
       'G_field direction, degrees'
  surface( log10( gabs+ small)) zoom(-50*r2,d-50*r2,100*r2,100*r2)
               interactive
end
```

In the graphs we make use of *zoom(xlow, ylow, delx, dely)*, which restricts the plot to a rectangular portion of the total domain. The two first arguments define the low-left corner and the last arguments the size of the plotting window.

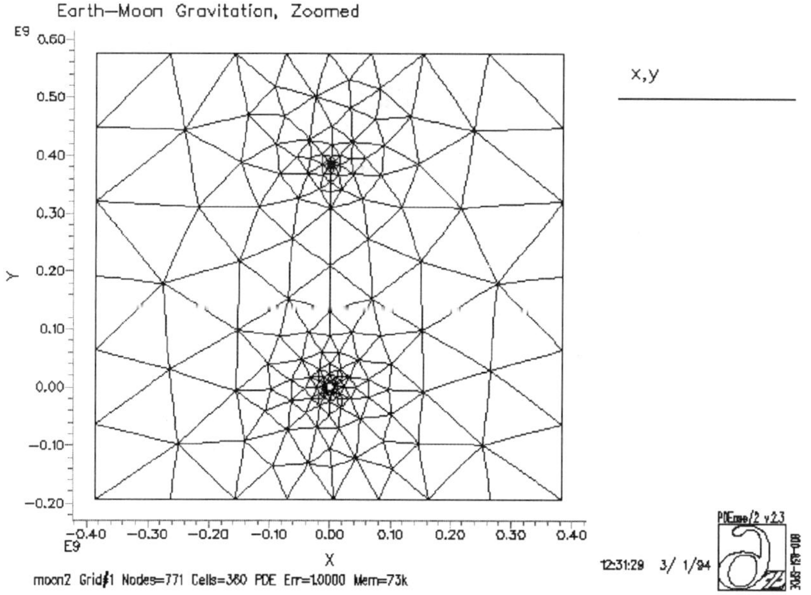

The above *grid* plot shows that PDEase divides the area into smaller triangles as you approach a relatively small feature, in this case the planets. This is useful since we are going to inspect the behavior of the gravitational field at close range.

The following figure shows the field direction close to the Moon. Evidently the perturbation would be easily observable at a distance of a few radii. It is worth noticing that the angular perturbation of the Moon's field is much larger at the corners towards the Earth than at the corners away from it, where the fields reinforce. You could verify this by means of a vector diagram, taking the changing strength of the Earth's field into account.

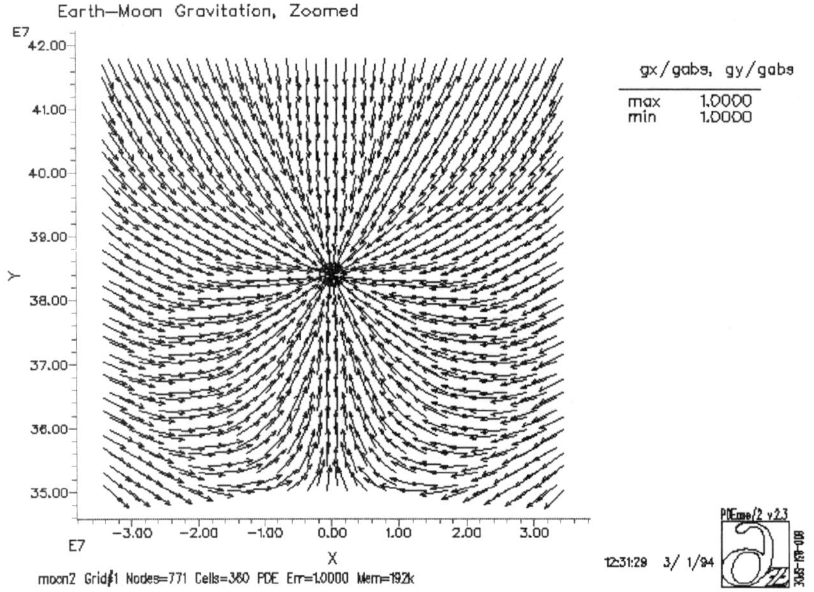

The angular perturbation can be seen even more clearly from the *elevation* plot on the next page, which displays the angle of the field as a function of radial distance from the Moon. Evidently the distortion is only about 0.1 degrees at the surface of the Moon. Thus, an astronaut should not find it difficult to remain on his feet.

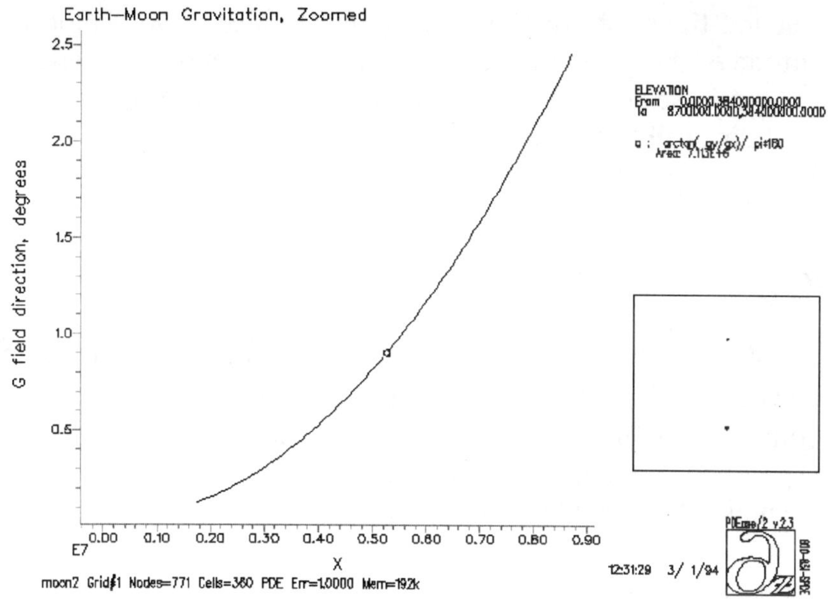

Earth–Moon Gravitation, Zoomed
G field direction, degrees
2.5
2.0
1.5
1.0
0.5
E7
0.00
0.10
0.20
0.30
0.40
0.50
0.60
0.70
0.80
0.90
X
moon2 Grid#1 Nodes=771 Cells=360 PDE Err=1.0000 Mem=192k
ELEVATION
From
To
12:31:29 3/ 1/94
PDEase/2 v2.3

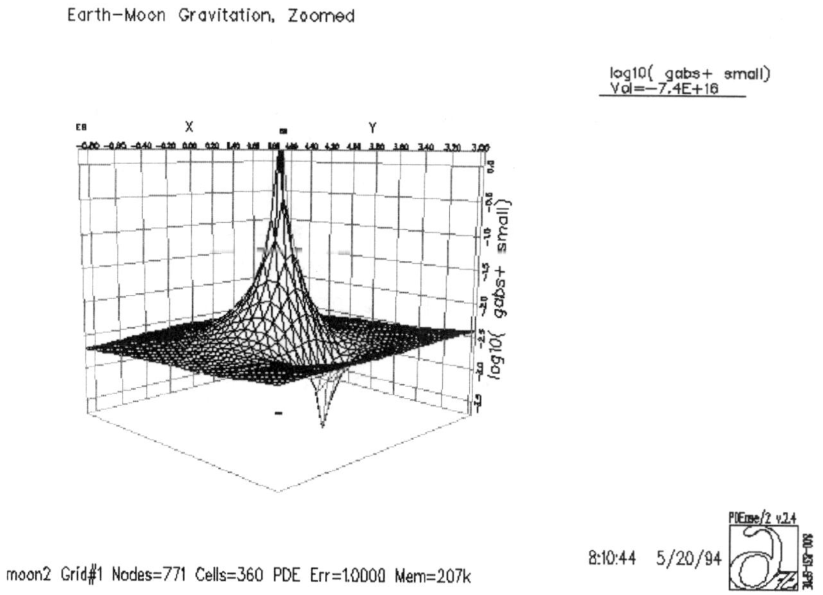

Earth–Moon Gravitation, Zoomed
log10(gabs+ small)
Vol=–7.4E+16
X
Y
log10(gabs+ small)
moon2 Grid#1 Nodes=771 Cells=360 PDE Err=1.0000 Mem=207k
8:10:44 5/20/94
PDEase/2 v2.4

The second figure above is a zoomed and rotated *surface* plot of the magnitude of the gravitational field close to the Moon. We have used a logarithmic scale for the field strength in order to see the Moon's near field as well as the sharp minimum (in fact a zero point) at some distance toward the Earth.

Divergence of g

Before leaving the gravitational field we should investigate whether the divergence $\nabla \cdot \mathbf{g} \equiv -\nabla^2 U$ is equal to zero. For this purpose we add the following definitions to the previous descriptor and make the corresponding contour plots. Here we use the underscore character (_) to make the variable names more readable.

```
...
laplace_xy= dxx(U)+dyy(U)
laplace_rz= (1/x)*dx[x*dx(U)]+ dyy(U)
...
```

To the reader who knows that the divergence of a gravitational field in free space vanishes, the plot of *laplace_xy* may come as a shock. We are now in three dimensions, however, although the plots are in a plane through the two planetary objects. If we had included the term *dzz(U)*, which is not allowed in the present versions of PDEase, we would in fact have obtained a vanishingly small value.

An alternative solution is to use cylindrical coordinates, ρ and z, without telling the program about this change of variables. We simply take x to mean ρ and y to mean z. The proper form for the Laplacian may be found in any book on electricity[3p705]. Plotting *laplace_rz* we recognize the irregular pattern of zero contours which indicates a vanishingly small function.

Towards the Center of the Earth

The gravitational field of course does not stop at the surface of the Earth, but penetrates through all of the planet. Newton already showed that the field at any radius below the surface is equivalent to the attraction from all the mass contained within that radius, concentrated at the center[2p78]. This does not mean that we may compute the *potential* by summation over internal masses only. What we can do is to derive the potential in this region from Newton's expression for the field strength, assuming the Earth's density to be constant. It is easy to show that the following expression for U yields the correct gravitational field on the inside, and that the potential takes the previous value at the Earth's radius r_1.

$$U = -\frac{GM_1}{2r_1}\left[3 - \frac{r^2}{r_1^2}\right]$$

We now study this potential using much the same descriptor as before, as suggested below.

```
...
definitions
  r1= 6.37e6     m1= 5.98e24                              { SI units }
  rad= sqrt(x**2+y**2)
  U= -6.67e-11* m1/(2*r1)*[ 3- (x**2+y**2)/r1**2]
  gx= -dx(U)     gy= -dy(U)          gabs= sqrt(gx**2+gy**2)
  laplace= (1/x**2)* dx[x**2*dx(U)]              { Spherical coordinates }
boundaries
                  region 1
  start(r1,0)
  arc to (0,r1) to (-r1,0) to (0,-r1) to finish
plots
  contour(U)              elevation( U) from (0,0) to (r1,0)
  elevation( gabs) from (0,0) to (r1,0)           surface( gabs)
  elevation( laplace) from (0,0) to (r1,0)
end
```

We have expressed the radial coordinate *rad* in terms of Cartesian x and y. The plots show that the force on an object in this region is proportional to the radius and directed towards the center. The *surface* plot below displays the magnitude of **g**.

In addition we introduce the Laplacian of U, using x temporarily as the radial coordinate in a polar (r, θ, ϕ) system. The *elevation* plot for y=0 demonstrates that the result is a constant value. This illustrates that the potential obeys the partial differential equation for gravitation

$$\nabla^2 U - 4\pi G\rho = 0$$

and you can easily verify that the value we obtain for the Laplacian is in accord with the uniform density assumed.

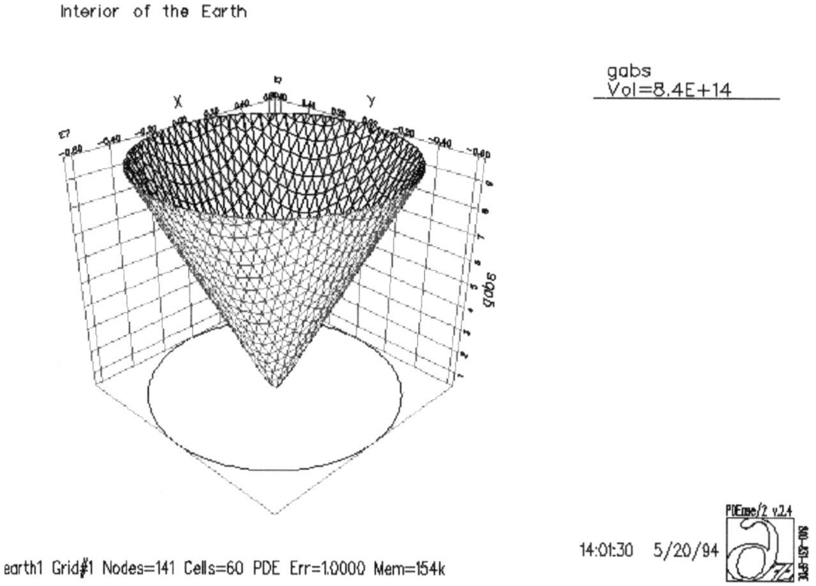

Exercises

❑ Make a *contour* plot of the z component of curl(**g**) in rectangular coordinates for the outer field of the Earth, putting the differential expression for $\text{curl}_z(\mathbf{g})$ directly into the plot parenthesis.

❑ Repeat the previous exercise with the field inside the earth. What does the result mean, and did you expect it?

Electric Point Charges

Using basic knowledge of the electrostatic potential[3p72] we may easily explore the field **E** around a set of point charges.

Field From Two Identical Charges

The only new feature in the *definitions* segment is the calculation of the angle of the field vector, *E_angle*. Since the program gives us access to the function *sign*(x), we may obtain not only the direction but also the sense of the field vector by means of arccos(x).

```
title
  'Equal Point Charges'
select
  contours= 10          surfacegrid= 30          paint= off
definitions                                                    { SI units }
  Lx= 1            Ly= 1            d= Ly/2
  q= 1                                           { Both charges positive }
  eps0= 8.85e-12      c= 1/(4*pi*eps0)
  U= c* [q/sqrt(x**2+(y+d)**2)+ q/sqrt(x**2+(y-d)**2)]
  Ex= -dx(U)      Ey= -dy(U)                          { Field components }
  Eabs= sqrt(Ex**2+Ey**2)                              { Magnitude of E }
  E_angle= sign(Ey)* arccos(Ex/Eabs)/pi*180
  small= 1e-15
boundaries
                    region 1
  start(-Lx,-Ly)
  line to (Lx,-Ly) to (Lx,Ly)  to (-Lx,Ly) to finish
plots
  contour( Ex)            contour( log10[ abs(Ex)+small])
  contour( Ey)            contour( log10[ abs(Ey)+small])
  contour( Eabs)          contour( log10[ Eabs+small])
  vector( Ex/Eabs, Ey/Eabs)          contour( E_angle)
end
```

As before, the linear plots convey much less information than logarithmic plots. An interesting feature concerning E_y is the oval zero curve which crosses a horizontal zero line. The logarithmic *contour* plots for E_x and E_y turn out to be rather complicated, whereas those of the magnitude are simpler, as seen in the figure below.

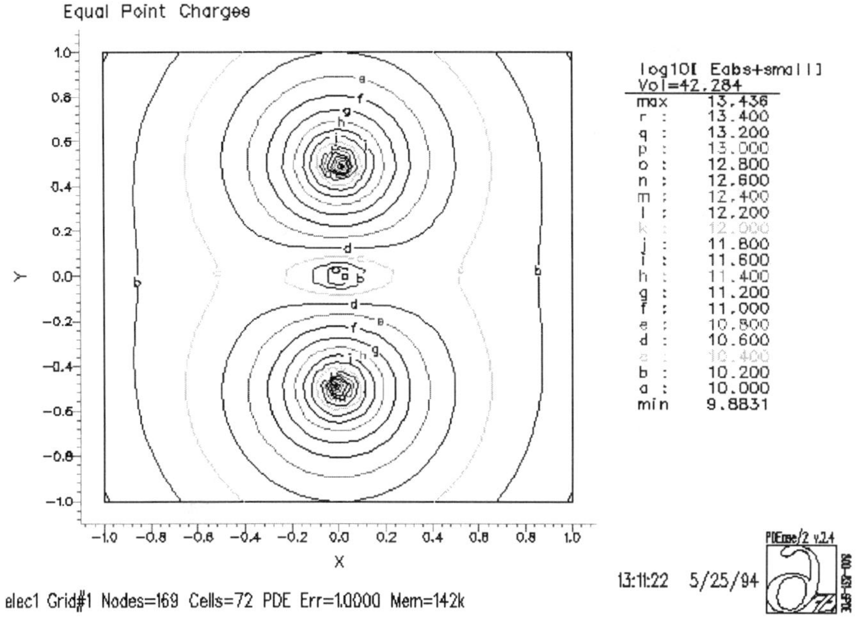

The *contour* plot of the angle of the field vector, shown on the next page, may be discussed in some detail. The contours corresponding to 180^0 have degenerated into bands, which may be an unintended feature of the software. Unfortunately there are few visible identifications, but the contours may still be assigned to the angles given in the table since we know that the contours must leave a point charge radially. Use of this fact and the alphabetic order permit us to assign angles to all the curves.

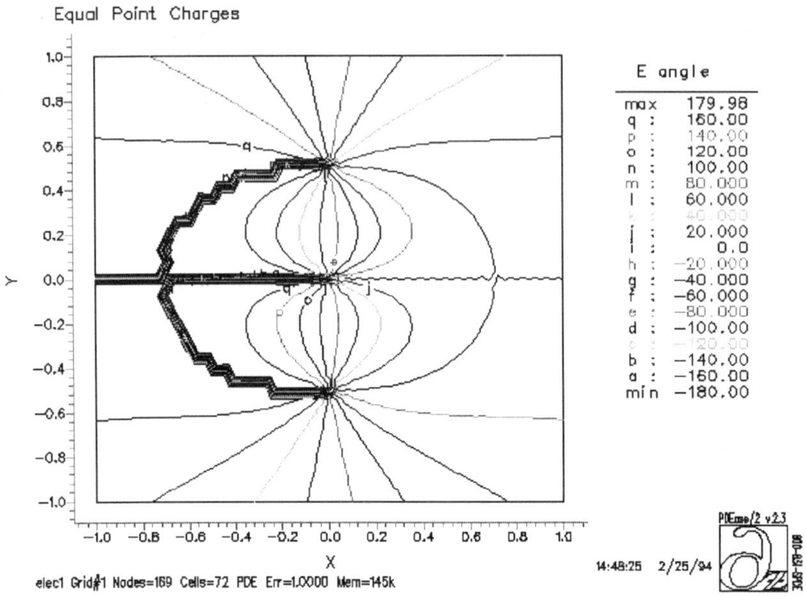

Charges of Opposite Signs

The fields of equal charges are rather similar to the gravitational field around planets. If there are charges of different sign, however, new and occasionally unexpected patterns appear. To study these effects you only need to modify a copy of the previous file by changing the sign of one of the charges (q) in the expression for the potential. The logarithmic *contour* plot of $|E_x|$ on the next page appears to be symmetric with respect to a horizontal line, since the change of sign is concealed by taking absolute values. Running this descriptor you will also find that E_y now exhibits hyperbola-like zero contours.

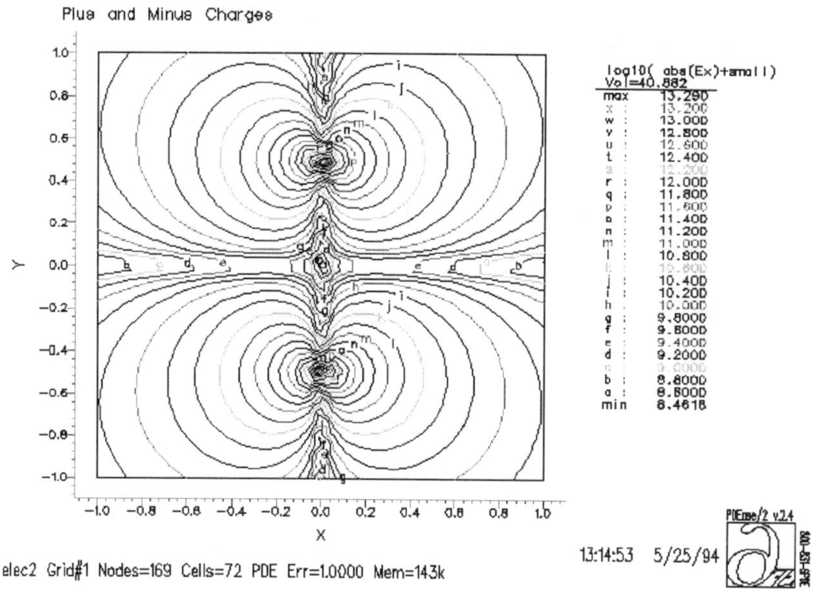

Dipole Field

We now proceed to an extreme case of the previous example: the electric dipole. This object is a combination of a positive and a negative charge ($\pm q$), at a distance of $2d$ from each other. The product $2dq$ is known as the dipole moment, and at distances much larger than d the resulting field is similar to that of an infinitesimally small dipole having the same moment. The electric dipoles occurring in the real world are of course of finite extension, but the expression for the potential and the field become simpler in the limit of vanishing d.

The precision of PDEase permits us to plot the dipole field directly from the point charge potentials we have used before. We just change the charge locations in the previous descriptor to d= 1e-6**Ly* and run again to obtain the following figure.

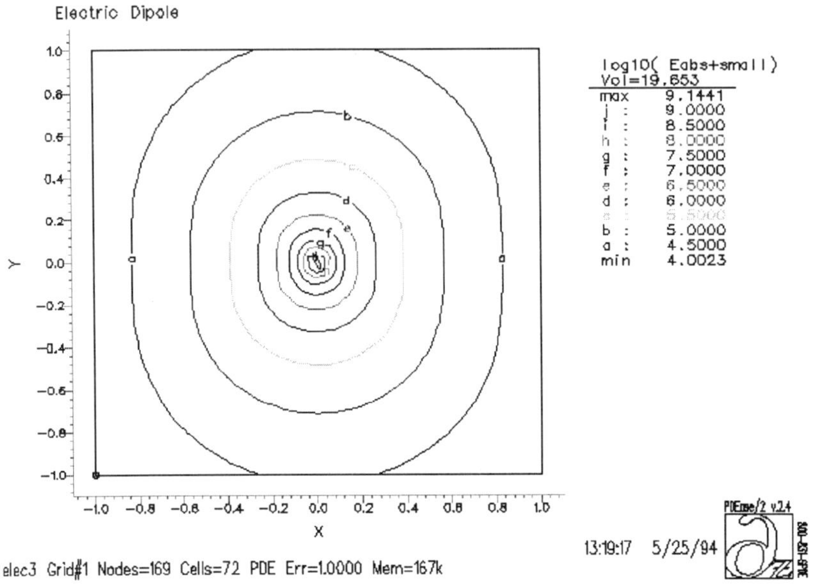

The above plot of the absolute value is remarkably simple. Looking back at the corresponding figure of the previous configuration we may imagine how the *contour* plot gradually transforms as we go from the near field to the far field.

The next figure shows the characteristic *vector* plot of the dipole field. Together with the plot of the absolute value it uniquely describes the field.

The plots in this chapter provide much food for thought. For example, it is instructive to relate the symmetries of the charge patterns to the symmetries or anti-symmetries of potentials and fields. Some figures also illustrate the limited accuracy of numerical methods, in this case two-dimensional interpolation.

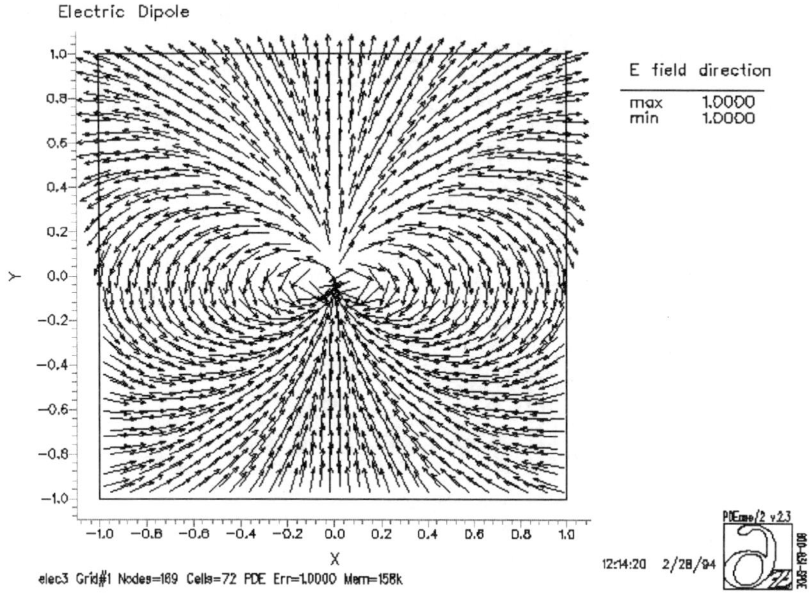

Quadrupole Fields

The total charge of a dipole is zero. A quadrupole is an object which not only has zero total charge, but also zero dipole moment. We may construct a quadrupole by placing two dipoles in opposition. As an example, we put four charges of equal magnitude on a line: a double, negative charge in the middle and two positive charges on either side at the same distance. The next two figures show a *vector* plot and a *contour* plot of the angle for this field. The latter is as simple as that for the dipole, in the sense that the contours are still radial.

Another procedure which produces a quadrupole is to place four charges of equal strength on a square, with positive charges on one diagonal, negative charges on the other one. The two following figures contain plots of the resulting field. As you see, the fields of a linear and a square quadrupole are not similar, and the *vector*

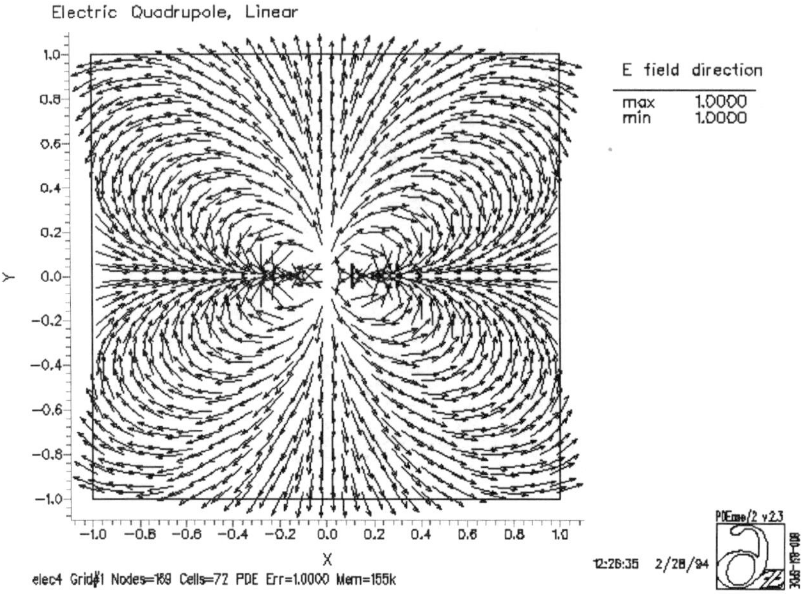
Electric Quadrupole, Linear
E field direction
max 1.0000
min 1.0000
Y
X
12:26:35 2/28/94
PDEase/2 v2.3
elec4 Grid#1 Nodes=169 Cells=72 PDE Err=1.0000 Mem=155k

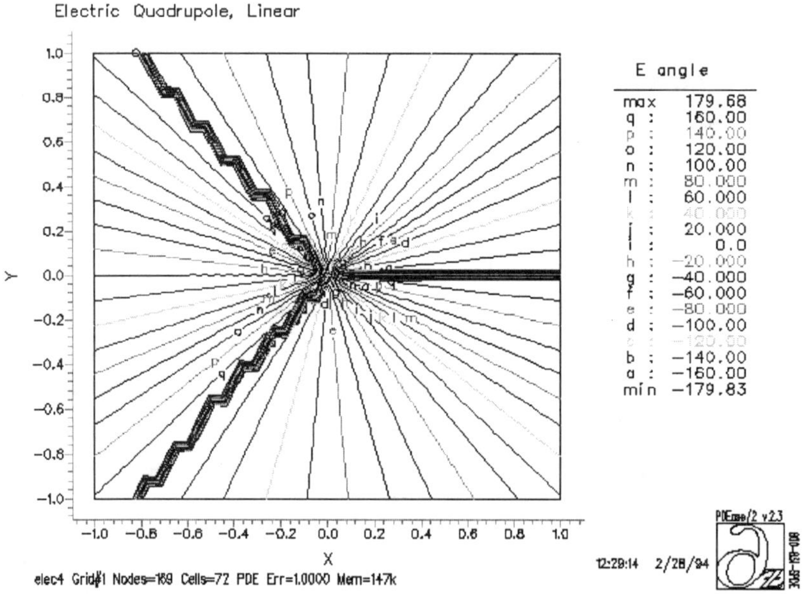
Electric Quadrupole, Linear
E angle
max 179.68
q : 160.00
p : 140.00
o : 120.00
n : 100.00
m : 80.000
l : 60.000
k : 40.000
j : 20.000
i : 0.0
h : −20.000
g : −40.000
f : −60.000
e : −80.000
d : −100.00
c : −120.00
b : −140.00
a : −160.00
min −179.83
Y
X
12:29:14 2/28/94
PDEase/2 v2.3
elec4 Grid#1 Nodes=169 Cells=72 PDE Err=1.0000 Mem=147k

plots show that one field cannot simply be rotated to fit the other one. This demonstrates that quadrupoles are far more complicated objects than dipoles. A dipole has a moment and a direction and transforms like a vector, whereas a quadrupole is a tensor and must be described by six distinct components.

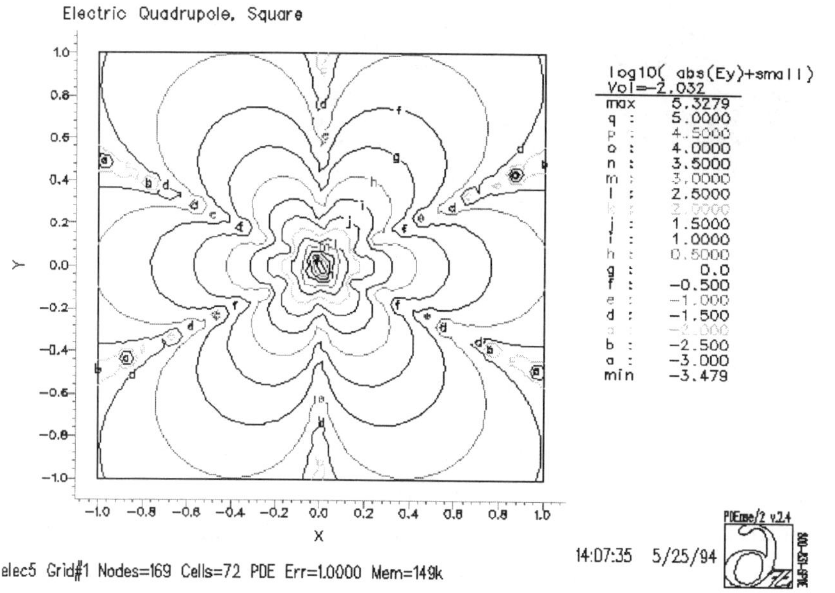

We may now ask ourselves if the Laplacian $\nabla^2 U$ and the perpendicular component $\mathrm{curl}_z(\mathbf{E})$ vanish for the above fields. As we found before for the gravitational field (p.32), we would need the three-dimensional form of $\nabla^2 U$ in Cartesian coordinates to obtain the expected result. For rotationally symmetric objects, such as dipoles and linear quadrupoles, however, we may use cylindrical coordinates to explore what comes out of a Laplacian operation.

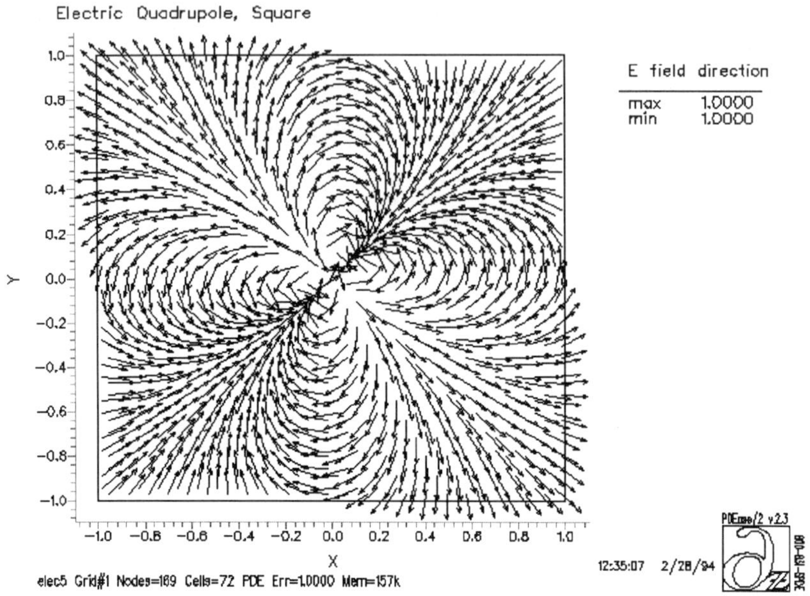

Electrically Charged Wires

The two-dimensional (2D) counterpart of the above charge configurations would be a set of thin, charged metal wires, all perpendicular to the *xy* plane. Let q_w denote the charge per unit length of wire and ρ the radial coordinate. The expression for the electrostatic potential of a wire through the origin then becomes

$$U_w = \frac{q_w \ln(1/\rho)}{2\pi\varepsilon_0} = \frac{q_w \ln(1/\sqrt{x^2+y^2})}{2\pi\varepsilon_0}$$

In order to study a 2D *dipole* field you only need to change the following lines in the *definitions* segment of the dipole descriptor.

```
...
c= 1/(2*pi*eps0)
U= c*q* (-ln[1/sqrt(x**2+(y+d)**2)]+ ln[1/sqrt(x**2+(y-d)**2)])
...
```

The 2D fields will look much the same as cross-sections of 3D dipole fields, but a closer inspection reveals certain differences. The contours of $|\mathbf{E}|$ in the far 2D field are now circular and the vectors are horizontal at the corners of a square domain. A plot of $\nabla^2 U$ clearly indicates a vanishing Laplacian, and the same is true of the *z*-component of curl, written as dx(-dy(U))-dy(-dx(U)).

Exercises

❑ The charge q=1 chosen in the first examples is arbitrary. Replace it with a more realistic value for charges in air.
❑ Make a surface plot of log10(abs(*Ex*)) and choose a suitable viewing angle before printing.
❑ Plot a dipole field with the two charges on a horizontal, rather than vertical, line. To what extent is the resulting field different?
❑ Type the expression for the dipole field components[3p84], into the *definitions* segment and plot the differences with respect to *Ex* and *Ey*.
❑ Make a *contour* plot of $\text{curl}_z(\mathbf{E})$ for one of the quadrupole fields.
❑ Test if the Laplacian of the dipole field is zero, using cylindrical coordinates.

The Laplace Equation

We are finally coming to the core of this book: deducing the field on the inside of an object from the governing partial differential equation (PDE) and certain knowledge about conditions on the boundary.

Known Values on Boundaries

The first example is the simplest one possible. We shall solve the Laplace equation, $\nabla^2 U = 0$, using known values of $U(x,y)$ on a rectangular boundary. The following descriptor for this problem makes use of a few new keywords, here emphasized by capitals.

First we specify the maximum relative error of the solution by the optional statement *errlim*= 1e-6. We can request this high accuracy, since we suspect that the problem will be an easy one.

```
title
  'Laplaces Equation'
select
  ERRLIM= 1e-6                                   { Requested relative accuracy }
  contours= 10          surfacegrid= 30            paint= off
VARIABLES
  U
definitions
  Lx= 1          Ly= 1
  U_ex= x**2- y**2
EQUATIONS
  dxx(U)+ dyy(U)= 0
boundaries
            region 1
  start(-Lx,-Ly)
  VALUE(U)= U_ex  line to (Lx,-Ly)                 { U known on boundaries }
  value(U)= U_ex  line to (Lx,Ly)
  value(U)= U_ex  line to (-Lx,Ly)
  value(U)= U_ex  line to finish
plots
```

```
  contour(U)            contour(U_ex)
  contour(U-U_ex)                          { Compare solution to exact value }
  surface(U) interactive
end
```

The next new keyword is *variables*, which heads the segment for the dependent variable, the one to solve for. Then comes the *equations* segment with the well-known Laplace expression. Finally, in the *boundaries* segment we specify values of the dependent variable U on all four lines of the rectangle by the statements *value*(*U*)= *U_ex*. We could have used almost any function for the boundary values, but the skeptical reader might want to convince himself that the solution emerging is correct. The best way of checking the numerical results is to use boundary values taken from a known, exact solution.

The chapter on harmonic functions (p.19) illustrated that the real as well as the imaginary part of any simple function $f(z)$ is harmonic, and hence a solution to our equation. Thus we specify $U = \mathrm{Re}(z^2) = x^2 - y^2$ for every point on the boundary. PDEase cannot "know" that *U_ex*= *x***2-*y***2, as specified under *definitions*, actually is the exact solution; it only uses the values on the boundary and the PDE. This means that our test is non-trivial.

In a second or two PDEase produces the following *contour* plot of the solution. The plots of U and *U_ex*, is indistinguishable to the naked eye. In anticipation we added a plot of *U-U_ex*, which presents the error of the numerical solution. As you see from the plot on the next page, the maximum error is less than 10^{-7} of the maximum value of U.

In this example we specified values on a rectangular boundary, but we could just as well have made it a triangular, circular or half-circular form. We could even have cut out areas inside the outer boundary, supplying values on these boundaries as well. Rigorous books on PDEs show that the Laplace equation has exactly one solution, if the values on the boundaries are known.

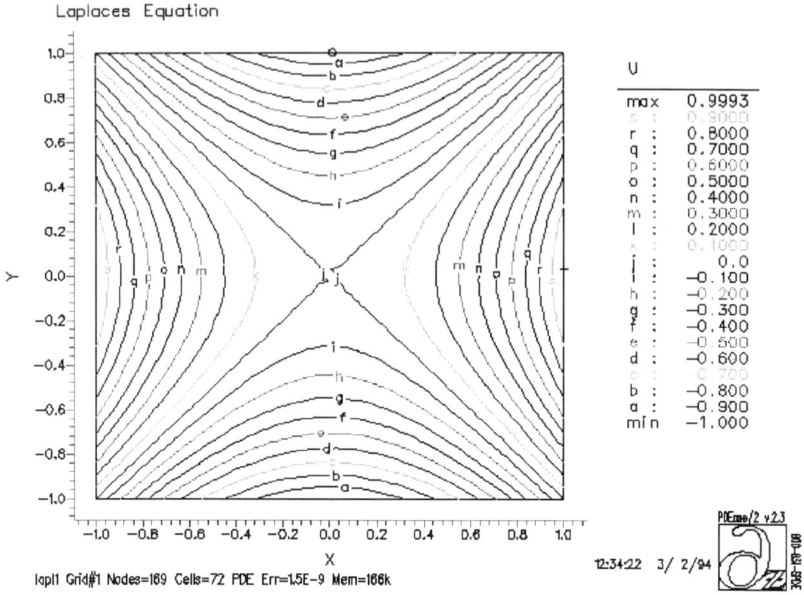
Laplaces Equation
1.0
0.8
0.6
0.4
0.2
0.0
-0.2
-0.4
-0.6
-0.8
-1.0
-1.0 -0.8 -0.6 -0.4 -0.2 0.0 0.2 0.4 0.6 0.8 1.0
X
Y
U
max 0.9993
s : 0.9000
r : 0.8000
q : 0.7000
p : 0.6000
o : 0.5000
n : 0.4000
m : 0.3000
l : 0.2000
k : 0.1000
j : 0.0
i : -0.100
h : -0.200
g : -0.300
f : -0.400
e : -0.500
d : -0.600
c : -0.700
b : -0.800
a : -0.900
min -1.000
lapl1 Grid#1 Nodes=169 Cells=72 PDE Err=1.5E-9 Mem=166k
12:34:22 3/ 2/94
PDEase/2 v2.3

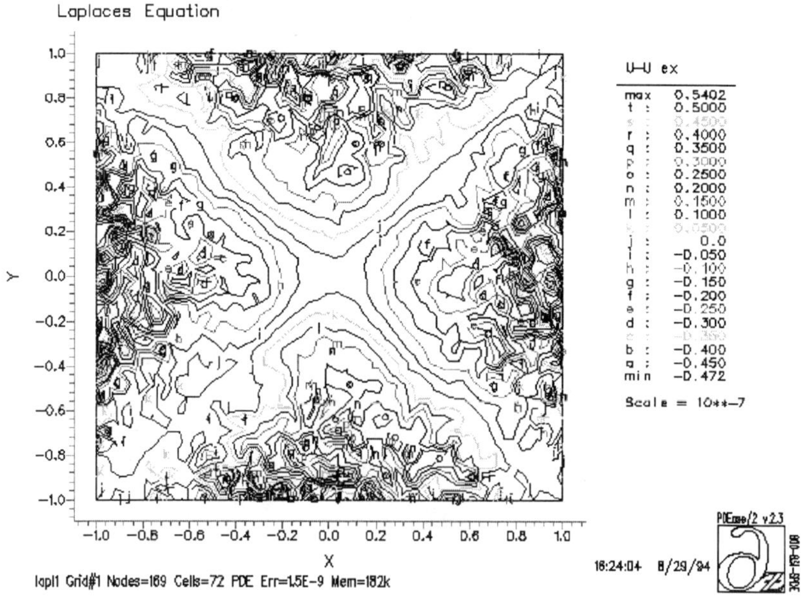
Laplaces Equation
1.0
0.8
0.6
0.4
0.2
0.0
-0.2
-0.4
-0.6
-0.8
-1.0
-1.0 -0.8 -0.6 -0.4 -0.2 0.0 0.2 0.4 0.6 0.8 1.0
X
Y
U-U ex
max 0.5402
t : 0.5000
s : 0.4500
r : 0.4000
q : 0.3500
p : 0.3000
o : 0.2500
n : 0.2000
m : 0.1500
l : 0.1000
k : 0.0500
j : 0.0
i : -0.050
h : -0.100
g : -0.150
f : -0.200
e : -0.250
d : -0.300
c : -0.350
b : -0.400
a : -0.450
min -0.472
Scale = 10**-7
lapl1 Grid#1 Nodes=169 Cells=72 PDE Err=1.5E-9 Mem=182k
16:24:04 8/29/94
PDEase/2 v2.3

Values and Derivatives on Boundaries

As an alternative to function values we could have specified values of the normal derivative, $\partial U / \partial n$, taken in a direction perpendicular to the boundary, and outwards from the solution domain. Let us do this in a new descriptor, where the *boundaries* segment reads as follows.

```
          region 1
start(-Lx,-Ly)
NATURAL(U)= 2*y  line to (Lx,-Ly)                { Outward derivative }
value(U)= U_ex  line to (Lx,Ly)
value(U)= U_ex  line to (-Lx,Ly)
value(U)= U_ex  line to finish
```

Here we have employed a new keyword, *natural*, which simply means outward derivative (The term *normal* would perhaps be more natural). For the boundary line at the bottom of the rectangle we must enter the downward derivative, i.e. $\partial U / \partial(-y) = 2y$. Running this descriptor we again find good agreement with the exact solution, although not as good as before.

Derivative Boundary Conditions Only

Now let us proceed to the extreme case, supplying only *natural* conditions on all of the four boundary lines, using the *boundary* descriptor below.

```
          region 1
start(-Lx,-Ly)
natural(U)= 2*y  line to (Lx,-Ly)          { Outward derivatives }
natural(U)= 2*x  line to (Lx,Ly)
natural(U)= -2*y  line to (-Lx,Ly)
natural(U)= -2*x  line to finish
```

The solution still becomes acceptable, although less accurate. This agreement may come as a surprise, if you thought (correctly) that a function cannot be uniquely determined by its derivatives.

To shed some light on this paradox we make another test, using *U_ex*= *x***2-*y***2+1, which obviously is harmonic as well (i.e. $\nabla^2 U\_ex = 0$). When you run that problem, however, you find that the error is no longer small, but equal to 1.000. In other words, the small error obtained when we first used *natural* boundary conditions everywhere was purely fortuitous, and was in fact due to the symmetry of the problem. This symmetry is evident from the surface plot, which also illustrates a well-known property of harmonic functions, i.e. that the highest and lowest points must reside on the boundary.

You may now guess that *natural* boundary conditions are sufficient except for a constant, and fortunately PDEase also allows us to assign a value to a single point. Using this feature we again modify the descriptor as follows.

```
           region 1
start(-Lx,-Ly)              POINT VALUE(U)= U_ex
natural(U)= 2*y   line to (Lx,-Ly)                    { Outward derivatives }
natural(U)= 2*x   line to (Lx,Ly)
natural(U)= -2*y  line to (-Lx,Ly)
natural(U)= -2*x  line to finish
```

The *point value* statement must be typed immediately after the parenthesis containing the coordinates. The value applies only to that point and takes no effect later in the descriptor. Running this problem we again obtain a solution with acceptably low error.

Solution Over a Quarter-Circle

We shall now solve the Laplace equation over the first quadrant of a circle, using $\mathrm{Im}(z^2)$ for the boundary values, as explained by the partial descriptor below. Here we must supply coordinates for the *center* of the *arc* in order to specify uniquely the boundary curve we want.

```
...
rad= 1
U_ex= 2*x*y
...
          region 1
start(0,0)
value(U)= U_ex   line to (rad,0)
value(U)= U_ex   arc(CENTER= 0,0) to (0,rad)
value(U)= U_ex   line to finish
```

Running the new descriptor you obtain the plot shown below. The error becomes somewhat larger than in the first example, as if the *arc* were a more difficult boundary to handle. The program also subdivides the triangles in the region close to the curved boundary. This is an example of adaptive grid refinement, which is one of the many intelligent features of PDEase.

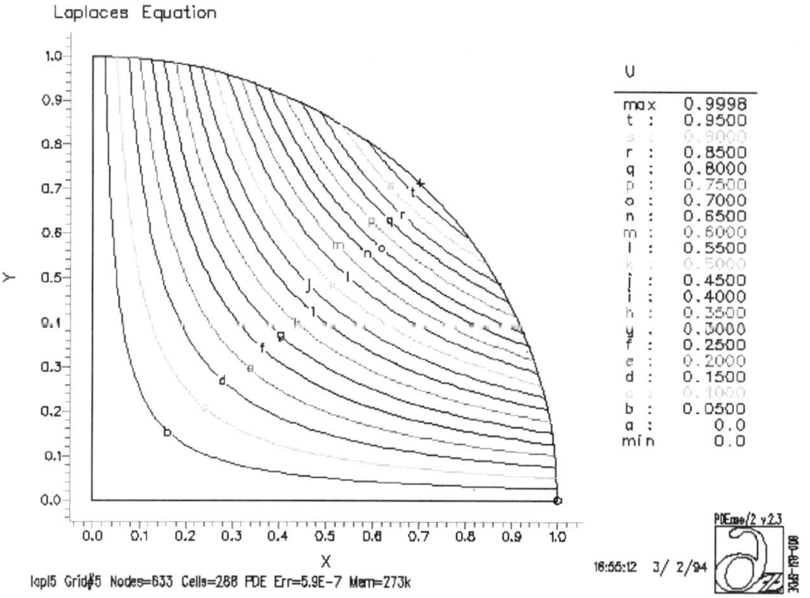

Every function we study yields additional confirmation that extreme values always lie on the boundary.

So far PDEase solved the problems with high accuracy using a modest number of triangular sub-regions, but we shall soon encounter cases where we must be content with larger errors. One way of reducing your ambition is to specify a larger value of *errlim*. A more flexible alternative is to interrupt the calculations at a reasonable number of triangles (cells) by pushing <Esc> and then <f> for finish. After terminating the procedure in course the program then stops, making the plots as usual.

Exercises

❑ Solve the Laplace equation over the first quadrant, as in the previous example, but supply normal derivatives instead on the straight boundary lines.

❑ Solve the Laplace equation over a rectangular region using natural boundary conditions from *U_ex*= *x*/(*x***2+*y***2) (which is the real part of 1/*z*) and a single point value. Choose your solution region such that it does not include the singular point at the origin.

Electrical Conduction

We are now ready to treat problems related to the real world. For a static (time-constant) electric field in the absence of net charge[3p153] we have $\mathbf{E} = -\nabla U$ and $\nabla \cdot \mathbf{E} = 0$, which combined yield the Laplace equation $\nabla^2 U = 0$. In an isotropically conducting material the electrical current density $\mathbf{j}$ is simply related to the field strength $\mathbf{E}$ by[3p203]

$$\mathbf{j} = \sigma \mathbf{E} = -\sigma \nabla U\,, \qquad \bullet$$

where σ is the electric conductivity. Having calculated $U(x,y)$ from the Laplace equation we just differentiate to obtain $\mathbf{j}$ everywhere.

Conduction in a Rectangular Plate

Let us apply our scheme to the simplest possible example. We apply a voltage over a rectangular region, so that the potential U of the lower surface is zero and that of the upper surface is 1.0 V. Although we have so far only considered a solution in the xy-plane, we now allow the object to have extension in the z direction. We always assume, however, that the dependent variable, U in this case, does not vary with respect to z. In practice, we could consider a conducting plate with insulated faces parallel to the xy-plane. The corresponding descriptor is shown below.

```
 ...
select
  errlim= 1e-5
  contours= 10          surfacegrid= 30        paint= off
variables
  U
definitions
  Lx= 0.5             Ly= 1
  Ex= -dx(U)          Ey= -dy(U)       Eabs= sqrt(Ex**2+Ey**2)
  cond= 5.99e7                                              { Copper }
```

```
  jx= cond*Ex        jy= cond*Ey      jabs= sqrt(jx**2+jy**2)
  U_ex= y/Ly* 1.0                              { Exact solution for 1.0 V }
equations
  dxx(U)+ dyy(U)= 0
boundaries
region 1
  start(-Lx,0)
  value(U)= 0         line to (Lx,0)
  natural(U)= 0       line to (Lx,Ly)                        { Insulated }
  value(U)= 1.0       line to (-Lx,Ly)
  natural(U)= 0       line to finish                         { Insulated }
plots
  contour( U)    contour( U_ex)     contour( U-U_ex)
  vector( Ex, Ey)
  contour( jx)   contour( jy)      contour( jabs)
  surface( U) interactive
end
```

When running the above file you discover that the solution satisfies the boundary conditions for U on the upper and lower surfaces and that the PDE error is acceptably small. You also find that the current density j_x through the insulated side is smaller than the vertical component by a factor of 10^5. We could not expect it to be exactly zero, however, since this solution is a numerical approximation.

Conduction in a Trapezoidal Plate

We now attempt to solve a problem which does not have an elementary solution. Let us deform the rectangle into a trapezoid by the following modifications.

```
  ...
definitions
  L1= 0.5          L2= 0.25        Ly= 1
  ...
boundaries
region 1
```

```
start(-L1,0)
value(U)= 0         line to (L1,0)
natural(U)= 0       line to (L2,Ly)                    { Insulated }
value(U)= 1.0       line to (-L2,Ly)
natural(U)= 0       line to finish                     { Insulated }
...
```

The solution takes a longer time than for the previous problem, and the density of triangular cells increases on each iteration at the corners. The program eventually finishes after having attained the required accuracy.

In this case we have no exact solution for comparison. How can we be sure that the solution is correct? First of all we should look for symmetries in the problem and verify that the solution obtained conforms to it. All the plots seem to have the expected left-right symmetry. From the plot of U below we see that the boundary conditions are satisfied on the upper and lower surfaces of the object. We also note that the contours of U meet the sloping boundaries at right angles, which demonstrates that the normal derivatives vanish.

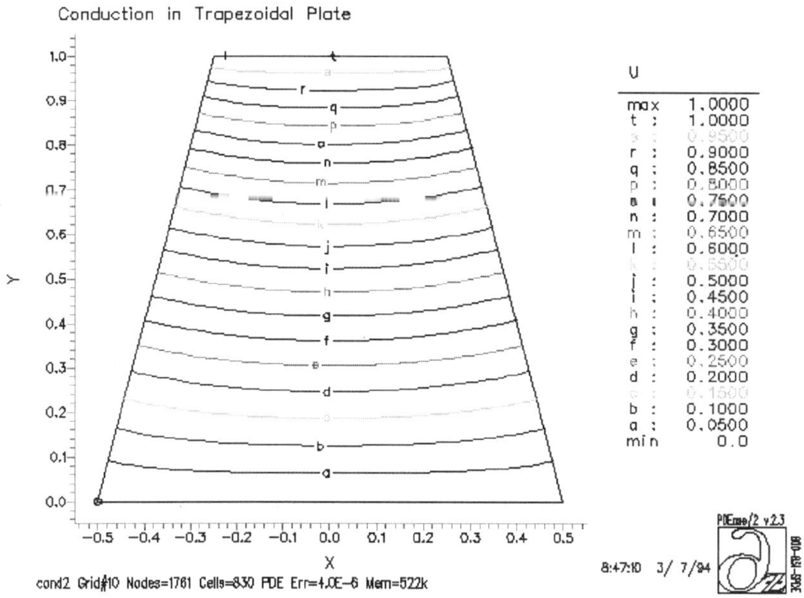

In the *vector* plot there are no arrows pointing outwards through the insulated surfaces.

We thus agree that the boundary conditions have been respected, but what about the PDE itself? PDEase internally checks for numerical errors at the end of each iteration, and the report in this case is *Err*=4.6E-6 at the bottom of the figure. This suggests that the solution should be good to about 1 part in 10^5.

Checking the Solution

If we still feel skeptical, we may plot the left side of the PDE as follows. First we define *eqn* to be $\nabla^2 U$ and then make a contour plot of the deviation from zero, usually called the residual.

The modifications we need for plotting the residual are given below. In addition we might check whether the current (per unit thickness of the plate) leaving the bottom surface equals that coming in through the top surface. This test we achieve by two *elevation* plots, which also report the integrated values (*Area*).

```
  ...
definitions
  ...
  eqn= dxx(U)+ dyy(U)
  ...
plots
  ...
  elevation( jy) from (-L1,0) to (L1,0)
  elevation( jy) from (-L2,Ly) to (L2,Ly)
  contour( eqn) as 'Residual'
  surface( jx) interactive
end
```

You will find that the plot of the residual exhibits irregular contours which we have learnt to associate with a vanishing function. The errors are evidently largest in the upper corners. The *elevation* curves show, however, that the integrated current densities over the top and bottom surfaces agree within 0.1 %, which seems satisfactory.

The plot below shows that the horizontal current density j_x is no longer zero, as it was in the rectangular geometry. We certainly expect the current entering at the top to fan out from the central line. From the plot of $|\mathbf{j}|$ you will also notice that the largest values occur at the upper corners. We may understand this by considering the contour plot of U on the preceding page. Since the contours close to the upper surface must be horizontal and all the contours must be perpendicular to the sloping sides, the curves come closer to each other at the top corners, which enhances the gradient.

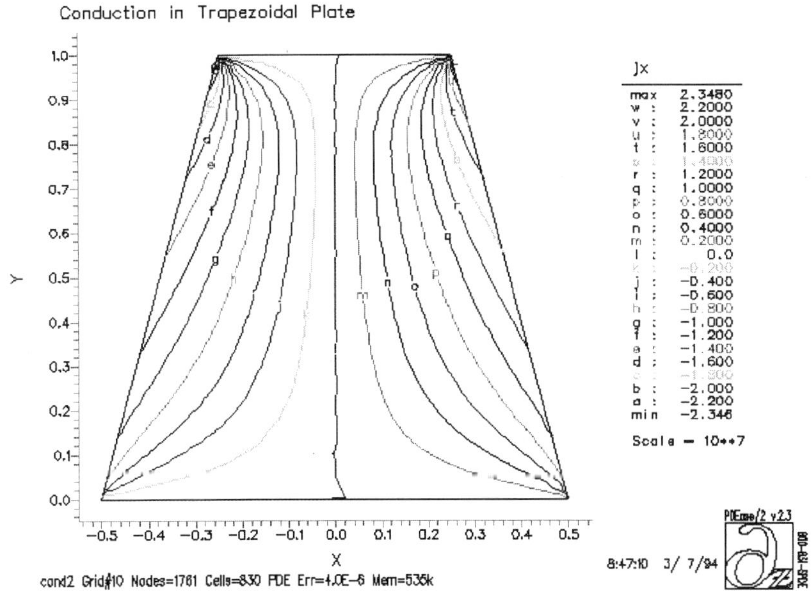

Radial Conduction in a Circular Foil

We shall now study the conduction through a thin copper foil, soldered perpendicularly to a tube of thick copper with a coaxial rod. A reasonable mathematical model would be a plate having constant potential levels on two concentric circles, corresponding to the edges of the tube and the central rod. Using the first descriptor (p.53) we modify it as follows.

```
  ...
  errlim= 1e-4
  ...
definitions
  rad1= 0.2          rad2= 1          U1= 1.0                { Radii and voltage }
  ...                                                          {Exact solution: }
  U_ex= U1* ln[ rad2/ sqrt(x**2+y**2)]/ ln[ rad2/rad1]
                    region 1
  start(rad2,0)
  value(U)= 0                                           { U=0 on outer circle }
  arc to (0,rad2) to (-rad2,0) to (0,-rad2) to finish
  start(rad1,0)
  value(U)= U1                                         { U=U1 on inner circle }
  arc to (0,-rad1) to (-rad1,0) to (0,rad1) to finish
```

This conduction problem has an elementary solution that we may use for *U_ex*. To obtain it we note (p.44) that $\ln(\rho)$ is a harmonic function in cylindrical coordinates, and we just have to determine the constants in $A + B\ln\sqrt{x^2 + y^2}$ to make it fit at the boundaries.

In the above descriptor each circular boundary consists of four sections, and there is a shorthand notation for assigning *value* or *natural* to each of these. You do not have to repeat *value*(*U*)=0, if it applies to several successive parts. If you should not specify any boundary conditions at all, the program assumes *natural*(*U*)=0 everywhere.

The plot of *U-U_ex* turns out to be about twice the requested error limit, the difficult region being near the circular cutout. The vector plot shows the expected radial field, but the *contour* plot of j_x below may be more of a surprise (try a surface plot as well).

The high symmetry of the present problem makes it possible to treat only part of the solution domain, e.g. only one quadrant. Since the field is radial, the radial line cuts must have *natural*(*U*)=0. This smaller domain requires only one fourth of the number of cells, and hence the calculations will be correspondingly shorter.

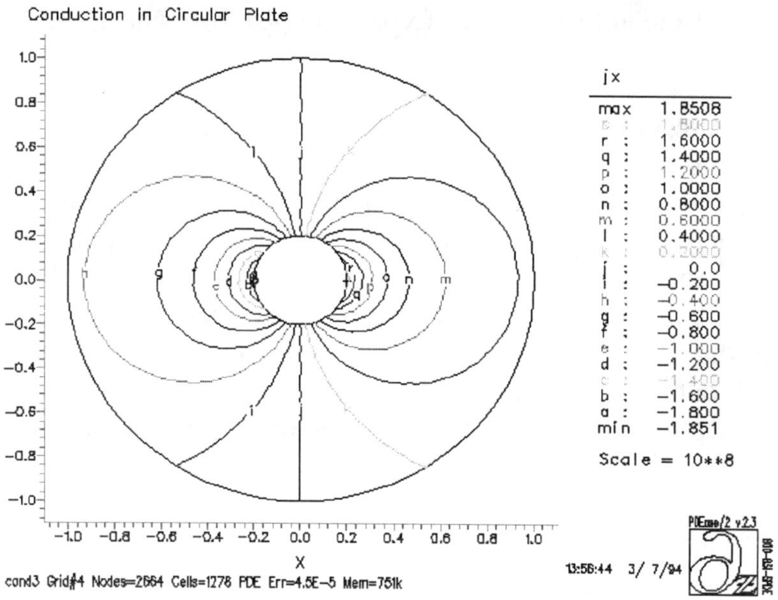

Constricted Rectangular Plate

In the next problem we study a rectangular plate with a constriction. Only a few changes of the first descriptor (p.53) are required, as suggested by the excerpt below.

```
  ...
  errlim= 1e-4
  ...
definitions
  Lx= 1.0            Ly= 2.0           d= 0.8
  ...
boundaries
            region 1
  start(-Lx,-Ly)
  value(U)= 0        line to (Lx,-Ly)
  natural(U)= 0                                        { Insulated }
     line to (Lx,-d) to (Lx-d,-d) to (Lx-d,d) to (Lx,d) to (Lx,Ly)
  value(U)= 1.0      line to (-Lx,Ly)
  natural(U)= 0                                        { Insulated }
```

```
line to (-Lx,d) to (-Lx+d,d) to (-Lx+d,-d) to (-Lx,-d) to finish
...
```

The plots resulting from the above descriptor demonstrate how the field deforms on going into a narrower or wider cross section. The following figure shows the effect on the magnitude of the current density. There are no contours in the narrow part of the plate, which means that the current density is virtually constant there. The last contour (*l*) extends somewhat into the constricted region, and the curves in the wider regions are strongly perturbed with respect to the horizontal contours expected close to the ends. These indications about the depth of perturbation are of great practical importance.

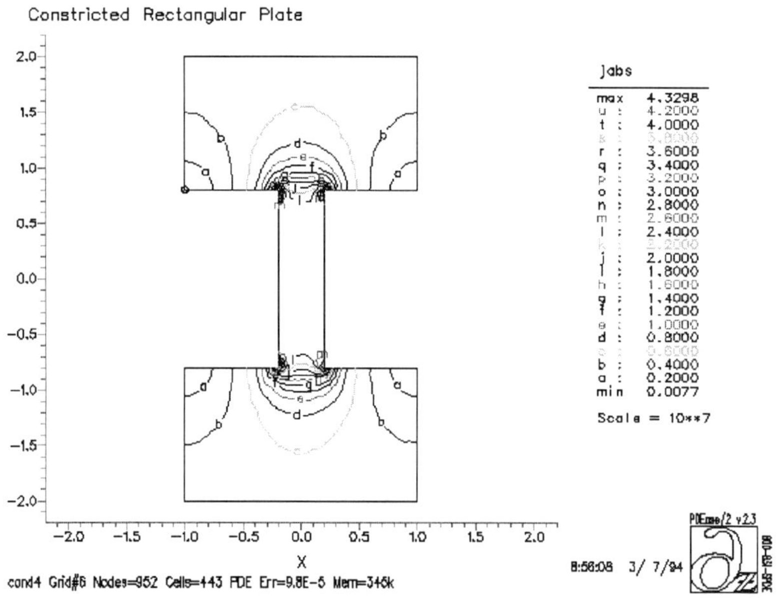

Plate Made of Two Different Materials

So far we have only been concerned with fields in homogeneous objects. A powerful feature of finite element analysis, however, is that the materials properties are allowed to vary over a domain, gradually as well as stepwise. In order to take advantage of this we must write the PDE in a form which contains the materials properties explicitly.

We have already used the relation $\mathbf{j} = \sigma\mathbf{E}$. In addition we need the equation of continuity,

$$\nabla \cdot \mathbf{j} + \frac{\partial \rho}{\partial t} = 0$$ ●

Here ρ is the charge density, and the equation simply means that charge cannot be destroyed. Since we shall be dealing with static phenomena, the time derivative is zero. Combining the two relations for **j** we obtain

$$\nabla \cdot \mathbf{j} = \nabla(-\sigma \nabla U) = 0$$ ●

where we shall keep the minus sign to make the parenthesis equivalent to **j**.

The first descriptor for a rectangular plate may serve as a starting point for the following one.

```
  ...
  errlim= 1e-4
  ...
  cond                                    { Conductivity parameter reserved }
  ...
  jx= cond*Ex        jy= cond*Ey      jabs= sqrt(jx**2+jy**2)
  j_angle= sign(jy)* arccos( jx/jabs)/ pi*180
equations
  dx[-cond*dx(U)]+ dy[-cond*dy(U)]= 0
boundaries
          region 1       cond= 5.99e7                          { Copper }
  start(-Lx,0)
  value(U)= 0        line to (Lx,0)
  natural(U)= 0      line to (Lx,Ly)                          { Insulated }
  value(U)= 1.0      line to (-Lx,Ly)
  natural(U)= 0      line to finish                           { Insulated }
          REGION 2       cond= 1.03e7                               { Iron }
  start(-Lx,0)
  line to (Lx,0) to (-Lx,Ly) to finish
  ...
  vector( jx/jabs, jy/jabs)
  contour( j_angle)
```

```
elevation( j_angle) from (-Lx,Ly/2) to (Lx,Ly/2)
elevation( jabs) from (-Lx,Ly/2) to (Lx,Ly/2)
...
```

Here we declare the conductivity parameter in the *definitions* segment, but wait until *boundaries* to give it a value. This is the way to handle an object consisting of different materials.

Another new feature is that we have introduced a triangular *region* 2, announcing that the material there is iron. The first region must always be the largest one, and on its border we specify boundary conditions. No boundary conditions are allowed on the sub-regions.

Running this descriptor we find the following contour plot of *U*. The potential is continuous across the interface between copper and iron, but the contours change direction. Since the current flow lines are perpendicular to these contours, the result of the *vector* plot is not surprising. It should be noted that a corresponding *vector* plot of **E** would have been similar to that of **j**.

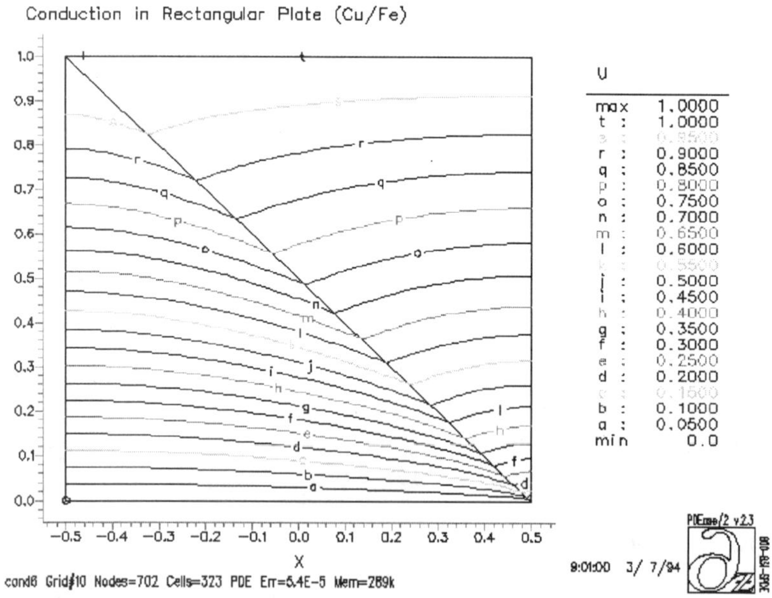

The *contour* plot of the angle contains quantitative information about the refraction of current lines, but unfortunately the plot is hard to read in this case. In order to obtain a better presentation of the change of flow direction at the interface, we request an elevation plot across the middle of the foil, as shown below.

Obviously the potential must be the same on either side of the diagonal, and the same is true for its rate of change in the diagonal direction. Hence, the field component E_p parallel to the diagonal must also be the same on either side. Using this component of common magnitude and the field angle we can calculate the normal components of **E** on either side of the diagonal, and from that the normal components of **j**. Finally we can test whether the latter components are equal, as they should be.

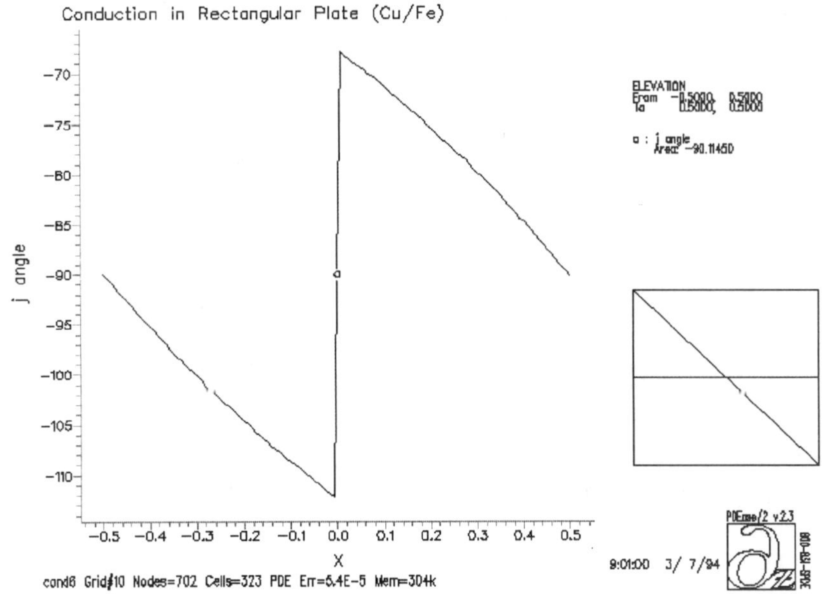

Using the *elevation* plot of *jabs* we may read off its values on both sides of the diagonal and then combine these with angles to compute values of E_p on either side. This provides another test of the numerical solution.

Line Integrals

Recent versions of PDEase provide means of calculating a line integral, using the *area* value present on an *elevation* plot. A line integral may be taken along a boundary or along a *feature*, which could be any curve defined (by *line* or *arc* or both) within the domain. To create a *feature* we add the following statements to the *boundaries* segment.

```
feature
start 'curve1' (0,0) line to (1,1)
```

This has the secondary effect of increasing the node density close to the curve, which may sometimes be exploited for its own sake. A device similar to that of naming a *feature* may also be applied to a *boundary*. To find the value of $\oint \mathbf{E}\cdot d\mathbf{l}$ we just define **E** as a vector (p.23) and include the plot command

```
elevation( tangential(E) ) on 'curve1'
```

Exercises

❑ Modify the trapezoid descriptor by adding a square to the shorter side. Compare the solutions.

❑ Use the symmetry of the circular Cu foil (p.57) to solve for one quadrant only.

❑ Change the circular plate problem by moving the cut-out by one hole diameter. Make maximum use of the remaining symmetry in order to shorten the calculations.

❑ Repeat the study of current flow through a constriction, exploiting the symmetry of the problem.

❑ Integrate $\mathbf{E}\cdot d\mathbf{l}$ along a vertical symmetry line through the constricted plate. Compare to the applied voltage.

Dielectrics in (*x*,*y*) Space

For dielectric materials, which become polarized when subjected to an external field, we must use relations[3p110] involving the electric flux density $\mathbf{D} = \varepsilon\mathbf{E} = \varepsilon_r\varepsilon_o\mathbf{E}$, where ε_r is the relative permittivity. The corresponding PDE[3p153] is of the form $\nabla\cdot\mathbf{D} = \rho_v$, where ρ_v is the volume charge density not arising from polarization. Using the potential U we may now write

$$\nabla(-\varepsilon_r\varepsilon_o\nabla U) - \rho_v = 0\ ,$$ ●

which is analogous to the PDE we used for electrical conduction.

Just as in the previous chapter we could now study the field in a rectangle or a trapezoid, subjected to a voltage. This would not be realistic, however, since there is nothing quite corresponding to "insulation" in an electric field. We may still study fields in 2D, however, if we consider the *xy* plane to be a section across the middle of a long specimen extending in the *z* direction.

Plate Capacitor in (x,y) Space

The following descriptor is for a cross-section of a parallel-plate capacitor, assumed to extend both ways along the *z*-axis. Here we reserve space for the field by *Lx* and *Ly* and cut out narrow rectangles corresponding to the metal plates. Then we assign potential values to these inner boundaries. Finally we define a sub-region for the dielectric material between the plates. The internal plate-to-plate distance is *d*. The *x*-component of **D** could not be denoted *Dx*, since PDEase would read this as *dx*, which is the reserved word for a derivative.

```
select
  errlim= 5e-4      contours= 10
  surfacegrid= 30    paint= off
variables
```

```
  U
definitions                                                    { SI units }
  Lx= 1     Ly= 1    delx= 0.5       d= 0.2     dely= 0.2*d
  Ex= -dx(U)              Ey= -dy(U)            U0= 1.0
  Eabs= sqrt(Ex**2+Ey**2)
  eps0= 8.854e-12
  eps                             { eps =epsr*eps0,  reserved for later }
  DEx= eps*Ex     DEy= eps*Ey
  Dabs= sqrt(DEx**2+DEy**2)
  small= 1e-15                                            { Small number }
equations
  dx( -eps* dx(U))+ dy( -eps* dy(U))= 0              { No volume charge }
boundaries
         region 1      eps= eps0                                { Vacuum }
  start(-Lx,-Ly)  natural(U)= 0                         { Outer boundary }
  line to (Lx,-Ly) to (Lx,Ly)  to (-Lx,Ly) to finish
  start(-delx/2,-d/2)                               { Lower plate, ground }
  value(U)= 0.0        line to (delx/2,-d/2)
  to (delx/2,-d/2-dely) to (-delx/2,-d/2-dely) to finish
  start(-delx/2,d/2+dely)
  value(U)= U0       line to (delx/2,d/2+dely)  { Potential on upper plate }
  to (delx/2,d/2)  to (-delx/2,d/2) to finish
         region 2      eps= 7.0*eps0                             { Glass }
  start(-delx/2,-d/2)  line to (delx/2,-d/2)
  to (delx/2,d/2) to (-delx/2,d/2) to finish
monitors
  contour(U)
plots
  contour(  U)            contour( Ex)              contour( Ey)
  elevation( Ey) from (-Lx,0) to (Lx,0)
  elevation( DEy) from (-Lx,0) to (Lx,0)
  contour( log10( Eabs+ small))          contour( log10(  Dabs+ small))
  vector( DEx/Dabs, DEy/Dabs) as 'Direction of D'
  surface( log10( Dabs+ small))  interactive
 end
```

The first figure below shows the resulting *contour* plot of *U*. The lines evidently are highly parallel in the dielectric between the plates, but our boundary conditions force the contours to meet the outer

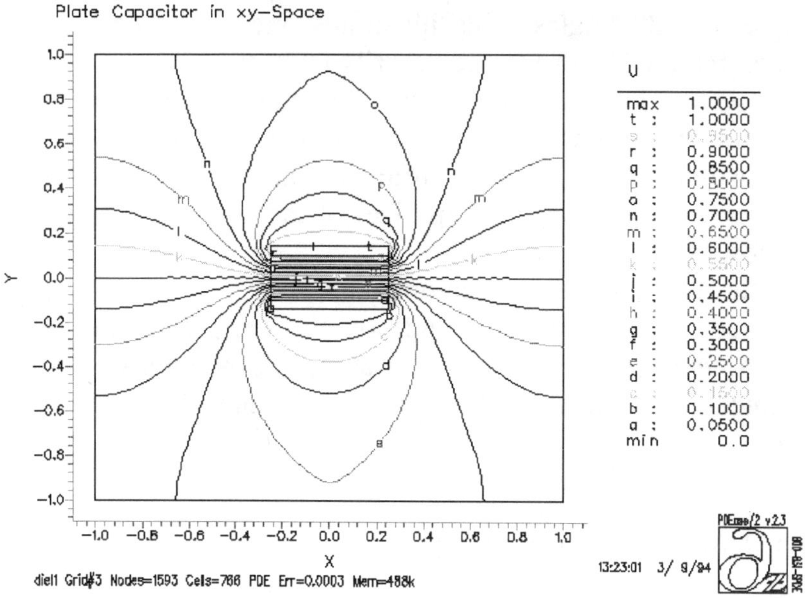

Plate Capacitor in xy-Space
U
max 1.0000
t : 1.0000
s : 0.9500
r : 0.9000
q : 0.8500
p : 0.8000
o : 0.7500
n : 0.7000
m : 0.6500
l : 0.6000
k : 0.5500
j : 0.5000
i : 0.4500
h : 0.4000
g : 0.3500
f : 0.3000
e : 0.2500
d : 0.2000
c : 0.1500
b : 0.1000
a : 0.0500
min 0.0
X
Y
diel1 Grid#3 Nodes=1593 Cells=786 PDE Err=0.0003 Mem=488k
13:23:01 3/ 9/94
PDEase/2 v2.3

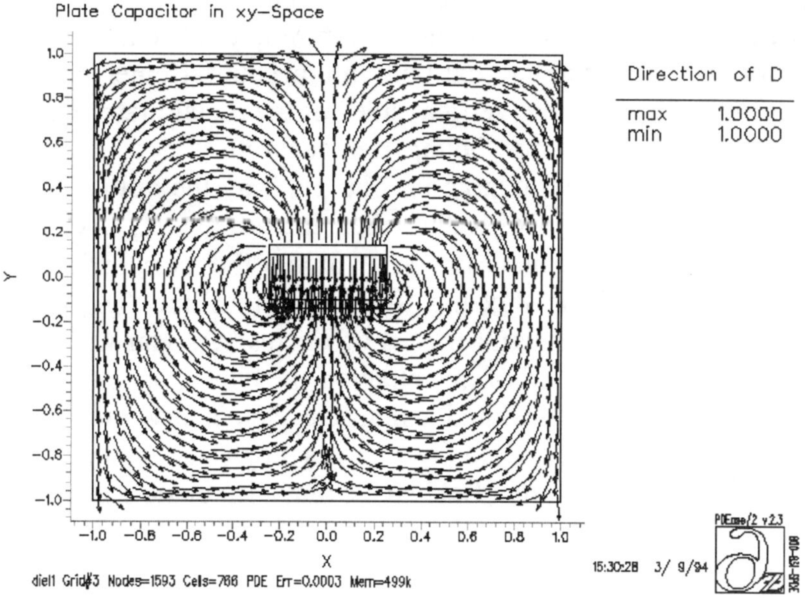

Plate Capacitor in xy-Space
Direction of D
max 1.0000
min 1.0000
X
Y
diel1 Grid#3 Nodes=1593 Cells=786 PDE Err=0.0003 Mem=499k
15:30:28 3/ 9/94
PDEase/2 v2.3

rectangle at right angles. This illustrates one of the problems with FEA, i.e. that the field space necessarily is limited.

The *vector* plot above illustrates how the far-field lines become flattened by the arbitrary rectangular limits we have imposed. We could reasonably expect this to influence the central part of the field to some extent. By trying various sizes of the outer boundary we may assess how large this effect is.

The *vector* plot also demonstrates that the field lines are constrained (by the boundary conditions) to remain within the region *Lx*Ly*. A few arrows appear to penetrate the boundary, but in fact they indicate the direction of the field at the end of the arrow, not at the tip.

Computing Capacitance

The flux of **D** leaves or enters the metal plates in the direction of the normal, and Gauss' law[3p87] states that $D_n = \rho$, the surface charge density. From the contour plot of E_y we estimate that D_y, hence the charge density, on the top surface of the capacitor may be as much as 3 % of that on the inside. The elementary model of the parallel-plate capacitor only takes the internal field into account. Remembering that there is also charge on the vertical sides of the metal plates we realize that the simplest model may be far from accurate.

If we divide the total charge on one plate by the voltage applied, we obtain the capacitance. Rather than integrating D_n over all four faces of a plate, we may integrate along the horizontal line of symmetry, since all the flux has to pass by that line. The *elevation* plot below yields the integral (*Area*) of interest. As you will discover in the exercises at the end of this chapter, the capacitance turns out to be higher than that calculated using the elementary formula. This discrepancy is mostly caused by charges neglected in the simplest model, but effects of limited field space may also be noticeable.

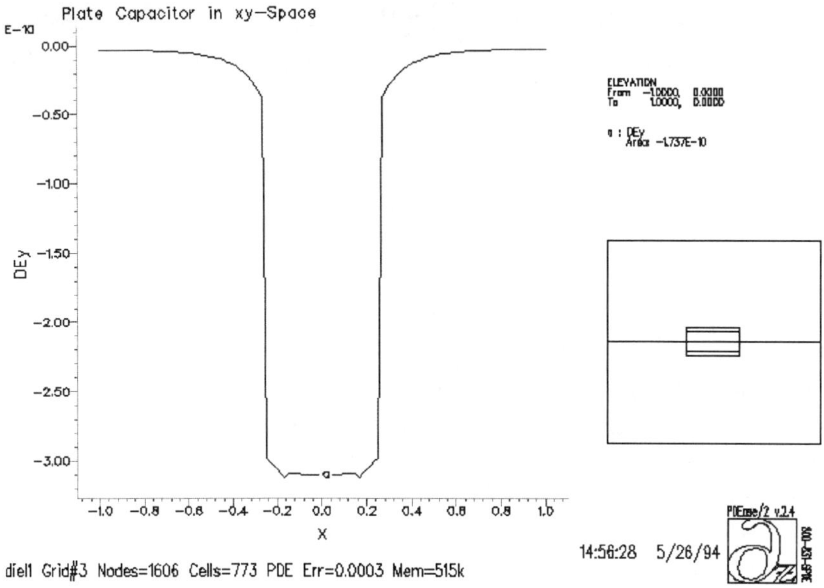

Electrostatic Energy

The expression for the energy[3p138] required to charge the above capacitor is $W = CU^2/2$. This quantity should equal the energy of the electrostatic field, given by

$$W = \frac{1}{2}\iiint \mathbf{D}\cdot\mathbf{E}\,dv\ ,$$

where the integral is to be taken over the volume of the solution domain. If we make a *contour* plot of the energy density, the program automatically gives us the value of the integral (*Vol*), assuming unit extension of the domain in the *z* direction. In order to test this equivalence we should add a definition and commands as follows.

```
  ...
  energy_d= 0.5* Eabs* Dabs
  ...
  contour( energy_d)                  surface( energy_d)
end
```

Running the extended descriptor we find that the two expressions for the energy agree to within a few percent. The cause of the discrepancy becomes apparent when we look at the *surface* plot, which shows that *energy_d* makes a sharp jump from a low value up to the maximum. This "discontinuity" increases the error of the integration routine, which otherwise should be very accurate.

Glass Rod in a Parallel Field

We now turn our attention to a circular rod of dielectric material (glass), exposed to an external field. We assume the field to have been homogeneous before the rod was placed perpendicularly to it. The list below contains the items to be changed in the last descriptor.

```
...
Lx= 1     Ly= 1          r0= 0.1*Lx     E0= 1.0                { Far field }
Ex= -dx(U)               Ey= -dy(U)
Eabs= sqrt(Ex**2+Ey**2)
eps0= 8.854e-12     epsr= 7.0        eps        D0= eps*E0
DEx= eps*Ex         DEy= eps*Ey      Dabs= sqrt(DEx**2+DEy**2)
U1_ex= -E0*x                                    { Exact solutions }
U2_ex= -E0*x*[1- (epsr-1)/(epsr+1)*r0**2/ (x**2+y**2)]
E2_ex= -dx(U2_ex)          small= 1e-15
...
     region 1            eps= eps0                         { Vacuum }
start(-Lx,-Ly)           natural(U)= 0    line to (0,-Ly)
point value(U)= 0        line to (Lx,-Ly)
natural(U)= D0           line to (Lx,Ly)
natural(U)= 0            line to (-Lx,Ly)
natural(U)= -D0          line to finish
        region 2         eps= epsr*eps0                    { Glass }
start(r0,0)
arc to  (0,r0  to  (-r0,0)  to  (0,-r0)  to finish
...
contour[ log10( abs(Ex-E2_ex)+ small)]
end
```

In this descriptor we introduce the radius *r0* of the rod. As pointed out earlier, the outer parentheses in the PDE contain components of **D**, and thus we should specify *natural*(*U*)= *D*0 at the right boundary in order to request a solution for a parallel and constant far field. The resulting *contour* plot of E_x shown below confirms the strength of the far field (curves *h*) at the left and right boundaries.

The descriptor includes an exact solution, although pertaining to a slightly different problem. It is possible to splice two cylindrical harmonics[3p183], one for the dielectric and another for the outside vacuum, so as to join continuously at the circular interface, also respecting the conditions on the field components. The solution for the vacuum region produces a parallel field at infinitely large *x*, however, whereas our descriptor imposes parallel directions on the left and right boundaries, hence at finite distance. In view of this difference we could never expect perfect agreement.

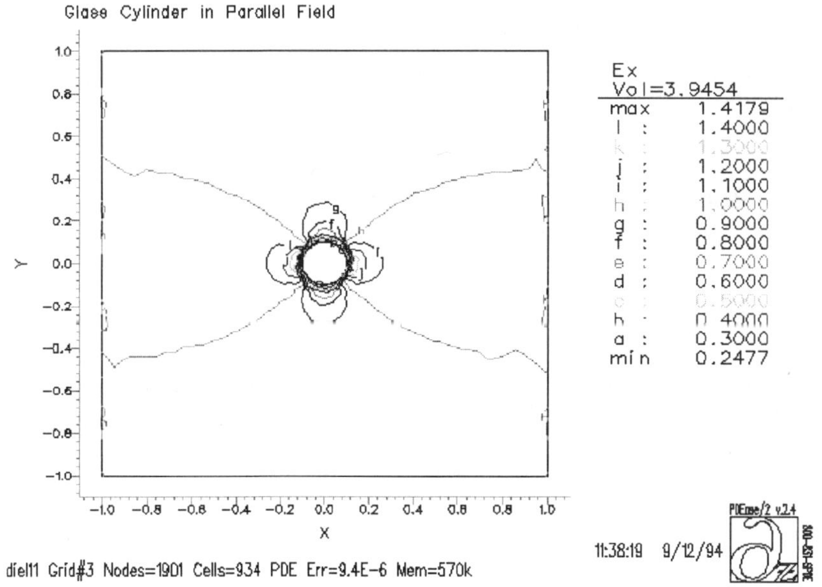

Plotting the discrepancy with respect to the exact expressions presents some practical problems. There are two "exact" solutions, one inside and one outside, and we can only show one at a time. If we choose the outer solution, *U2_ex* or its derivative *E2_ex*, which perhaps is more interesting, the difference is bound to be large inside the dielectric rod, and the number of contours become sparse in the outside region, since the difference is relatively small. To avoid this problem we display the logarithm of the absolute difference between the FEA solution and the harmonic function. This difference turns out to be about 1 % in the vacuum region, and choosing a smaller *errlim* does not improve the agreement. The discrepancy is mostly due to the different boundary conditions.

Solution With Value Boundary Conditions

We may inspect the solution in even more detail, if we bring our boundary conditions into complete agreement, as indicated by the following excerpt from a modified descriptor.

```
...
        region 1                    eps= eps0                    { Vacuum }
start(-Lx,-Ly)
value(U)= U2_ex   line to (Lx,-Ly)  to (Lx,Ly)  to (-Lx,Ly) to finish
...
contour[ ustep( x**2+y**2- 1.1*r0**2)*(Ex-E2_ex)]  as 'Error in Ex'
...
```

Now we have copied the boundary values from the harmonic solution and have every reason to expect excellent agreement. In order to display the error more clearly we use a new device to eliminate the large values associated with the dielectric region and its immediate vicinity. We simply multiply the difference to be plotted by a discontinuous function, *ustep*(x), which is unity for $x>1$ and else zero. Using this step function we arrange to exclude the high values inside a circle around the dielectric rod. As shown in the following plot we now obtain irregular contours of zero value over most of the region.

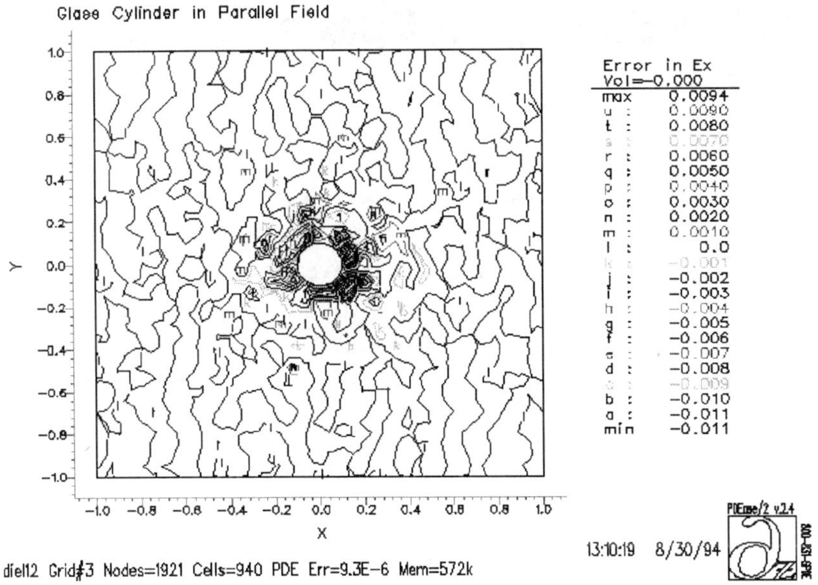

Exercises

❑ What will the fields look like, if you replace the glass in the capacitor (p.65) with vacuum, or a dielectric with very high ε?

❑ Investigate how the capacitance value changes if you extend the domain around the capacitor (p.65) to 2×2 m^2, then to 10×10 m^2, using zoom to inspect the fields of the original domain.

❑ Make two *elevation* plots which will permit you to calculate the charge on the interior of the capacitor and the remainder of the charge.

❑ Show by plots that a very large value of ε in a dielectric rod simulates a metallic cylinder.

Steady Fields in (ρ,z) Space

The present version of PDEase only handles problems in two spatial dimensions. If the geometry, the materials and the boundary conditions all are axially symmetric, however, any cross-section through the axis of symmetry would be equivalent. This permits us to solve a 3D problem in two cylindrical coordinates (ρ,z) and present the results in a plane.

Electrical Conduction in a Cone

Our first example of an axially symmetric object is a truncated, circular cone of Si with metallic electrodes of high conductivity connected to the flat ends, such that we may consider the ends to have constant potential. A potential difference of 1.0 V is applied, and we wish to calculate the current distribution and the resistance between the ends. The following descriptor specifies that problem.

```
...
select
  errlim= 1e-4                    contours= 10
  VECTORGRID= 10                  surfacegrid= 30       paint= off
  ALIAS(x)= 'Radius'              alias(y)= 'Z'
variables
  U
definitions
  mm= 1e-3     r0= 1.0*mm       r1= 10*mm        h= 10*mm
  cond= 1.0e-3                                                  { Silicon }
  Er= -dx(U)                Ez= -dy(U)
  jr= cond*Er               jz= cond*Ez          jabs= sqrt(jr**2+jz**2)
  small= 1e-15
equations
  (1/x)* dx( -x*cond*dx(U))+ dy( -cond*dy(U))= 0
boundaries
          region 1
```

```
  start(0,0)
  value(U) = 0.0          line to (r0,0)
  natural(U)= 0           line to (r1,h)
  value(U)= 1.0           line to (0,h)
  natural(U)= 0           line to finish
monitors
  contour(U)
plots
  contour( U)             contour( Er)        contour( Ez)
  contour[ log10( abs(Ez)+ small)]
  vector( jr/jabs, jz/jabs) as 'Direction of j'
  elevation( 2*pi*x*jz)  from (0,0) to (r0,0) as 'Current'
  elevation( 2*pi*x*jz)  from (0,h) to (r1,h) as 'Current'
end
```

In the select segment of the above list we encounter the new keyword *alias*, which provides an optional means of changing notations in the plot labels. PDEase always works internally with the

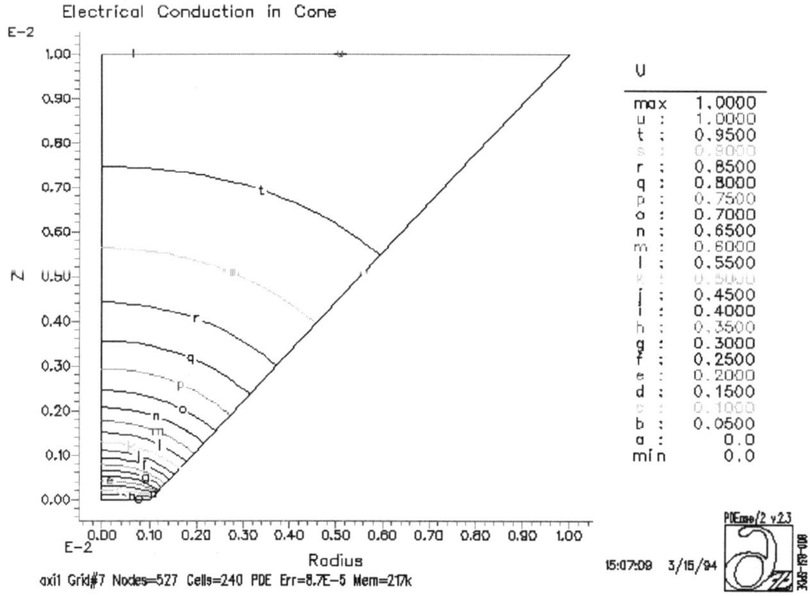

space coordinates x and y, even if we specify the problem in terms of cylindrical coordinates. In order to avoid confusion when interpreting the results, we may thus introduce more familiar labels.

In ρ,z coordinates the gradient expression becomes similar[2p705], but $\nabla \cdot \mathbf{E}$ takes a different form, as seen in the first term of the equations segment.

The contour plot of U above indicates that the axially symmetric surfaces of constant potential are nearly spherical, except close to the plane ends. We could use this fact to generate an approximate solution over the region close to the axis.

In order to calculate the current traversing an end surface one needs to integrate j_z over a circular area, consisting of annular elements of size $2\pi\rho dr$. Here we make use of the integration routine included in the *elevation* plot, which yields the area under the curve. From the current obtained from the next plot you easily calculate the resistance.

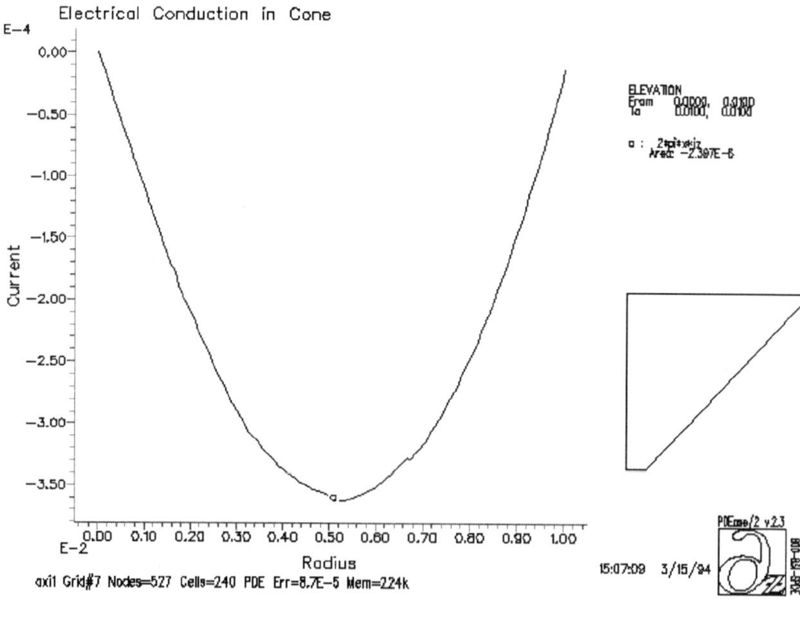

Plate Capacitor in (ρ,z) Space

In the previous chapter we studied a plate capacitor in 2D. We may now extend the calculations to a capacitor having two circular plates, using the following descriptor.

```
  ...
select
  errlim= 5e-4            contours= 10
  surfacegrid= 30         paint= off
  alias(x)= 'Radius'      alias(y)= 'Z'
variables
  U
definitions
  Lx= 2     Ly= 1     r0= 0.7          d= 0.2     dely= 0.2*d
  Er= -dx(U)       Ez= -dy(U)          Eabs= sqrt(Er**2+Ez**2)
  eps0= 8.854e-12
  eps                                              { Reserved for later }
  DEr= eps*Er     DEz= eps*Ez     Dabs= sqrt(DEr**2+DEz**2)
  small= 1e-20
equations
  (1/x)* dx( -x*eps*dx(U))+ dy( -eps*dy(U))= 0
boundaries
               region 1          eps= eps0                    { Vacuum }
  start(0,-Ly)  natural(U)= 0
  line to (Lx,-Ly) to (Lx,Ly)  to (0,Ly) to (0,d/2+dely)
  value(U)= 1.0                                         { Upper plate }
  line to (r0,d/2+dely) to (r0,d/2) to (0, d/2)
  natural(U)= 0
  line to (0,-d/2)
  value(U)= 0.0                                         { Lower plate }
  line to (r0,-d/2) to (r0,-d/2-dely) to (0,-d/2-dely)
  natural(U)= 0
  line to finish
               region 2     eps= 7.0* eps0                     { Glass }
  start(0,-d/2)  line to (r0,-d/2)
  to (r0,d/2) to (0,d/2) to finish
monitors
  contour(U)
plots
  contour( U)      contour( Er)       contour( Ez)
```

```
  elevation( Ez) from (0,0) to (Lx,0)
  elevation( 2*pi*x* DEz)  from (0,0) to (Lx,0)
  contour( log10( Eabs+ small))            contour(log10( Dabs+ small))
  vector( DEr/Dabs, DEz/Dabs)  as 'Direction of D'
  surface( log10( Dabs+ small))  interactive
end
```

In this problem we are dealing with a cross-section of the field, starting at the axis of symmetry. The ρ-axis is horizontal and the z-axis vertical. The surfaces of constant potential cut the plane of the figure along the curves in the plot below. In other words, you may generate the potential surfaces by rotating the contours around the vertical axis.

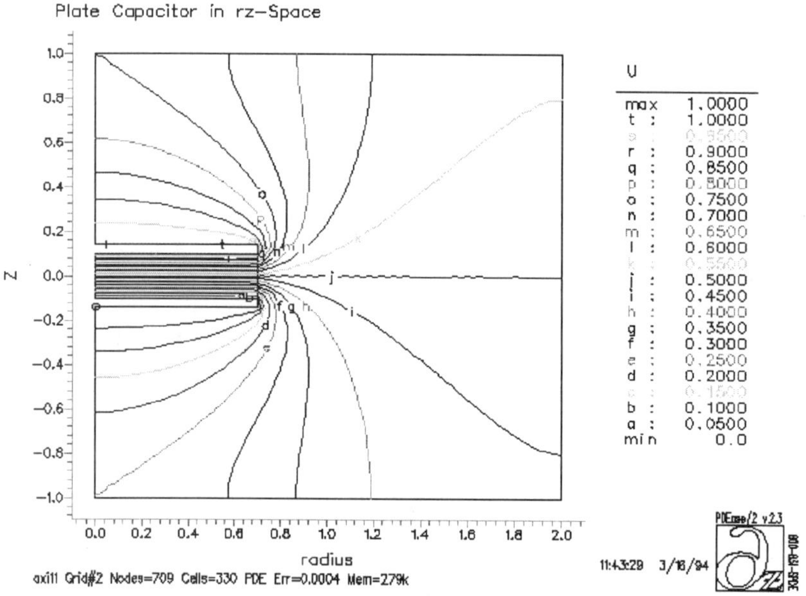

The following *elevation* plot displays the flux of D_z through annular area elements, the integral being equal to the total charge on a capacitor plate. From this charge and the voltage you directly obtain the capacitance, to be compared with what you obtain from the elementary capacitor formula.

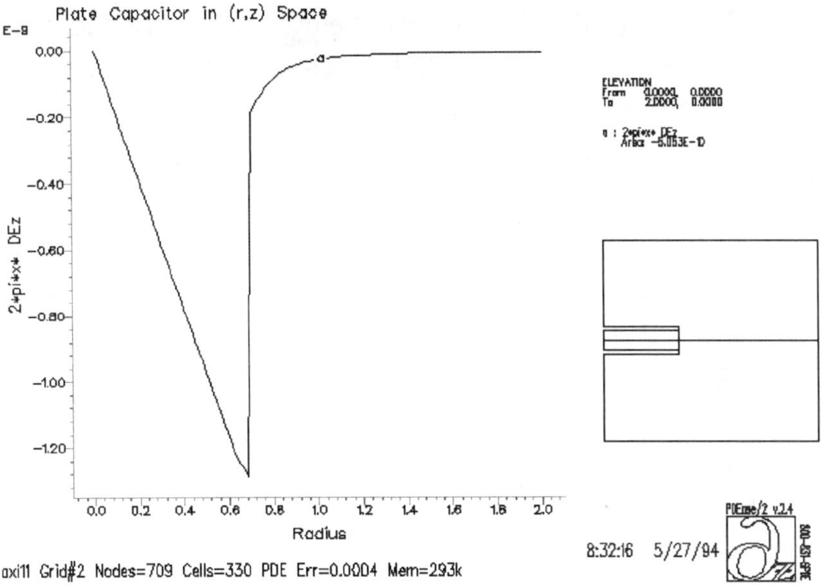

Two Metallic Spheres in Glass

The cylindrical coordinate system also permits us to treat problems with spheres on a common axis. The following list is for two metallic spheres, immersed in glass. The spheres are kept at potentials of 0.5 V and -0.5 V.

```
Lx= 1.5          r0= 0.08* Lx     Lc= 2*r0        Ly= 1       U0= 0.5
...
                 region 1   eps= 7.0*eps0                     { Glass }
start(0,-Ly)
natural(U)= 0     line to (Lx,-Ly) to (Lx,Ly)  to (0,Ly) to (0,Lc+r0)
value(U)= U0      arc to (r0,Lc) to (0,Lc-r0)
                                                              { Sphere }
natural(U)= 0     line to (0,-Lc+r0)
value(U)= -U0     arc to (r0,-Lc) to (0,-Lc-r0)               { Sphere }
natural(U)= 0     line to finish
...
elevation(2*pi*x* DEz) from (0,0) to (Lx,0)                   { Flux of D }
```

The resulting plot of $|\mathbf{D}|$ is shown below. Evidently the field strength at the upper, lower and right boundaries is only about 1/100 of that between the spheres, and hence we can feel assured that the outer constraint on the field does not seriously perturb the charge. Of course we may test, by direct computation, the effect of extending the boundaries.

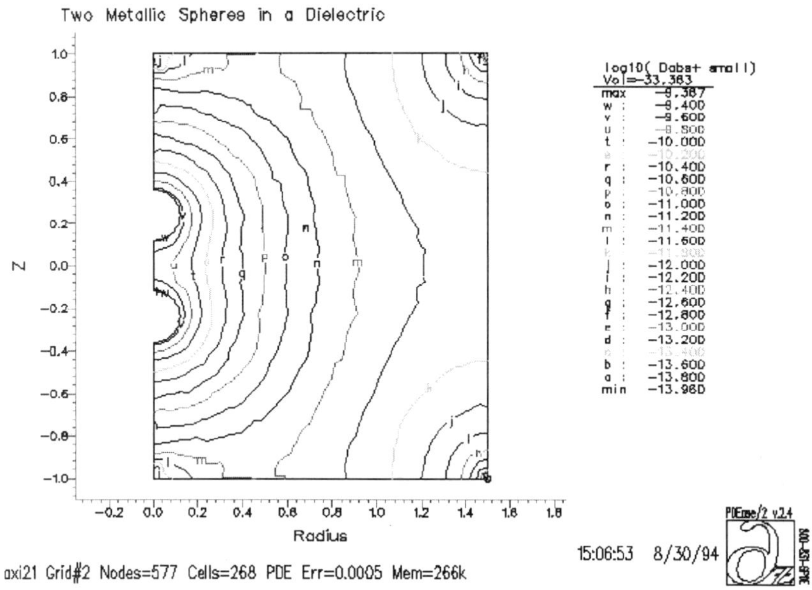

Glass Sphere in a Parallel Field

The next application concerns a dielectric sphere in an (initially) parallel field of strength *E*0. We create this field by imposing voltages of -*U*0 and *U*0 to the horizontal boundaries as shown in the partial descriptor below.

```
...
errlim= 1e-5     vectorgrid= 20
...
Lx= 1.5          r0= 0.1          Ly= 1.0
Er= -dx(U)       Ez= -dy(U)       Eabs= sqrt(Er**2+Ez**2)
```

```
eps0= 8.854e-12      epsr         epsr2= 7            eps= epsr*eps0
U0= 1.0           E0= -2*U0/(2*Ly)        DE0= eps*E0           { Far field }
DEr= eps*Er    DEz= eps*Ez     Dabs= sqrt(DEr**2+DEz**2)
rad= sqrt(x**2+y**2)          outer= ustep(rad-1.05*r0)
U1_ex= -3*E0/(epsr+2)*y                                    { Exact solutions }
U2_ex= -E0*y*[1- (epsr2-1)/(epsr2+2)* ( r0/rad)**3]
E2r_ex= -dx(U2_ex)                   E2z_ex= -dy(U2_ex)
E2abs_ex= sqrt(E2r_ex**2+ E2z_ex**2)
...
          region 1     epsr= 1                                       { Vacuum }
start(0,-Ly)
value(U)= -U0          line to (Lx,-Ly)
natural(U)= 0          line to (Lx,Ly)
value(U)= U0           line to (0,Ly)
natural(U)= 0          line to finish
          region 2     epsr= 7.0                                     { Glass }
start(0,r0)            line to (0,-r0)
arc to (r0,0) to finish
...
contour( U)
vector( Er/Eabs, Ez/Eabs) as 'Direction of E'
vector( Er/Eabs, Ez/Eabs) zoom(0,0, 2*r0,2*r0)

surface( Eabs/E0)  zoom(0,0, 3*r0,3*r0) interactive
surface[ outer* (Eabs-E2abs_ex)/E0]  zoom(0,0, 3*r0,3*r0) interactive
contour[ outer* (Eabs-E2abs_ex)/E0]  zoom(0,0, 3*r0,3*r0)
end
```

According to the linear *contour* plot and the *vector* plot of U the far field appears to be negligibly perturbed by the presence of the sphere. In fact, the influence of the dielectric is definitely smaller at the outer boundary than in the case of the dielectric rod (p.70).

The zoomed *vector* plot shows that the field is parallel inside the sphere, which is also predicted by theory. The zoomed *surface* plot of *Eabs*/*E*0, shown suitable rotated below, brings out the perturbation of the field near the sphere, as well as the constant value inside the volume.

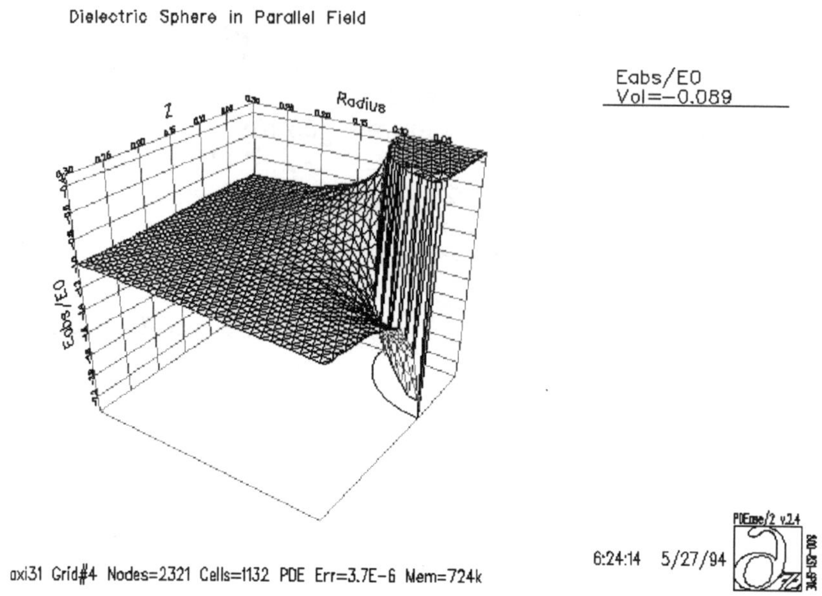

There is an exact solution to a very similar problem[3p191], that of a dielectric sphere in a field which is parallel and of constant strength as $z \to \pm\infty$. As we see from the descriptor there are two solutions, *U1_ex* inside the sphere and *U2_ex* outside. The outside field rapidly approaches a linear term in *z*, corresponding to a homogeneous field. In order to compare our solution to the exact expression for the outer field we exclude the volume of the sphere, and its noisy proximity, by the factor *outer*. This factor is unity outside a radius 10% larger than the radius of the sphere, and when plotting the error we hence obtain more closely spaced *contour* levels.

The following figure is a *contour* plot of the relative error in the outside field. There are evidently many irregular zero contours, which indicate that the error there is smaller than 0.1%. Larger values appear as we come closer to the interface. The accuracy of the solution may of course be further improved by choosing a smaller value of *errlim*. We might also use a solution domain of larger size and zoom the plots to fit the original boundaries.

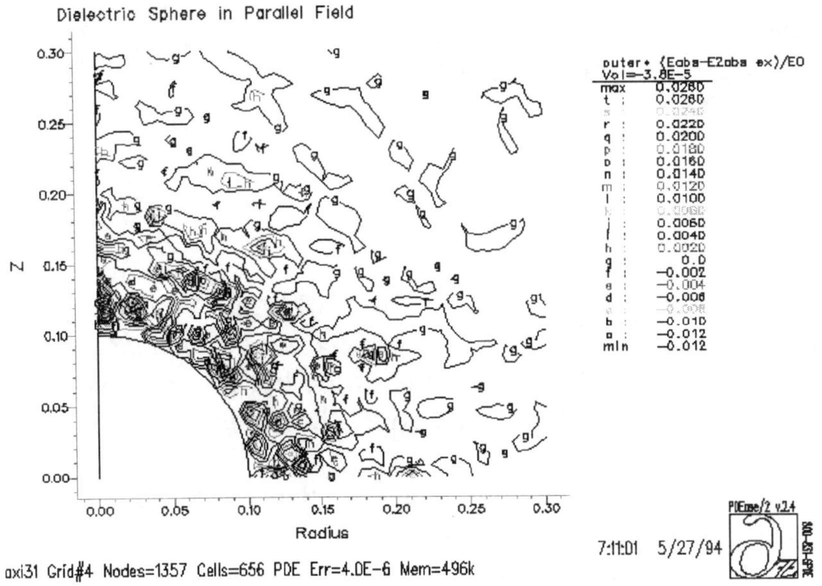

The Gauss Integral

The following relation[3p324] holds for the field **D** and the charge Q enclosed by the surface S

$$\oiint_S \mathbf{D} \cdot d\mathbf{s} = Q .$$

Here the vector $d\mathbf{s}$ is directed along the outward normal of the surface and its magnitude is equal to the area of the element. Recent versions of PDEase permit us to calculate such integrals over cylindrical surfaces of various shapes. We may define the field vector by

```
D= -eps*grad( U)
```

and then request an elevation plot over the surface S, given as a *feature* (p.64) in the plane of the screen, i.e.

```
elevation( normal( D)) on 'curve'
```

in (x,y) space or

```
elevation( 2*pi*x*normal( D)) on 'curve'
```

in (ρ,z) space.

Exercises

❑ Modify the cone descriptor using *r*0= 0.01*mm. You may then find that PDEase no longer can create the necessary triangular grid. If you create a region 2 of height 0.5 mm at the tip, however, without specifying any new materials properties, the program will finish normally.

❑ You found that the charge on the circular-plate capacitor is higher than expected from elementary theory. What does the elementary model leave out, and how can you calculate that part separately using an elevation plot?

❑ Compare the two equivalent expressions for electrostatic energy in the case of the circular-plate capacitor.

❑ Find the capacitance between the two spheres already treated.

❑ Use the Gauss integral to evaluate the charge on one of the two metal spheres. Define D as a vector, and create a spherical surface, 10% larger than that of the ball but with the same center. Make an elevation plot of 2*pi*x*normal(D) and compare to the integral of 2*pi*x*DEz which appears in the original descriptor.

❑ Exploit the symmetry of the problem of the dielectric sphere to solve only for the upper half of the domain (for positive *z*). Show that the dielectric ball behaves like a metallic one as *epsr* becomes very large.

❑ Write a descriptor for a charged metallic sphere in a dielectric medium, specifying a boundary condition for **D**. Compare the solution to the exact expression.

Magnetostatics in (*x*,*y*) Space

From Maxwell's equations[3p323] we know that

$$\nabla \times \mathbf{H} = \mathbf{J} + \frac{\partial \mathbf{D}}{\partial t},$$ •

which in the static case simply becomes $\nabla \times \mathbf{H} = \mathbf{J}$. Combining this relation with $\mathbf{B} = \mu\mathbf{H} = \mu_r \mu_o \mathbf{H}$ we also obtain

$$\nabla \times \left(\frac{\mathbf{B}}{\mu} \right) = \mathbf{J}$$ •

or, in more detail,

$$\nabla \times \left(\frac{\mathbf{B}}{\mu} \right) = \begin{vmatrix} \mathbf{i} & \mathbf{j} & \mathbf{k} \\ \frac{\partial}{\partial x} & \frac{\partial}{\partial y} & \frac{\partial}{\partial z} \\ \frac{B_x}{\mu} & \frac{B_y}{\mu} & \frac{B_z}{\mu} \end{vmatrix} = \begin{Bmatrix} \frac{\partial}{\partial y}\left(\frac{B_z}{\mu} \right) - \frac{\partial}{\partial z}\left(\frac{B_y}{\mu} \right) \\ \frac{\partial}{\partial z}\left(\frac{B_x}{\mu} \right) - \frac{\partial}{\partial x}\left(\frac{B_z}{\mu} \right) \\ \frac{\partial}{\partial x}\left(\frac{B_y}{\mu} \right) - \frac{\partial}{\partial y}\left(\frac{B_x}{\mu} \right) \end{Bmatrix} = \mathbf{J}$$

where we have kept the factor $1/\mu$ inside the parentheses, considering that it may vary with the spatial coordinates.

Electromagnetic theory[3p232] shows that a vector function $\mathbf{A}(\mathbf{r})$ exists, such that

$$\mathbf{B} = \begin{Bmatrix} B_x \\ B_y \\ B_z \end{Bmatrix} = \nabla \times \mathbf{A} = \begin{vmatrix} \mathbf{i} & \mathbf{j} & \mathbf{k} \\ \frac{\partial}{\partial x} & \frac{\partial}{\partial y} & \frac{\partial}{\partial z} \\ A_x & A_y & A_z \end{vmatrix} = \begin{Bmatrix} \frac{\partial A_z}{\partial y} - \frac{\partial A_y}{\partial z} \\ \frac{\partial A_x}{\partial z} - \frac{\partial A_z}{\partial x} \\ \frac{\partial A_y}{\partial x} - \frac{\partial A_x}{\partial y} \end{Bmatrix}.$$ •

We wish to study fields **B** with directions confined to the *xy* plane. As seen from the third vector equation above, A_x and A_y may produce a perpendicular component, B_z, and hence we keep only one component of the vector potential, i.e. A_z. The expression then reduces to

$$\begin{Bmatrix} B_x \\ B_y \\ B_z \end{Bmatrix} = \begin{Bmatrix} \dfrac{\partial A_z}{\partial y} \\ -\dfrac{\partial A_z}{\partial x} \\ 0 \end{Bmatrix}$$

Putting these expressions for the components of **B** into the equation for **J**, we have the result

$$\begin{vmatrix} \mathbf{i} & \mathbf{j} & \mathbf{k} \\ \dfrac{\partial}{\partial x} & \dfrac{\partial}{\partial y} & \dfrac{\partial}{\partial z} \\ \dfrac{1}{\mu}\dfrac{\partial A_z}{\partial y} & -\dfrac{1}{\mu}\dfrac{\partial A_z}{\partial x} & 0 \end{vmatrix} = \begin{Bmatrix} -\dfrac{\partial}{\partial z}\left(-\dfrac{1}{\mu}\dfrac{\partial A_z}{\partial x}\right) \\ \dfrac{\partial}{\partial z}\left(\dfrac{1}{\mu}\dfrac{\partial A_z}{\partial y}\right) \\ \dfrac{\partial}{\partial x}\left(-\dfrac{1}{\mu}\dfrac{\partial A_z}{\partial x}\right) - \dfrac{\partial}{\partial y}\left(\dfrac{1}{\mu}\dfrac{\partial A_z}{\partial y}\right) \end{Bmatrix} = \begin{Bmatrix} J_x \\ J_y \\ J_z \end{Bmatrix}$$

The first two vector components may be taken to be zero, if we assume that nothing varies along the *z*-axis. The remaining component yields the following PDE

$$\frac{\partial}{\partial x}\left(\frac{1}{\mu}\frac{\partial A_z}{\partial x}\right) + \frac{\partial}{\partial y}\left(\frac{1}{\mu}\frac{\partial A_z}{\partial y}\right) + J_z = 0 \qquad \bullet$$

After having solved this auxiliary equation for A_z we easily compute the magnetic induction from $\mathbf{B} = \nabla \times \mathbf{A}$.

Magnetic Field Around Wire

Our first application will be to the magnetic field in a plane perpendicular to a long wire of circular cross-section, carrying a uniform current density $J_z = 1.0$. We assume the permeability to be that of free space (μ_o). In the descriptor below, the components of **B** are denoted *Bxx* and *Byy*, for the simple reason that *By* would be equivalent to "*by*", which is a reserved word in PDEase.

If your PC system has a 486DX or Pentium processor you could now try removing the plot restrictions *contourgrid*, *vectorgrid* and *surfacegrid* to see more details.

```
  ...
select
  errlim= 1e-3          contours= 10          paint= off
variables
  Az                                    { Magnetic vector potential }
definitions
  mu0= 4*pi*1e-7    mu= mu0      J                { Current density }
  r0= 0.2          r1= 1.0
  rad= sqrt(x**2+ y**2)
  Bxx= dy(Az)       Byy= -dx(Az)          { x and y components of B }
  Babs= sqrt(Bxx**2+ Byy**2)                      { Magnitude of B }
  J0= 1.0                                      { Exact solution ... }
  B1_ex= mu/(2*pi)* (J0*pi* rad**2)/rad
  B2_ex= mu/(2*pi)* (J0*pi* r0**2)/rad
equations
  dx( (1/mu)* dx(Az))+ dy( (1/mu)* dy(Az))+ J = 0
boundaries
        region 1         J= 0
  start(-r1,0)     value(Az) = 0
  arc to (0,-r1) to (r1,0) to (0,r1) to finish
        region 2         J= 1.0                   { Interior of wire }
  start (-r0,0)
  arc to (0,-r0) to (r0,0) to (0,r0) to finish
monitors
  contour(Az)
plots
  contour( Az)  as 'Vector potential Az'
  contour( Babs)
  vector( Bxx, Byy)
  vector( Bxx, Byy)  zoom(-r0,-r0,2*r0,2*r0)
  elevation( Babs, B1_ex)  from (0,0) to (r0,0)
  elevation( Babs, B2_ex)  from (r0,0) to (r1,0)
  surface( Az)   as 'Vector potential Az'
end
```

The condition imposed at the outer boundary, *value*(Az)=0, should perhaps be explained. In view of the relation $\mathbf{B} = \nabla \times \mathbf{A}$, the component B_y may be written $-\partial(A_z)/\partial x$. Consequently, if the vector potential is constant along the x-axis, the normal component B_y becomes zero. This remains true at any point of the outer circle, if we place an x-axis tangentially to the boundary at that point. Hence, a constant value on the boundary gives us a vanishing normal component of **B**.

In this simple case the vector potential has only one component, and the figure below shows a *surface* plot of it. The region of the wire is marked by a horizontal circle near the maximum.

The figure on the next page shows the direction of **B** and also gives a rough indication about the field strength. The field is perpendicular to the radius everywhere and takes its largest value at the edge of the wire.

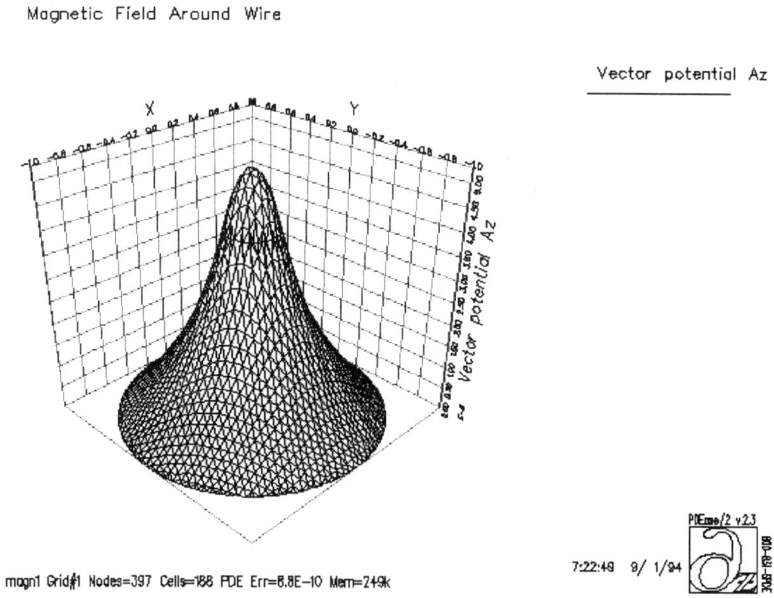

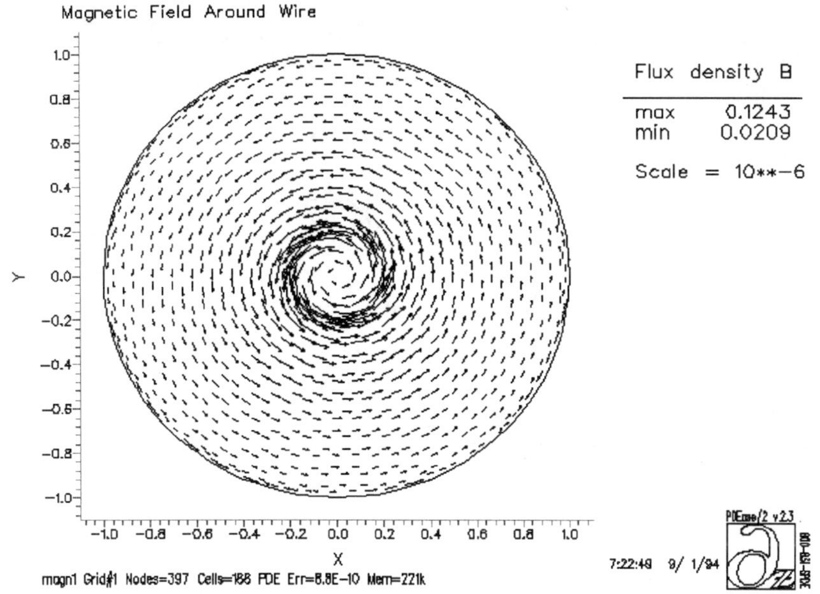

The simple problem of a straight wire has an exact solution which we obtain by the equation[3p228]

$$\oint \mathbf{B} \cdot d\mathbf{s} = \mu_0 I$$

where I means the current enclosed by the contour of integration. Making use of the symmetry of rotation we integrate **B** along concentric circles. The expression for the current becomes $I = J_z \pi r^2$ inside the wire, whereas it takes the constant value $I = J_z \pi r_o^2$ for field points outside. Since there are two different expressions, *B1_ex* and *B2_ex*, for the two regions of space, we compare the solutions in separate *elevation* plots.

Field Around Two Wires

We next consider the case of two wires, conducting current in opposite directions, as in the following descriptor.

```
...
definitions
  mu0= 4*pi*1e-7      mu= mu0          J                    { Current density }
  r0= 0.1             dr0= 1.5*r0     r1= 1.0
  ...
              region 1            J= 0
  start(-r1,0)      value(Az) = 0
  arc to (0,-r1) to (r1,0) to (0,r1) to finish
          region 2            J= -1.0
  start (-dr0,-r0)
  arc to (-dr0+r0,0) to (-dr0,r0) to (-dr0-r0,0) to finish
              region 3            J= 1.0
  start (dr0,-r0)
  arc to (dr0+r0,0) to (dr0,r0) to (dr0-r0,0) to finish
  ...
  contour( Babs)  zoom(-2*r0,-2*r0,4*r0,4*r0)
  vector( Bxx/Babs, Byy/Babs)  as 'Flux density B'
  vector( Bxx/Babs, Byy/Babs)  zoom(-2*r0,-2*r0,4*r0,4*r0)
end
```

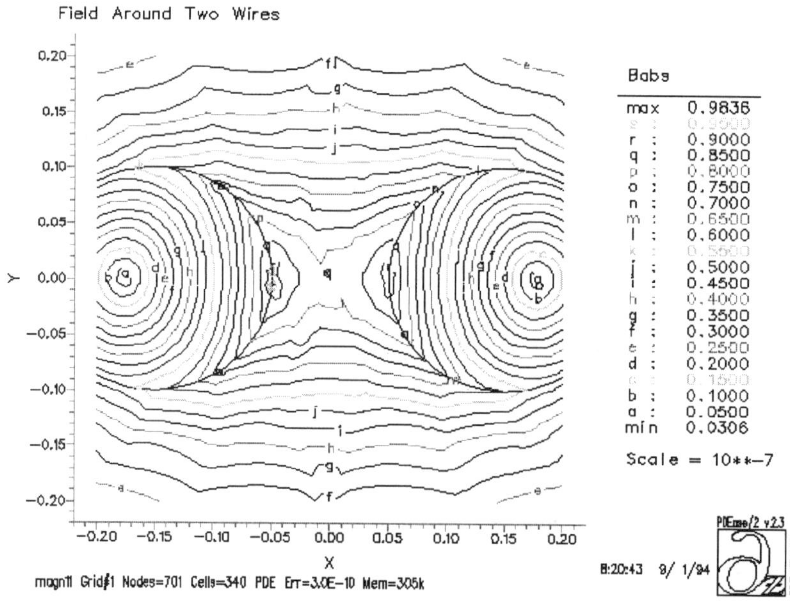

The zoomed *contour* plot above reveals the location of the largest magnitude of **B** at the inner edges of the wires.

Two Wires With a Ferromagnetic Shield

We now install a tube of ferromagnetic material around the two wires, as shown in the descriptor excerpt below.

```
...
r01= 0.5          r02= 0.55        r1= 1.0
...
small= 1e-30
...
region 4   J= 0    mu= 2500                                  { Ferrite }
start (0,-r02)
arc to (r02,0) to (0,r02) to (-r02,0) to finish
start (0,-r01)
arc to (-r01,0) to (0,r01) to (r01,0) to finish
...
contour( log10(Babs+ small))
contour( log10(Babs+ small)) zoom(-0.3,0.45,0.6,0.15)
elevation( Babs) from (-r1,0) to (r1,0)
surface( Babs) interactive          end
```

The following *surface* plot indicates the magnitude of the field both inside and outside the magnetic shield. Changing the viewpoint you may explore the variation inside the shielding tube in more detail.

In order to take a closer look at *Babs* in the various regions we create an *elevation* plot, also shown below. Clearly the field is extremely weak outside the shield, and this is further evidenced by the logarithmic plots. Accordingly, we should be warned not to take the field pattern pertaining to the outside region too seriously.

The zoomed plot presents some interesting details about the field distribution at the top edge of the tube.

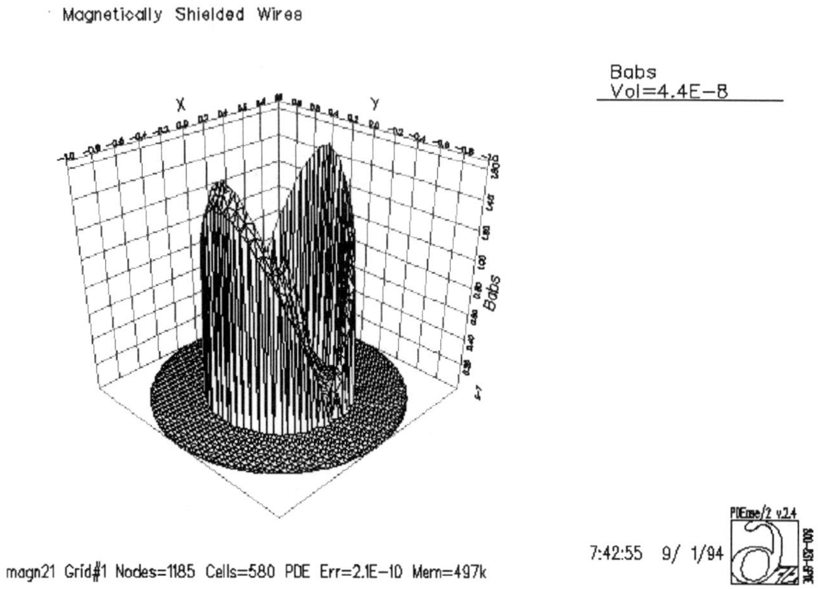
Magnetically Shielded Wires
Babs
Vol=4.4E-8
X
Y
Babs
7:42:55 9/ 1/94
PDEase/2 v2.4
magn21 Grid#1 Nodes=1185 Cells=580 PDE Err=2.1E-10 Mem=497k

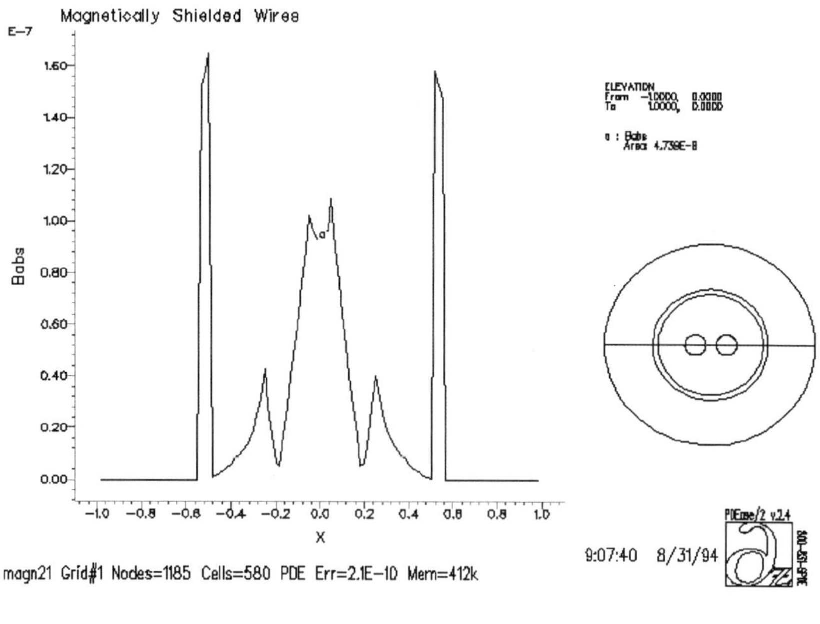
Magnetically Shielded Wires
E-7
1.60
1.40
1.20
1.00
0.80
0.60
0.40
0.20
0.00
Babs
-1.0 -0.8 -0.6 -0.4 -0.2 0.0 0.2 0.4 0.6 0.8 1.0
X
ELEVATION
From -1.0000, 0.0000
To 1.0000, 0.0000
a : Babs
Area 4.739E-8
9:07:40 8/31/94
PDEase/2 v2.4
magn21 Grid#1 Nodes=1185 Cells=580 PDE Err=2.1E-10 Mem=412k

Exercises

❑ Make contour plots of div(**B**) and $curl_z$(**B**) using the exact expression *B2_ex* in the descriptor for a single wire.
❑ Modify the descriptor for the two wires by putting *r*0=0.01 to investigate the far field. Compare with the 2D electric dipole (p.39).
❑ Zoom on regions outside the magnetic shield to find the residual magnitude of **B**.

Magnetostatics in (ρ,z) Space

In cylindrical coordinates (ρ, φ, z) the expression for the field $\mathbf{B}$ (p.85) becomes[3p705]

$$\mathbf{B} \equiv \begin{Bmatrix} B_\rho \\ B_\varphi \\ B_z \end{Bmatrix} = \nabla \times \mathbf{A} = \begin{vmatrix} \frac{1}{\rho}\mathbf{e}_\rho & \mathbf{e}_\varphi & \frac{1}{\rho}\mathbf{e}_z \\ \frac{\partial}{\partial \rho} & \frac{\partial}{\partial \varphi} & \frac{\partial}{\partial z} \\ A_\rho & \rho A_\varphi & A_z \end{vmatrix} = \begin{Bmatrix} \frac{1}{\rho}\frac{\partial A_z}{\partial \varphi} - \frac{\partial A_\varphi}{\partial z} \\ \frac{\partial A_\rho}{\partial z} - \frac{\partial A_z}{\partial \rho} \\ \frac{1}{\rho}\frac{\partial(\rho A_\varphi)}{\partial \rho} - \frac{1}{\rho}\frac{\partial A_\rho}{\partial \varphi} \end{Bmatrix}$$

and similarly for the relation between $\mathbf{B}$ and $\mathbf{J}$

$$\nabla \times \left(\frac{\mathbf{B}}{\mu}\right) = \begin{vmatrix} \frac{1}{\rho}\mathbf{e}_\rho & \mathbf{e}_\varphi & \frac{1}{\rho}\mathbf{e}_z \\ \frac{\partial}{\partial \rho} & \frac{\partial}{\partial \varphi} & \frac{\partial}{\partial z} \\ \frac{B_\rho}{\mu} & \rho\frac{B_\varphi}{\mu} & \frac{B_z}{\mu} \end{vmatrix} = \mathbf{J}$$

Substituting the expression for $\mathbf{B}$ into the equation for $\mathbf{J}$ we obtain

$$\nabla \times \left(\frac{\mathbf{B}}{\mu}\right) =$$

$$\begin{vmatrix} \frac{1}{\rho}\mathbf{e}_\rho & \mathbf{e}_\varphi & \frac{1}{\rho}\mathbf{e}_z \\ \frac{\partial}{\partial \rho} & \frac{\partial}{\partial \varphi} & \frac{\partial}{\partial z} \\ \frac{1}{\mu}\left(\frac{1}{\rho}\frac{\partial A_z}{\partial \varphi} - \frac{\partial A_\varphi}{\partial z}\right) & \frac{\rho}{\mu}\left(\frac{\partial A_\rho}{\partial z} - \frac{\partial A_z}{\partial \rho}\right) & \frac{1}{\mu}\left(\frac{1}{\rho}\frac{\partial(\rho A_\varphi)}{\partial \rho} - \frac{1}{\rho}\frac{\partial A_\rho}{\partial \varphi}\right) \end{vmatrix} = \mathbf{J}$$

or after expansion of the determinant

$$\begin{Bmatrix} \dfrac{1}{\rho}\dfrac{\partial}{\partial\varphi}\left(\dfrac{1}{\mu\rho}\dfrac{\partial(\rho A_\varphi)}{\partial\rho}-\dfrac{1}{\mu\rho}\dfrac{\partial A_\rho}{\partial\varphi}\right)-\dfrac{1}{\rho}\dfrac{\partial}{\partial z}\left(\dfrac{\rho}{\mu}\dfrac{\partial A_\rho}{\partial z}-\dfrac{\rho}{\mu}\dfrac{\partial A_z}{\partial\rho}\right) \\ \dfrac{\partial}{\partial z}\left(\dfrac{1}{\mu\rho}\dfrac{\partial A_z}{\partial\varphi}-\dfrac{1}{\mu}\dfrac{\partial A_\varphi}{\partial z}\right)-\dfrac{\partial}{\partial\rho}\left(\dfrac{1}{\mu\rho}\dfrac{\partial(\rho A_\varphi)}{\partial\rho}-\dfrac{1}{\mu\rho}\dfrac{\partial A_\rho}{\partial\varphi}\right) \\ \dfrac{1}{\rho}\dfrac{\partial}{\partial\rho}\left(\dfrac{\rho}{\mu}\dfrac{\partial A_\rho}{\partial z}-\dfrac{\rho}{\mu}\dfrac{\partial A_z}{\partial r}\right)-\dfrac{1}{\rho}\dfrac{\partial}{\partial\varphi}\left(\dfrac{1}{\mu\rho}\dfrac{\partial A_z}{\partial\varphi}-\dfrac{1}{\mu}\dfrac{\partial A_\varphi}{\partial z}\right) \end{Bmatrix} = \begin{Bmatrix} J_\rho \\ J_\varphi \\ J_z \end{Bmatrix}$$

In the case of axial symmetry, $A_\rho = A_z = 0$, this equation takes the special form

$$\begin{Bmatrix} \dfrac{1}{\rho}\dfrac{\partial}{\partial\varphi}\left(\dfrac{1}{\mu\rho}\dfrac{\partial(\rho A_\varphi)}{\partial\rho}\right) \\ \dfrac{\partial}{\partial z}\left(-\dfrac{1}{\mu}\dfrac{\partial A_\varphi}{\partial z}\right)-\dfrac{\partial}{\partial\rho}\left(\dfrac{1}{\mu\rho}\dfrac{\partial(\rho A_\varphi)}{\partial\rho}\right) \\ -\dfrac{1}{\rho}\dfrac{\partial}{\partial\varphi}\left(-\dfrac{1}{\mu}\dfrac{\partial A_\varphi}{\partial z}\right) \end{Bmatrix} = \begin{Bmatrix} J_\rho \\ J_\varphi \\ J_z \end{Bmatrix}$$

which shows that rotational symmetry about the z-axis requires $J_\rho = J_z = 0$. The relation between the only non-zero components now becomes

$$\frac{\partial}{\partial z}\left(\frac{1}{\mu}\frac{\partial A_\varphi}{\partial z}\right)+\frac{\partial}{\partial\rho}\left(\frac{1}{\mu\rho}\frac{\partial(\rho A_\varphi)}{\partial\rho}\right)+J_\varphi = 0, \qquad \bullet$$

which is similar to the Poisson equation we already used.

Simple Magnet Coil

Is there a practical device where an electrical current always runs in the direction of increasing or decreasing φ ? One single circular turn of metal wire would be completely symmetric with respect to an axis perpendicular to the circle and going through its center, but how could we drive an electrical current through it? A good approximation to a single turn is a number of layers of wire, wound on the surface of a circular cylinder. In the ρ,z plane we shall only see the cross section of this coil.

In the descriptor below we have chosen *value*(*A_phi*)=0 for the outer boundary. Just as in the previous chapter we wish the normal field component B_n across the boundary to vanish, and this we achieve by imposing a constant value for the potential.

```
...                                                        { magn31.pde }
select
  errlim= 1e-3            contours= 20
  contourgrid= 30         vectorgrid= 30
  surfacegrid= 30         paint= off
  alias(x)= 'Radius'      alias(y)= 'Z'
variables
  A_phi
definitions
  r1= 0.1                 r2= 0.2                 r3= 2.0
  L12= 0.2                L3= 1.0
  mu0= 4*pi*1e-7          mu= mu0                 J
  Br= -dy(A_phi)                   Bz= (1/x)* dx(x*A_phi)
  Babs= sqrt(Br**2+Bz**2)
  small= 1e-30
equations
  dy[ (1/mu)*dy(A_phi)]+ dx[1/(mu*x)* dx(x*A_phi)]+ J= 0
boundaries
          region 1      J= 0
  start(0,-L3)            value(A_phi)= 0
  line to (r3,-L3)   to (r3,L3)   to (0,L3) to finish
          region 2      J= 1                       { Coil cross-section }
  start(r1,-L12)
  line to (r2,-L12) to (r2,L12) to (r1,L12) to finish
monitors
  contour(A_phi)
plots
  contour( A_phi)
  contour( Babs)                contour( log10(Babs+ small))
  contour( Br)                  contour( Bz)
  vector( Br/Babs, Bz/Babs)
end
```

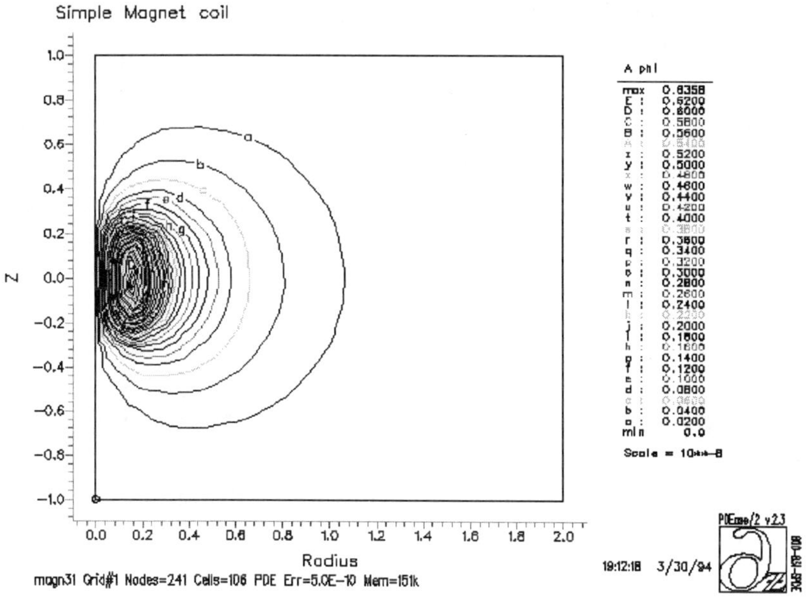
Simple Magnet coil
Z
Radius
A phi
max 0.6358
min 0.0
Scale = 10**-8
19:12:18 3/30/94
PDEase/2 v2.3
magn31 Grid#1 Nodes=241 Cells=106 PDE Err=5.0E-10 Mem=151k

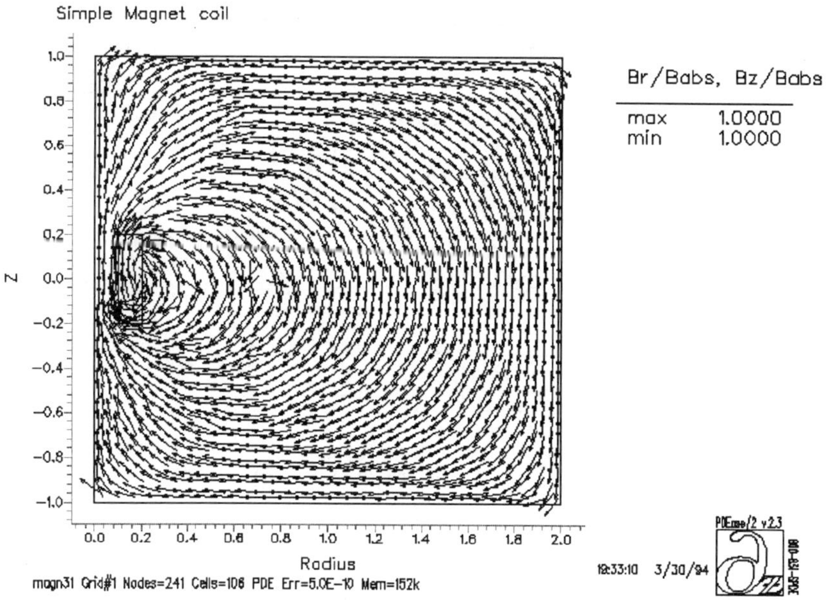
Simple Magnet coil
Z
Radius
Br/Babs, Bz/Babs
max 1.0000
min 1.0000
19:33:10 3/30/94
PDEase/2 v2.3
magn31 Grid#1 Nodes=241 Cells=106 PDE Err=5.0E-10 Mem=152k

As seen in the first of the preceding figures the contours of the vector potential never meet the boundary, and the contours crowd inside the coil, where the field is strong.

The last figure above shows that the field **B** circulates in a similar manner, never crossing the outer boundary. The radial field vanishes on the axis, since it must have the same strength in both directions, in view of the symmetry of rotation.

Helmholtz Coil

We next explore the field inside a special device, known as the Helmholtz coil. It consists of two identical, simple coils on the same axis and with a mean spacing equal to the mean radius. Most of the previous descriptor may be kept, and the following list contains the few necessary changes.

Here we use a new keyword, *initial values*, for the first time. This problem is somewhat more complicated than that of the simple coil, and it appears that PDEase needs to know at least the order of magnitude of the solution in order to work stably.

```
  ...                                                    { magn41.pde }
  errlim= 1e-4             contours= 20
  ...
  r1= 0.3          r2= 0.5       r3= 2.0
  L12= 0.1         L3= 1.0       D= 0.2
  ...
INITIAL VALUES
  A_phi= 1e-7
equations
  ...
           region 2                 J= 1
  start(r1,D-L12)
  line to (r2,D-L12) to (r2,D+L12) to (r1,D+L12) to finish
           region 3                 J= 1
  start(r1,-D-L12)
  line to (r2,-D-L12) to (r2,-D+L12) to (r1,-D+L12) to finish
  ...
  elevation(Babs) from (0,0) to (0.3,0)          end
```

The Helmholtz coil is designed to yield a magnetic field of highly constant direction and magnitude in a region close to the center. The plot below illustrates this by the *q* and *r* contours, large spacing indicating small variation.

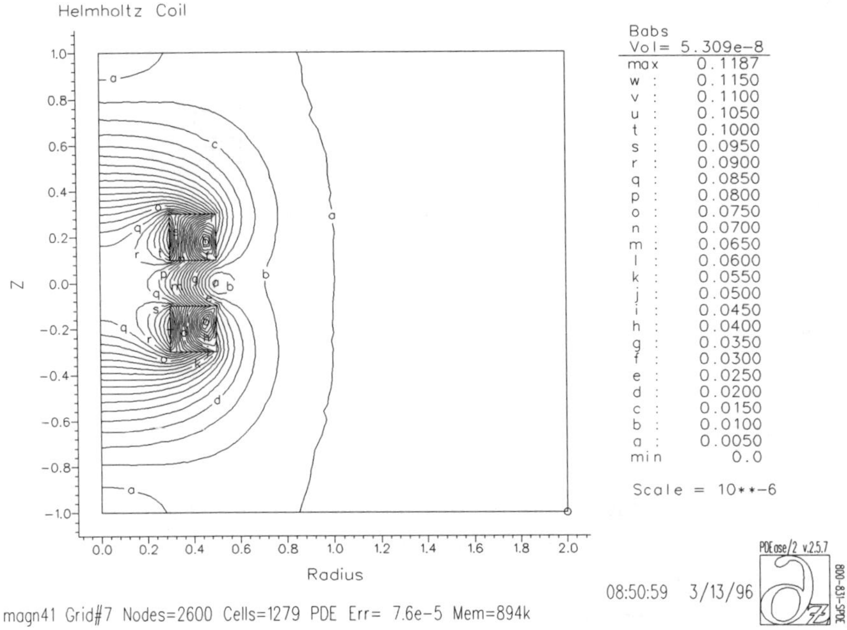

We may inspect the field in the central region more closely by elevation plots of *Br*, *Bz* and *Babs*, taken along the axis or along a radius. The next figure shows how small the radial variation of $|\mathbf{B}|$ is close to the axis.

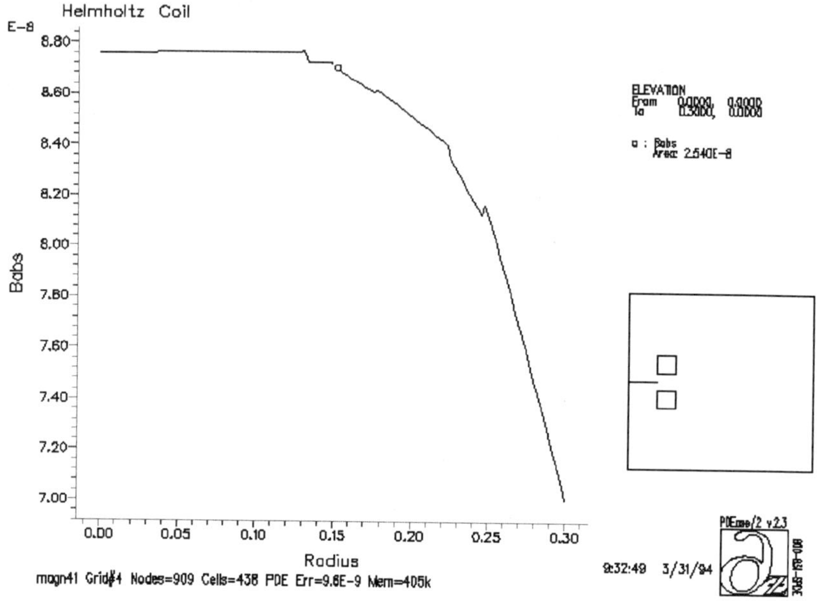

Exercises

❑ Reduce the dimensions of the simple coil by a factor of ten to explore the far field, which for large factors approaches that of a magnetic dipole.

❑ Inspect the homogeneity of the Helmholtz field by several radial and axial elevation plots.

❑ Change the spacing between the Helmholtz coils to smaller and larger values to find out how the constancy of the central field is influenced.

Steady Heat Conduction in (x,y) Space

A steady distribution of temperature in an object means that the heat leaving any volume element equals the quantity produced. The corresponding mathematical statement is

$$\nabla(-\lambda \nabla T) - h = 0 \, , \qquad \bullet$$

where λ is the thermal conductivity, T the temperature and h the amount of heat generated per unit volume. Internal heating may for instance be caused by an electric current flowing through the material, by radiation, or by a chemical reaction. The expression in parenthesis is equal to the heat flux density, i.e. the heat power transferred per unit area.

The above PDE is similar to that for the potential in a dielectric (p.65), the heat production term replacing the volume charge, etc.. Electric conduction is also a closely analogous process. Hence many of the previous problems may be transcribed into descriptors for heat conduction. In order to avoid near-repetition, however, we shall choose different problems in this chapter.

Two Hot-Water Tubes

The following example concerns the flow of heat in a cylinder of insulating material, in a soil environment, containing two water tubes of different temperatures. We cannot use the notation T for temperature, since PDEase would confuse this with t, which is the system variable for time.

The heat flux density and its components, *fluxd_x* and *fluxd_y*, are analogous to the electric current densities (p.53). An important derived quantity is the power loss from the heat-carrying tubes to the surrounding soil. We may calculate this power by elevation plots of the outward component of flux density, which the program automatically integrates over an area of unit extension in the z

direction. Due to the symmetry of the problem, we only need to integrate over three sides, doubling the integral obtained for the horizontal plane.

```
  ...
select
  errlim= 2e-4
  contours= 10          surfacegrid= 30        paint= off
variables
  temp                                              { SI units: m, K }
definitions
  r0= 0.1          d= 0.15          r1= 0.5
  Lx= 0.3          Ly= 0.2
  k= 0.03                            { Thermal conductivity of insulation }
  fluxd_x= -k*dx(temp)      fluxd_y= -k*dy(temp)
  fluxd= sqrt( fluxd_x**2+ fluxd_y**2 )
  f_angle= sign(fluxd_y)* arccos(fluxd_x/fluxd)/pi*180
equations
  dx[-k*dx(temp)]+ dy[-k*dy(temp)]= 0
boundaries
        region 1
  value(temp)= 0                       { Frozen soil, Celsius temperature }
  start(0,-r1)
  arc to (r1,0) to (0,r1) to (-r1,0) to finish
  value(temp)= 80                           { Cutout for hot water tube 1 }
  start(d-r0,0)
  arc to (d,r0) to (d+r0,0) to (d,-r0) to finish
  value(temp)= 50                           { Cutout for hot water tube 2 }
  start(-d-r0,0)
  arc to (-d,r0) to (-d+r0,0) to (-d,-r0) to finish
monitors
  contour(temp)
plots
  contour(temp)
  vector(fluxd_x/fluxd, fluxd_y/fluxd)      contour(f_angle)
  contour(fluxd_x)       contour(fluxd_y)         contour(fluxd)
  elevation(fluxd_y) from (-Lx,Ly) to (Lx,Ly)
  elevation(-fluxd_x) from (-Lx,-Ly) to (-Lx,Ly)
  elevation(fluxd_x) from (Lx,-Ly) to (Lx,Ly)
  surface(temp) interactive
end
```

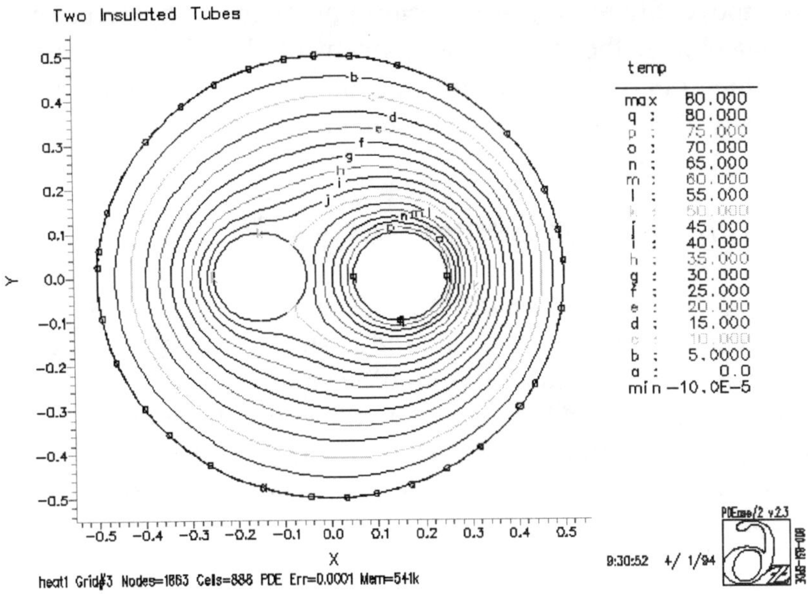
Two Insulated Tubes
temp
max 80.000
q : 80.000
p : 75.000
o : 70.000
n : 65.000
m : 60.000
l : 55.000
k : 50.000
j : 45.000
i : 40.000
h : 35.000
g : 30.000
f : 25.000
e : 20.000
d : 15.000
c : 10.000
b : 5.0000
a : 0.0
min -10.0E-5
Y
X
9:30:52 4/ 1/94
PDEase/2 v2.3
heat1 Grid#3 Nodes=1863 Cells=888 PDE Err=0.0001 Mem=541k

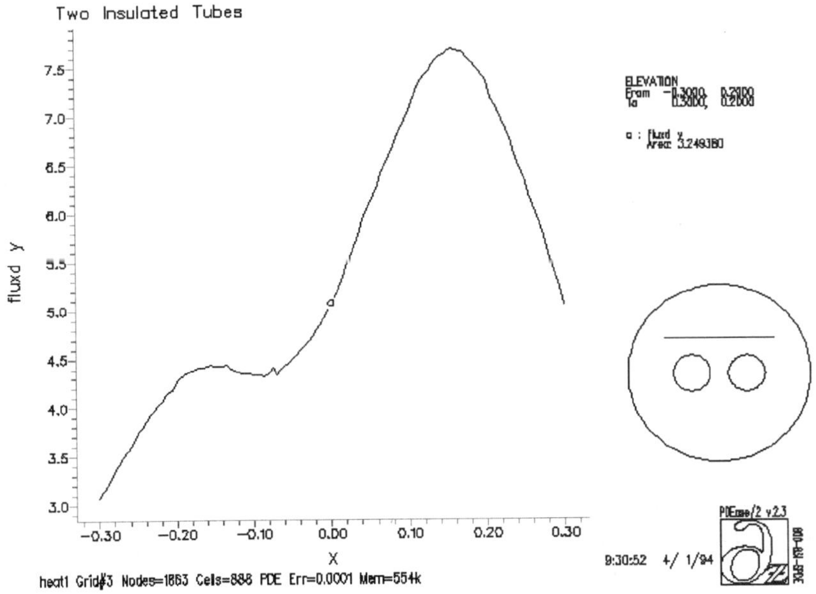
Two Insulated Tubes
ELEVATION
From -0.3000, 0.2000
To 0.3000, 0.2000
a : fluxd y
Area: 3.249380
fluxd y
X
9:30:52 4/ 1/94
PDEase/2 v2.3
heat1 Grid#3 Nodes=1863 Cells=888 PDE Err=0.0001 Mem=554k

The above figures are a *contour* plot of the temperature and an *elevation* plot of the vertical component of flux density, taken over one of the four lines enclosing the hot tubes. The total heat flux leaving any enclosing rectangle is equal to that leaving the outer tube.

Uniformly Heated Semi-Circular Rod

In the next problem a semi-circular rod of stainless steel is uniformly heated by an electric current. The rod is surrounded by granite, which in turn is in contact with a mixture of ice and water.

Here we need to include the volume heating term in the PDE, the power per unit volume being zero in the granite and 1e5 in the steel. The elevation plots from the previous descriptor no longer apply and can be erased.

```
...
Lx= 0.1              Ly= 0.1                  r0= 0.05
...
dx[-k*dx(temp)]+ dy[-k*dy(temp)]- heat= 0
...
     region 1            k= 3.5                                      { Granite }
heat= 0
start (-Lx,-Ly)              value(temp)= 0
line to (Lx,-Ly)  to (Lx,Ly) to (-Lx,Ly)  to finish
               region 2            k= 45                       { Stainless steel }
heat= 1e5
start (-r0,0) line to (r0,0)
arc to (0,r0) to finish
...
```

The first of the following figures shows a temperature maximum within the region of the steel rod. The second plot yields contours of constant heat flux density, which is zero at a point inside the rod, where the temperature takes its maximum. The largest flux density evidently occurs at the edges of the rod, and you could view this feature more clearly by means of a zoomed *surface* plot.

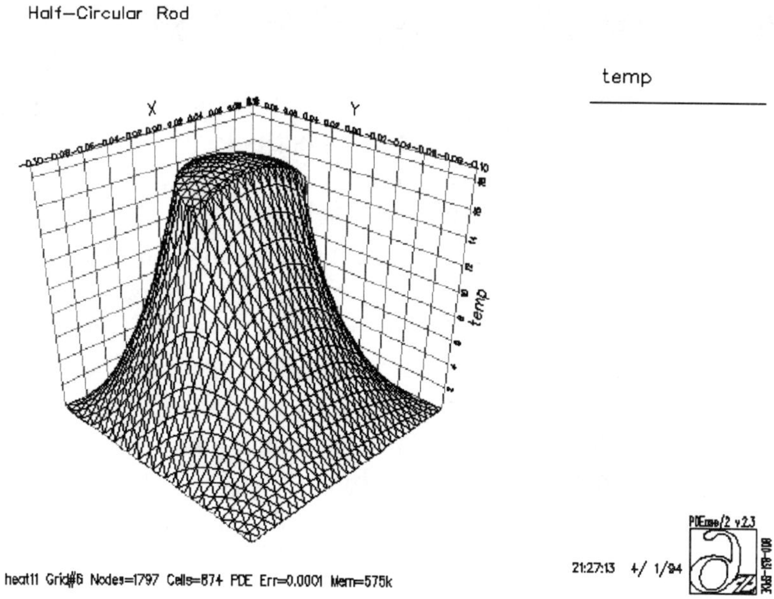
Half-Circular Rod
temp
X
Y
temp
heat11 Grid#6 Nodes=1797 Cells=874 PDE Err=0.0001 Mem=575k
21:27:13 4/ 1/94
PDEase/2 v2.3

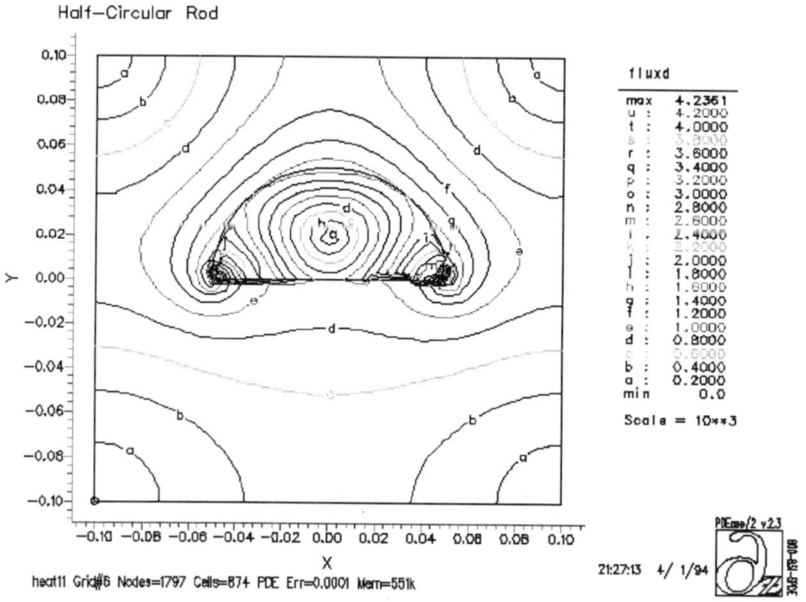
Half-Circular Rod
fluxd
max 4.2361
u : 4.2000
t : 4.0000
r : 3.6000
q : 3.4000
p : 3.2000
o : 3.0000
n : 2.8000
m : 2.6000
l : 2.4000
j : 2.0000
i : 1.8000
h : 1.6000
g : 1.4000
f : 1.2000
e : 1.0000
d : 0.8000
b : 0.4000
a : 0.2000
min 0.0
Scale = 10**3
X
Y
heat11 Grid#6 Nodes=1797 Cells=874 PDE Err=0.0001 Mem=551k
21:27:13 4/ 1/94
PDEase/2 v2.3

Heat Conduction in Anisotropic Wood

In the next descriptor we find the first example of conduction in an anisotropic material. The thermal conductivity of pine wood is considerably higher for flow in the direction of the fibers than across. We consider a bar of wood, cut such that the fibers are in the y direction (vertical). On the top surface we mount an electric heater in the form of a strip of width $2w$, extending along the z axis. The bottom surface of the wooden block is assumed to be at zero temperature, all other surfaces being well insulated, including that side of the heater which is not in contact the wood.

Since the power dissipated in the heater can only flow into the specimen, we represent the heater by a downward flux density of 100 at the upper boundary.

```
...
Lx= 0.1            Ly= 0.1            w= 0.01
kx= 0.14           ky= 0.35                    { Thermal conductivities }
heat= 0                                             { No volume heating }
fluxd_x= -kx*dx(temp)     fluxd_y= -ky*dy(temp)
...
dx[-kx*dx(temp)]+ dy[-ky*dy(temp)]- heat= 0
...
            region 1
start (-Lx,-Ly)     value(temp)= 0   line to (Lx,-Ly)
natural(temp)= 0     line to (Lx,Ly) to (w,Ly)
natural(temp)= -100.0  line to (-w,Ly)
natural(temp)= 0 line to (-Lx,Ly) to finish
...
```

The temperature plot below exhibits contours which are vertically elongated in the upper part of the figure. The finite extension of the strip tends to stretch the contours *horizontally*, so the anisotropic effect becomes attenuated. Nevertheless we clearly see the signs of enhanced flow along the fibers. In spite of the concentrated heating on part of the upper surface, the flow becomes almost parallel in the lower part of the specimen.

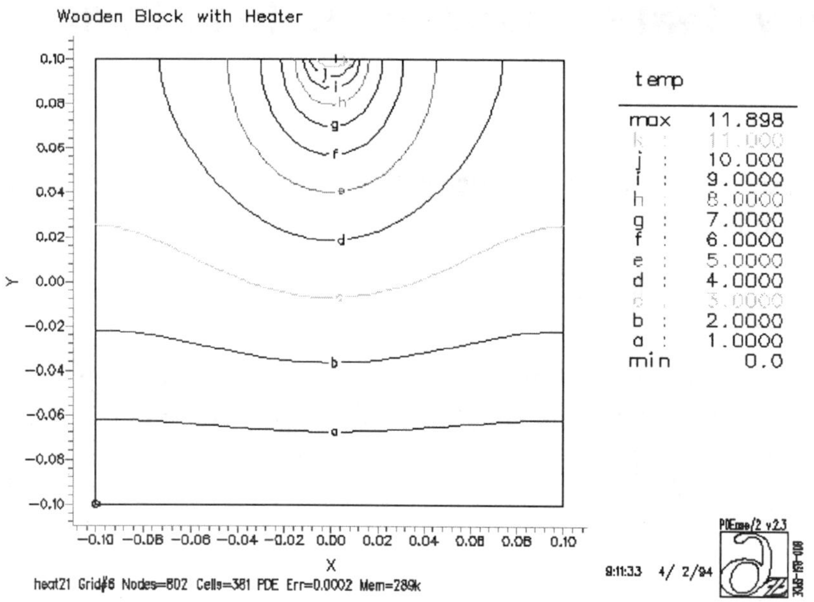

Exercises

❑ In the case of the two insulated tubes, make two more *elevation* plots to find out how much heat is lost by each of the two tubes.

❑ Compare the heat dissipated in the steel rod with the flux leaving the rod.

❑ Combine the half-cylinder of steel with an magnesium oxide bar of triangular cross-section, having a thermal conductivity of 55. Let the flat surface be common to the two rods and place the remaining corner of the triangle at ($r0$,-$r0$).

❑ Find the power (per unit length in the z direction) transferred to the bottom surface of the wooden block, and compare with the power from the strip. In addition, make an elevation plot just below the upper face.

❑ Solve the problem of the heated pine wood rod under other conditions. Assume first that the fibers are in z direction, i.e. that kx= ky= 0.14. Then assume anisotropic conditions again, but with horizontal fibers.

Steady Heat Conduction in (ρ,z) Space

In the case of full axial symmetry the PDE for heat conduction becomes, in the cylindrical coordinate system[3p705],

$$\frac{1}{\rho}\frac{\partial}{\partial\rho}\left(-\rho\lambda\frac{\partial T}{\partial\rho}\right)+\frac{\partial}{\partial z}\left(-\lambda\frac{\partial T}{\partial z}\right)-h=0\ ,$$ •

where λ is the thermal conductivity. The radial and axial components of heat flux density are, by analogy with electrical conduction,

$$f_\rho=-\lambda\frac{\partial T}{\partial\rho}\ \text{and}\ f_z=-\lambda\frac{\partial T}{\partial z}\ .$$ •

Radial Heat Conduction in a Hollow Cylinder

As a first test of PDEase in this situation, let us study a simple case of radial heat conduction where T is independent of z. Any such problem may also be treated in (x,y) space for verification.

The following descriptor applies to a hollow cylinder of magnesium oxide, uniformly heated from the inside by an electrical power. The outer cylindrical surface is kept at 300 K by water cooling, and the flat ends are well insulated to ensure radial flow.

The heating power is 10^4 W per unit of axial length, and from this we calculate the flux density entering radially from the central bore. This flux density is opposite to the direction of the normal and we must thus input the negative of it in the *natural* boundary condition.

This problem has an exact solution which we may construct using cylindrical harmonics (p.44) in the form $A+B\ln(\rho)$. The boundary conditions determine the constants A and B.

```
...
errlim= 2e-5
contours= 10          surfacegrid= 30          paint= off
```

```
  alias(x)= 'Radius'            alias(y)= 'Z'
variables
  temp
definitions
  mm= 1e-3      r1= 10*mm        r2= 50*mm        L= 40*mm
  temp2= 300    k= 55                              { Magnesium oxide }
  power_m= 1e4    heat_fd= power_m/(2*pi*r1)
  fluxd_r= -k*dx(temp)      fluxd_z= -k*dy(temp)
  fluxd= sqrt( fluxd_r**2+ fluxd_z**2 )
  f_angle= sign(fluxd_z)* arccos(fluxd_r/fluxd)/pi*180
  tempx= r1*heat_fd/k* ln(r2/x)+ temp2
equations
  (1/x)* dx( -x*k*dx(temp))+ dy( -k*dy(temp))= 0
boundaries
        region 1
  start(r1,0)
  natural(temp) = 0            line to (r2,0)
  value(temp)= temp2           line to (r2,L)
  natural(temp)= 0             line to (r1,L)
  natural(temp)= -heat_fd            line to finish
monitors
  contour(temp)
plots
  contour(temp)   contour(temp-tempx)  as 'Temperature error'
  vector(fluxd_r/fluxd, fluxd_z/fluxd)        contour(f_angle)
  contour(fluxd_r)             contour(fluxd_z)
  contour(fluxd)               surface(temp) interactive
end
```

The following plot shows the temperature contours, which are crowded close to the axis, as could be expected. The plot of the difference with respect to the exact solution yields irregular contours at about 10^{-4} of the total temperature drop. The angle plot demonstrates the high degree of parallelism in the heat flow.

There is an important point to notice concerning *natural* boundary conditions: we are not allowed to multiply the PDE by a factor (other than unity). For example, if we run the previous descriptor with the following modification,

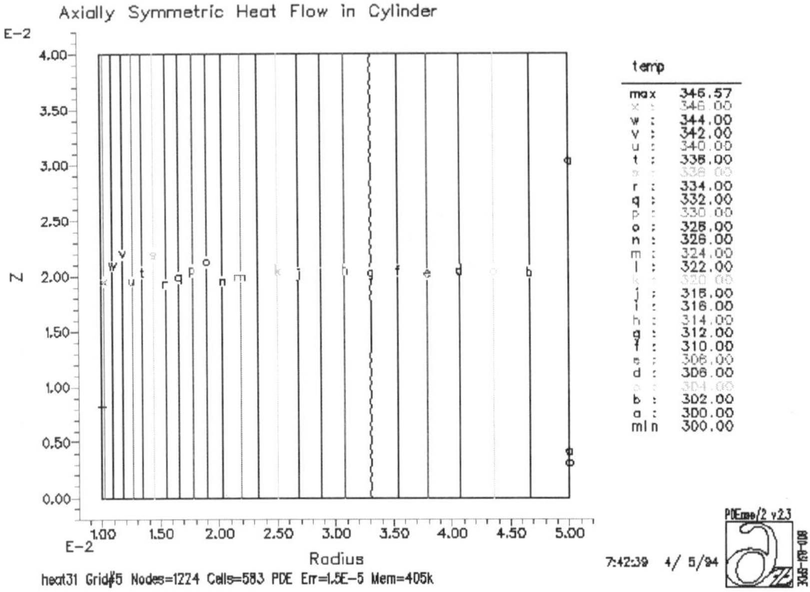


```
...
  2* [ (1/x)* dx( -x*k*dx(temp))+ dy( -k*dy(temp)) ]= 0
  ...
```

the error no longer becomes negligible, although the new equation appears to be mathematically equivalent. This surprising feature is caused by the way PDEase transforms the equation in order to take the normal derivative into account, a subject which we shall discuss in the appendix on FEA. We shall have reasons to multiply PDEs by a function in later chapters. In such cases we always multiply the *natural* boundary conditions by the same factor. If only *value* boundary conditions occur, however, multiplying all terms of the equation by the same factor does not change the solution.

Cooling by Forced Convection

So far we have always specified a constant value or a function of (x,y) as the *natural* boundary condition. We are also allowed, however, to use *natural* conditions containing the dependent variable (T in this case). This is required in problems involving steady convection or radiation at the boundary. If a fluid is driven at constant speed by means of a pump or a fan past a hot object, the latter is cooled at a heat flux density proportional to the difference between the temperature of the object and the initial temperature of the fluid. This proportionality, which is more of a rule of thumb than a law of nature, is known as Newton's law of cooling.

To apply this simple law of cooling we modify the first descriptor in this chapter as follows. We heat from the inside over the lower half of the bore, and assume the outer cylindrical surface to be cooled in proportion to the temperature difference to the coolant. The temperature of the fluid we take to be 300 K and assume a cooling coefficient of 1e3.

```
...
errlim= 1e-4
...
coef= 1e3                                          { Coefficient of cooling }
...
          region 1
start(r1,0)
natural(temp)= 0                 line to (r2,0)
natural(temp)= coef* (temp-300)       line to (r2,L)
natural(temp)= 0                 line to (r1,L)  to  (r1,L/2)
natural(temp)= -heat_fd          line to finish
...
```

You should obtain a temperature plot similar to the one shown below. By adding two *elevation* plots of *fluxd_r*, one close to the bore and the other close to the outer surface, you could check (by the integrals) that the solution satisfies conservation of energy. You could also plot the left side of the PDE to see if the contours are irregular, as expected for a vanishing residual.

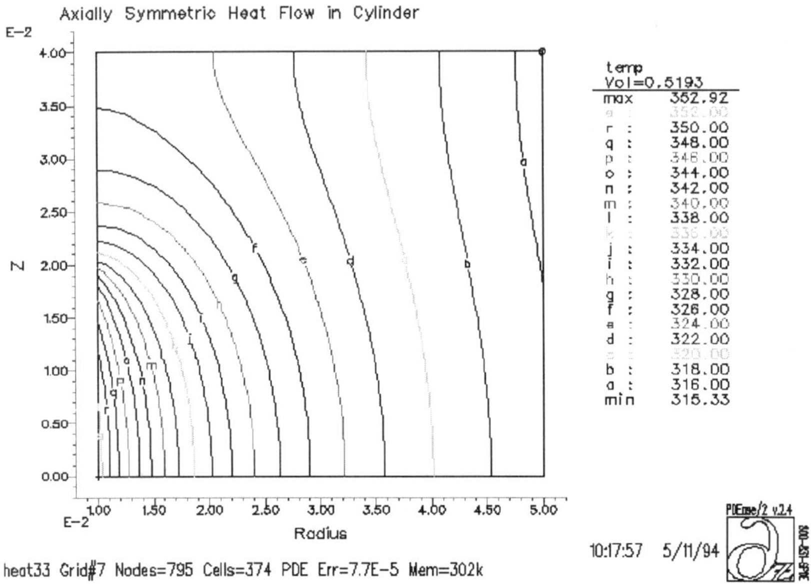

Continuously Varying Thermal Conductivity

In the above examples the thermal conductivity was constant throughout the object. We may introduce a space-varying conductivity, however, varying both in the axial and in the radial direction. Shaking a powder, for instance, would make the heavy grains segregate from the lighter ones, thus creating a density gradient. Centrifuging about the vertical axis would produce a corresponding gradient in the radial direction. After compacting under high pressure we would be left with a solid with varying density, but with axial symmetry. This case should, however, be distinguished from that of anisotropic conductivity, since the conductivity is the same in all directions within a given grain.

The next excerpt of a descriptor pertains to a rotationally symmetric object, having a thermal conductivity (k) which varies both radially and axially. The boundary conditions apply to forced cooling as before.

```
...
errlim= 5e-5
...
k= 55* exp[20*(x-y)]                    { Specimen shaken, not stirred }
...
contour(k)
surface(fluxd_z) interactive
```

We notice that PDEase solves the above problem without difficulty. The figure below is a plot of the radial flux density. You could easily test if the equation and the boundary conditions are satisfied, the latter by plotting, say, *fluxd_r*/(*temp*-300).

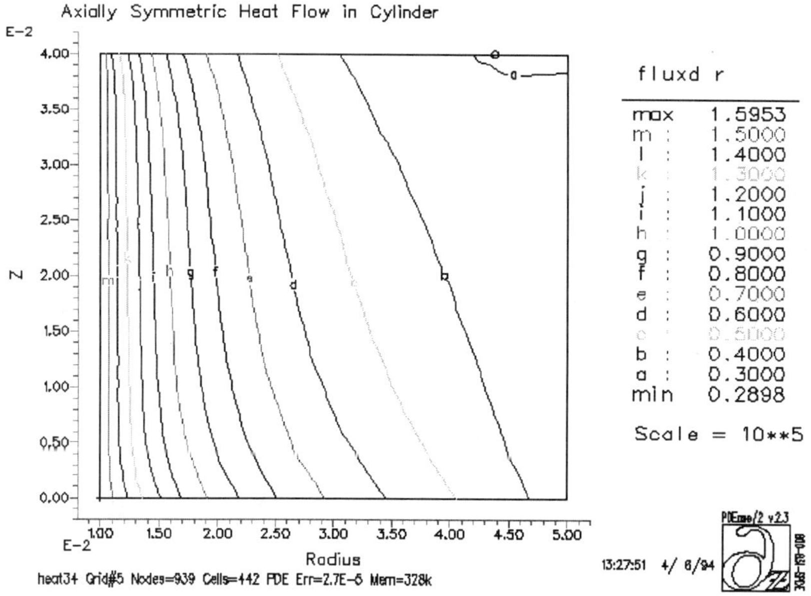

Steel Tube With Cooling Flanges

We shall now proceed to a more practical problem: a series of circular cooling flanges on a steel tube. The flanges are distributed at equal distances along the tube, and all are at about the same temperature, such that there is no axial heat flow at mid-distance between two adjacent flanges. Thus we may limit our study to one flange and a half-length of tube on either side of it.

The following partial descriptor and *contour* plot illustrate the geometry. Let us assume cooling by air flow at a speed yielding a coefficient of cooling of 20.

```
...
errlim= 1e-4
...
cm= 1e-2    r1= 10*cm      r2= 12*cm      r3= 20*cm
L= 5*cm    dL= 2*cm
k= 45                                                    { Steel }
coef= 20                                  { Coefficient of cooling }
small= 1e-4
...
            region 1
start (r1,0)
natural(temp)= 0               line to (r2,0)
natural(temp)= coef*(temp-300)
     line to (r2,L) to (r3,L) to (r3,L+dL) to (r2,L+dL) to (r2,2*L+dL)
natural(temp)= 0               line to (r1,2*L+dL)
value(temp)= 373               line to finish
...
elevation(2*pi*x* fluxd_r) from (r1+small, 0) to (r1+small, 2*L+dL)
elevation(2*pi*x* fluxd_r) from (r2-small, 0) to (r2-small, 2*L+dL)
surface(temp) interactive
...
```

The plot below shows the temperature distribution. The *elevation* plots permit us to compare the power (integrated flux density) entering and leaving the tube.

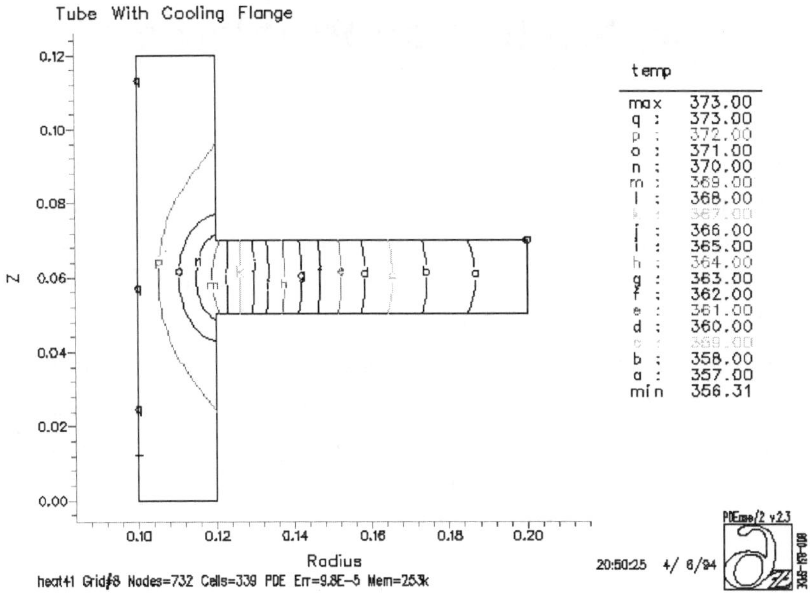

Exercises

❑ Modify the first descriptor (simple, hollow cylinder) to specify 400K on the inner surface and 300 K on the outer one. Calculate the power transferred using *elevation* plots.

❑ Change the first descriptor again to specify pine wood (p.106). Which fiber orientations can you handle?

❑ In the last example the cooling efficiency depends on the radius of the flange. How would you plan a calculation of the maximum ratio of cooling power to weight per tube section?

Non- Linear Conduction Problems

The PDEs we have considered so far have been linear, even if some coefficients have been non-constant, such as in the PDE $\nabla(-\lambda(x,y)\nabla T)-h(x,y)=0$. If the corresponding homogeneous equation $\nabla(-\lambda(x,y)\nabla T)=0$ has two solutions, T_1 and T_2, then $T=T_1+T_2$ will also be a solution. If the conductivity λ should depend on T, however, the PDE is known as non-linear, which means that it will no longer be possible to generate new solutions by summing, and very few exact solutions have been found. Fortunately, PDEase usually permits us to solve non-linear equations.

Heat Conduction Through a Large Temperature Difference

The descriptor below is for a plate of MgO, thick enough to let us neglect the radiation from the surfaces in comparison with the power transmitted through the material. There is a hole in the center of the plate. The edges are at very different temperatures, 300 and 1000 K, and since the thermal conductivity λ is known to be proportional to $1/T$ for this material, the PDE will evidently be non-linear.

The only unusual descriptor feature is the *initial values* segment, where we give a constant value as a guide in searching for a solution. Depending on the case, this hint may or may not be necessary. If the equation is nearly linear, PDEase may well solve it without a hint, but the search will perhaps take more time.

In non-linear problems it is particularly interesting to study the convergence of the consecutive trial functions to the final solution. For this purpose we include an *elevation* plot in the monitors segment, which permits us to inspect the progress at a glance.

```
...
errlim= 5e-5
```

```
  contours= 10                    paint= off
variables
  temp
definitions
  Lx= 0.6          Ly= 0.4              r0= 0.1
  k= 15000/temp                                    { Thermal conductivity versus T }
  fluxd_x= -k*dx(temp)      fluxd_y= -k*dy(temp)
  fluxd= sqrt( fluxd_x**2+ fluxd_y**2 )
  f_angle= sign(fluxd_y)* arccos(fluxd_x/fluxd)/pi*180
initial values
  temp= 700
equations
  dx[-k*dx(temp)]+ dy[-k*dy(temp)]= 0
boundaries
            region 1
  start (0,0)
  natural(temp)= 0              line to (Lx,0)
  value(temp)= 300              line to (Lx,Ly)
  natural(temp)= 0              line to (0,Ly)
  value(temp)= 1000             line to finish
  start (Lx/2-r0,Ly/2)
  natural(temp)= 0
  arc to (Lx/2,Ly/2+r0)  to (Lx/2+r0,Ly/2)  to (Lx/2,Ly/2-r0) to finish
monitors
  elevation(temp) from (0,0) to (Lx,0)
plots
  contour(temp)                 elevation(temp) from (0,0) to (Lx,0)
  vector(fluxd_x/fluxd,  fluxd_y/fluxd) as 'Heat flow'
  contour(f_angle)
  contour(fluxd_x)      contour(fluxd_y)          contour(fluxd)
  surface(temp) interactive
end
```

The first figure below shows how the temperature contours crowd in the hot region to the left, where the resistivity ($1/k$) is high, and also where the hole creates constrictions. The temperature plot suggests asymmetry of the horizontal gradient $\partial T / \partial x$, and a separate plot of that quantity will provide a striking demonstration.

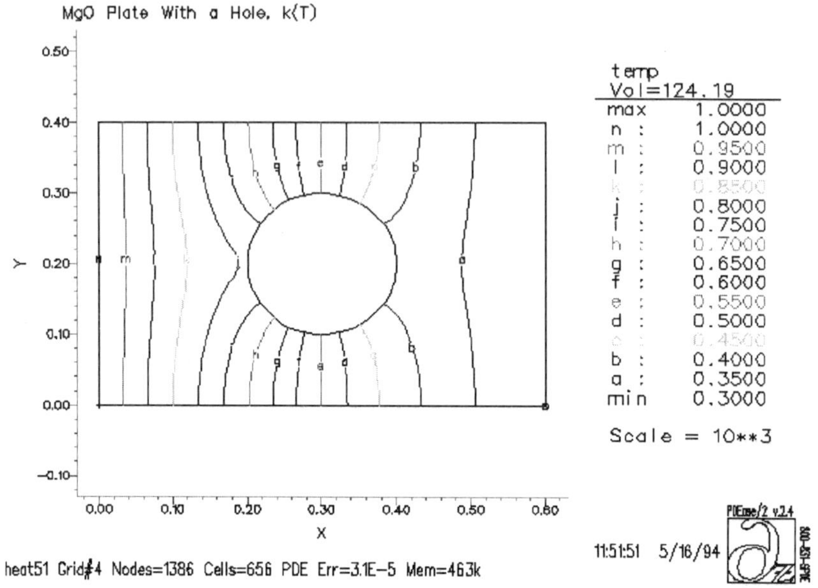

MgO Plate With a Hole, k(T)
temp
Vol=124.19
max 1.0000
n : 1.0000
m : 0.9500
l : 0.9000
k : 0.8500
j : 0.8000
i : 0.7500
h : 0.7000
g : 0.6500
f : 0.6000
e : 0.5500
d : 0.5000
c : 0.4500
b : 0.4000
a : 0.3500
min 0.3000
Scale = 10**3
Y
X
11:51:51 5/16/94
PDEase/2 v2.4
heat51 Grid#4 Nodes=1386 Cells=656 PDE Err=3.1E-5 Mem=463k

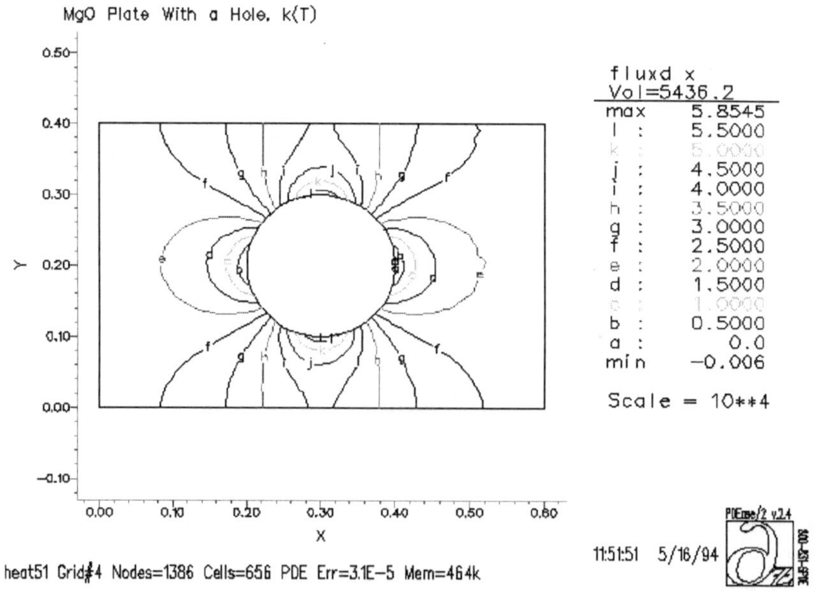

MgO Plate With a Hole, k(T)
fluxd x
Vol=5436.2
max 5.8545
l : 5.5000
k : 5.0000
j : 4.5000
i : 4.0000
h : 3.5000
g : 3.0000
f : 2.5000
e : 2.0000
d : 1.5000
c : 1.0000
b : 0.5000
a : 0.0
min -0.006
Scale = 10**4
Y
X
11:51:51 5/16/94
PDEase/2 v2.4
heat51 Grid#4 Nodes=1386 Cells=656 PDE Err=3.1E-5 Mem=464k

The second figure on the preceding page is a *contour* plot of the x component of the heat flux density. You may be surprised at the symmetry of this plot, having seen the asymmetry of the temperature. We may understand this behavior if we write the PDE in the form $\nabla \cdot \mathbf{f} = 0$, with $\mathbf{f} = -\lambda \nabla T$. Now it seems obvious that the field **f** could not be influenced by the non-linearity, since the corresponding PDE and its boundary conditions are independent of λ.

Steel Foil Emitting Infrared Radiation

In the previous case conduction through the material dominated any radiative loss from the surfaces. We shall now consider the case of a steel sheet, where the heat loss by infrared radiation becomes considerable in comparison with the heat conducted through the foil. To simplify, we assume the thermal conductivity of steel to be independent of temperature, which is approximately true. The program would solve the problem even if λ were a function of T, but we prefer to study one effect at a time.

The following list contains the new features to include in the preceding descriptor, after eliminating the circular cut-out. The PDE now has a source term specifying the radiative heat loss per unit volume. Strictly speaking, the heat loss occurs at the front and rear surfaces, but in the present 2D version of PDEase we do not have access to the normal derivatives in the z direction. Hence we assume the sheet to be thin enough so that the temperature variation across the foil may be neglected.

```
...
errlim= 1e-4
...
Lx= 0.6                 Ly=0.4                  d= 1e-3
k= 45                   { Steel }
equ1= dx[-k*dx(temp)]+ dy[-k*dy(temp)]
equ2= 2*0.3*  5.67e-8* (300**4-temp**4)* (1/d)
...
dx[-k*dx(temp)]+ dy[-k*dy(temp)]-
                    2* 0.3*  5.67e-8* (300**4-temp**4)* (1/d)= 0
```

```
                   { 2 faces, emissivity, S-B constant }
monitors
  elevation(temp) from (0,0) to (Lx,0)
plots
  elevation(equ1, equ2) from (0,0) to (Lx,0)
  contour(equ2) as 'IR power'
  ...
```

The power radiated per unit area is given by the formula $f = \varepsilon\sigma T^4$, where ε is the emissivity (for which we take the value 0.3) and σ the Stefan-Boltzmann constant. The foil loses heat at this rate, but it also gains heat by radiation from the surroundings, which we assume to be at 300 K. The net power so obtained must now be expressed in terms of power per unit area. Since we concentrate this surface power into a thickness d, we must divide by d to express it as power per unit

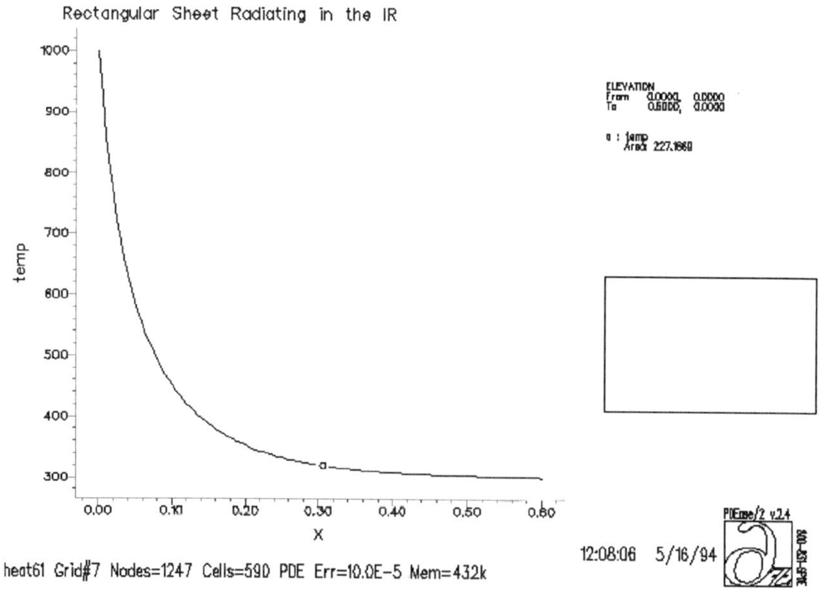

volume. Finally, we multiply by 2 in order to take both faces into account. Since the PDE now has two terms which are required to balance, we may compare these two expressions in a single *elevation* plot. The latter expression is equal to the power lost by radiation.

The above curve shows the final temperature distribution, obtained after a number of iterations. The *contour* plot of the IR emission shows that most of the power entering the sheet at high temperature is lost by radiation. The two curves corresponding to the terms of the PDE are in reasonable agreement.

Exercises

❑ Remove the circular cut-out from the first descriptor. Run the resulting, simpler problem and verify that the x-component of flux density is constant for all values of x, as required.

❑ Extend the preceding exercise by plotting *elevation* curves of the left side of the PDE. First take a plot in the x direction along the middle line. Read the integral (*Area*) value and compute the average. Then try other locations of the line.

❑ Modify the descriptor for the steel sheet to include a circular cut-out as in the first descriptor.

Simultaneous Electric and Thermal Conduction

An electric current always heats the material through which it passes. If the heating is non-uniform, which is often the case, temperature differences will also occur, and we have conduction of heat as well as conduction of charge. The charge distribution is still governed by the equation $\nabla(-\sigma\nabla U)=0$, if there is no source of charge inside the material. The resulting current density, $\mathbf{j}=-\sigma\nabla U$ for an isotropic material, causes dissipation of heat at the rate $\mathbf{j}\cdot\mathbf{E}=\sigma(\nabla U)^2$. The latter will appear as a source term in the PDE for heat conduction:

$$\nabla(-\lambda\nabla T)-\sigma(\nabla U)^2=0.$$

To calculate the potential and temperature distributions we thus have to face a system of simultaneous PDEs

$$\begin{cases}\nabla(-\sigma\nabla U)=0\\ \nabla(-\lambda\nabla T)-\sigma(\nabla U)^2=0\end{cases}$$

This problem is obviously non-linear, already because of the gradient squared in the second equation. If the electrical conductivity had been independent of T, we could have solved the first equation with respect to U, then used that solution in the second equation. This approximation is rarely valid, however. For a pure metal, such as copper, the electric conductivity is roughly proportional to $1/T$ over a large range of temperatures. Hence, both equations are non-linear, and the equations have to be solved simultaneously.

PDEase helps us to solve a system of equations almost as easily as if it were a single equation. It is wise, however, to supply initial values for the dependent variables, if at all possible. If you have a rough idea about the shape of the solutions, give an expression for it. At least you could indicate an order of magnitude, or a typical value for the solution. If no initial values are available, PDEase assumes the value zero, which may or may not be a good starting point.

There are as many dependent variables (solutions) as there are equations. For each of the dependent variables you must supply boundary values or normal derivatives. If not, PDEase will assume *natural*()= 0, which may be quite different from what you intended.

As the number of equations increase, the number of possible plots multiply. One soon finds it necessary to be restrictive in the choice of graphical presentations, in order not to drown in figures.

Copper Block, Heated by an Electric Current

In the first example we have chosen a block of copper, to which a voltage is applied on two opposite faces. We shall ignore the practical problems of keeping constant potentials over each of the two faces. Similarly, we assume a constant temperature of 300 K on these faces without describing how it could be achieved. For this first application we choose the simplest case possible.

Since we are using a 2D version of finite element analysis, it is understood that there is no variation in the invisible z direction, hence no heat loss. As regards the top and bottom faces, which are insulated, we may simply specify that there is no heat loss.

In the descriptor below we have specified $U1/2$ as the initial value for the potential. We could have done better by giving a linear function between the two extreme potential levels, 0 and $U1$. For the temperature we guess at a value just above that at the ends of the specimen.

Again we order an *elevation* plot of the temperature in order to watch the progress of the solution.

```
select
  errlim= 1e-5          contours= 10          paint= off
variables
  U  temp
definitions
  Lx= 0.6        Ly=0.4          d= 1
  temp0= 300     U1= 0.1
  cond0= 1/1.67e-8    cond= cond0*temp0/temp               { Copper }
  k=400
```

```
  fluxd_x= -k*dx(temp)        fluxd_y= -k*dy(temp)
  fluxd= sqrt( fluxd_x**2+ fluxd_y**2 )
  f_angle= sign(fluxd_y)* arccos(fluxd_x/fluxd)/pi*180
  j_x= -cond*dx(U)      j_y= -cond*dy(U)
  jabs= sqrt( j_x**2+ j_y**2 )
initial values
  U= U1/2          temp= 350
equations
  dx[-cond* dx(U)]+ dy[-cond* dy(U)]= 0
  dx[-k*dx(temp)]+ dy[-k*dy(temp)]- cond* [dx(U)]**2= 0
boundaries
        region 1
  start (0,0)
  natural(U)= 0    natural(temp)= 0       line to (Lx,0)
  value(U)= U1     value(temp)= 300        line to (Lx,Ly)
  natural(U)= 0    natural(temp)= 0       line to (0,Ly)
  value(U)= 0      value(temp)= 300        line to finish
monitors
  elevation(temp) from (0,Ly/2) to (Lx,Ly/2)
plots
  contour(temp)          elevation(temp) from (0,Ly/2) to (Lx,Ly/2)
  vector(fluxd_x/fluxd, fluxd_y/fluxd) as 'Heat flow'
  contour(f_angle)
  contour(fluxd_x)       contour(fluxd_y)         contour(fluxd)
  contour(U)
  elevation(j_x) from (0,Ly/2) to (Lx,Ly/2)
  elevation( dx(U) ) from (0,Ly/2) to (Lx,Ly/2)
  vector(j_x/jabs, j_y/jabs) as 'Current'
  surface(temp) interactive
end
```

The *surface* plot on the next page shows the temperature variation in the object. It turns out that the final solution for the temperature becomes symmetrical with respect to the middle of the specimen, which is what we expect. If the program had not worked properly, it could have been of any shape. We also note that the main component of current density, j_x, is constant across the object, still according to expectations. The *elevation* plot of the electric field strength shows a maximum at mid-distance, which is reasonable considering that the electrical resistivity is largest there. Finally, we find that U and T do not vary significantly along the y direction, which had not been specified explicitly but is a consequence of the boundary conditions and the equations.

We have thus obtained plausible solutions to a non-linear system of PDEs, involving electrical as well as thermal variables. If we wish to verify the solutions more in detail, we may compare the first and last terms in the second equation.

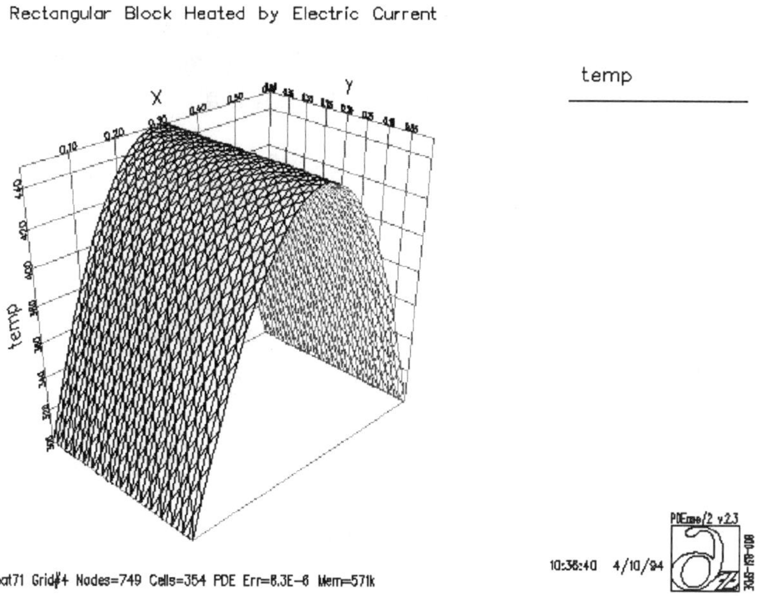

Electrically Heated Copper Foil, Losing Heat by Radiation

Let us now replace the copper block by a thin foil of the same material, emitting infrared radiation in the positive and negative z directions. In the same manner as in the previous chapter (p.119) we lump the radiative loss from the front and rear sides into a volume heating term, neglecting radiation from the very small top and bottom surfaces. We need to complement the previous descriptor as indicated below.

This is a case of strong non-linearity, where the program may work unstably, in particular at high radiative loss. For this reason we use a device provided by PDEase, which permits us to introduce the non-linearity gradually. We first announce that we intend to employ 10 *stages*. After having made this declaration we may use *stage* as an integer variable, which automatically steps from 1 to 10. There are several ways of exploiting this variable, but here we employ it to increase the applied voltage $U1$, increasing the maximum temperature and the radiative loss each time. The program thus solves the equations for the first value of $U1$, then steps *stage* to the next higher integer value. The program uses the first solution as an "initial value" for the next stage of the calculation. This is an important tool to grasp for in difficult situations.

```
select
  STAGES= 10            errlim= 1e-3
  ...
  Lx= 0.6               Ly=0.4          d= 1e-4
  temp0= 300            U1= STAGE* 0.1
  k= 400                                              { Copper }
  ...
equations
  dx[-cond* dx(U)]+ dy[-cond* dy(U)]= 0
  dx[-k*dx(temp)]+ dy[-k*dy(temp)]- cond* [dx(U)]**2-
                  0.3* 5.67e-8* (300**4-temp**4)* (1/d)* 2= 0
                  { Emissivity= 0.3, power per m**3, both sides }
boundaries
        region 1
  start (0,0)
  natural(U)= 0         natural(temp)= 0        line to (Lx,0)
  value(U)= U1          value(temp)= temp0      line to (Lx,Ly)
  natural(U)= 0         natural(temp)= 0        line to (0,Ly)
  value(U)= 0           value(temp)= temp0      line to finish
monitors
  elevation(temp)  from (0,Ly/2) to (Lx,Ly/2)
plots
  contour(temp)         elevation(temp) from (0,Ly/2) to (Lx,Ly/2)
end
```

Apart from helping the program to cope with non-linearity, the use of stages gives us a whole series of results for different applied voltages. Since the number of figures thus will be multiplied by ten in the present case, we must reduce the number of plots to a bare minimum. We keep the contour plot of T in order to read the

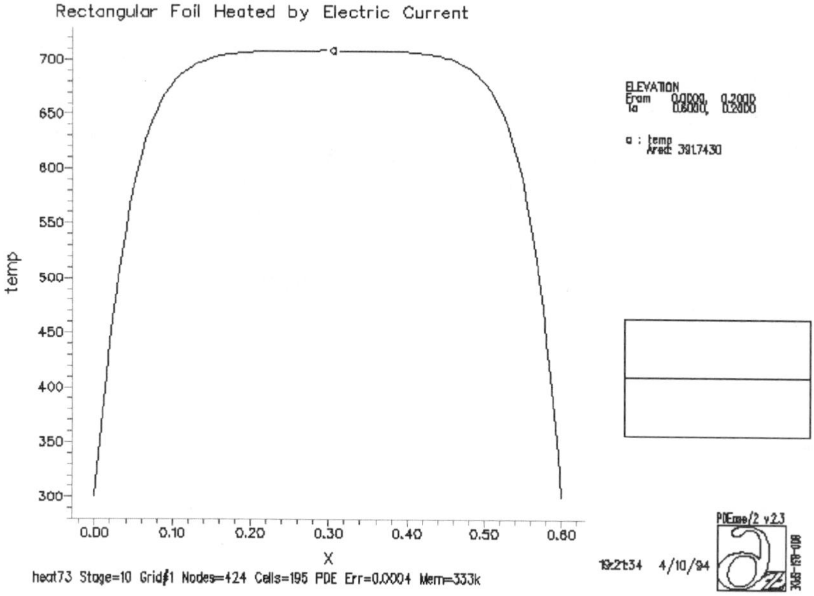

maximum value and to check that the solution is invariable with respect to y. In addition we like to store elevation plots of T to see the change of shape as the power increases.

The figure above shows the last plot, the one for the highest voltage. The curve is virtually flat over the middle part, since the electrical heat dissipation balances the loss by radiation. Hence, there is practically no conduction of heat out of this central region of the foil.

Semicircular Sheet, Heated by Electric Current

Finally we investigate the temperature distribution in a thin, semicircular strip of copper, heated by a current, taking radiation into account. We limit the number of stages to five and increase the voltage by 0.5 V for each stage. Because of the special shape of the object we prefer *contour* and surface plots to *elevation* plots.

```
 ...
 stages= 5                errlim= 1e-3
 ...
 r1= 0.5                  r2= 1.0              d= 1e-4
 k= 400                   U1= stage* 0.5
 ...
 start (-r2,0)
   value(U)=  0           value(temp)=300      line to (-r1,0)
   natural(U)= 0          natural(temp)= 0
      arc to (0,r1) to (r1,0)
   value(U)= U1           value(temp)= 300     line to (r2,0)
   natural(U)= 0          natural(temp)= 0
      arc to (0,r2) to finish
 ...
 contour(temp)            surface(temp)
end
```

The next figure shows the temperature *surface* plot for the final *stage* of this example. The dependent variables now vary in both directions, and the temperature is considerably higher on the inner edge of the strip. We may of course test the solution set by various plots suggested in earlier chapters.

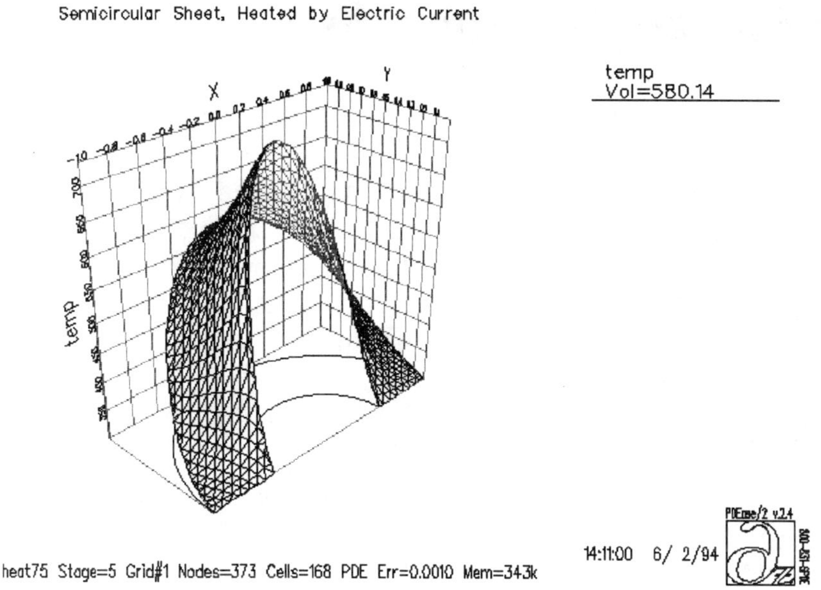

Exercises

❑ Use *elevation* plots to compare the first, second and third terms in the second equation of the first example (p.124).

❑ Change the second descriptor for the copper foil from rectangular to trapezoidal. Let the width at the left be 0.4 as before, and that at the right 0.8.

❑ Calculate the current flowing through the semicircular strip and the resistance in the last example.

❑ Find the temperature distribution of the semicircular strip in the case of negligible radiation and $U1 = 0.2$.

Time-Sinusoidal Problems

Let us now consider the simplest case of time-dependence, i.e. where the boundary value or the source term oscillates around an average. For instance, the heating power per unit volume could vary with time according to $h = h_0 \operatorname{Re}(\exp(i\omega t))$, where h_0 is a time-constant amplitude and Re(z) means the real part of the complex quantity z.

We assume the solution to be

$$f = f_s + f_0 \cos(\omega t) \equiv f_s + f_0 \operatorname{Re}((\exp(i\omega t)) ,$$

where f_s is a static (time-average) value and f_0 the amplitude of oscillation. In this situation, a sinusoidal solution usually exists, and the system will approach that solution while transient terms vanish.

The equations for periodic problems need not involve the time explicitly. For example, let us consider the equation of heat conduction[4p8]

$$\nabla(-\lambda \nabla T) - h + \rho c_p \frac{\partial T}{\partial t} = 0 ,$$ ●

where ρ is the mass density and c_p the specific heat capacity. This PDE just expresses the conservation of energy. We now try a solution of the form

$$T = T_o(\mathbf{r})\exp(i\omega t) = \left[T_r(\mathbf{r}) + iT_i(\mathbf{r})\right]\exp(i\omega t) ,$$

where the real and imaginary parts are functions of the spatial variables. Substituting into the PDE we obtain two equations, one for the real and one for the imaginary term, as follows

$$\begin{cases} \nabla(-\lambda \nabla T_r) - h_0 - \omega \rho c_p T_i = 0 \\ \nabla(-\lambda \nabla T_i) + \omega \rho c_p T_r = 0 \end{cases}$$ ●

Here we have assumed that the materials properties (λ, ρ and c_p) are independent of T, although they may vary with the space coordinates, even in a discontinuous fashion. For sufficiently small temperature oscillations, this would be a good approximation. The system of PDEs is then linear, i.e., this problem is simpler than those we solved in the previous chapter. The interpretation of the solution is completely analogous to that used with vibrating structures and alternating current circuits[1p97]. The actual temperature increment is given by the real part of the complex solution (including the time factor).

Steel Block With Temperature Oscillations

As a first example we choose the simplest possible situation: a steel block, insulated except at the front face, where the temperature oscillates around 300 K at the same amplitude and phase everywhere. There are two variables, the real and imaginary parts, which must be declared.

The input temperature oscillation, applied on the left side, is chosen as real. Oscillations at any point of the object hence refer to this phase angle (0). Since the attenuation of the temperature wave is extremely strong we need to present logarithmic values, defined in the definitions segment.

There is an exact solution to a related problem, which we may use for comparison. This solution is valid for the same boundary conditions, except that the right boundary is at infinite distance. In view of the strong attenuation this difference may not always be significant. The exact solution[4p65] is

$T_{exact} = a_0 \exp(-s_q x[1+i])$ with $s_q = \sqrt{\omega \rho c_p / (2\lambda)}$,

which yields, for the real part,

$T_{r,exact} = a_0 \exp(-s_q x) \cos(-s_q x)$

and similarly for the imaginary part. You may easily verify that these functions satisfy the PDEs and the boundary conditions.

```
select
  errlim= 1e-3                         contours= 10
variables
  tempr      tempi
definitions
```

```
 Lx= 0.2           Ly= 0.1
 k= 45                                   { Thermal conductivity }
 rcp= 3e6                                { Density * Specific_heat_capacity }
 heat= 0                                 { No volume heating }
 a0= 1                                   { Re(tempr) at boundary }
 omega= 0.25                             { Angular frequency }
 small= 1e-12
     log_tempr= log10( abs(tempr)+ small )
     log_tempi= log10( abs(tempi)+ small )
 tempa= sqrt(tempr**2+ tempi**2)               { Amplitude of oscillations }
     log_tempa= log10( abs(tempa)+ small )
 phase= sign(tempi)* arccos(tempr/tempa)/pi*180
 sq= sqrt( omega* rcp/ [ 2*k])                          { Exact solutions ...}
     tempr_ex= a0* cos(-sq*x)* exp(-sq*x)
     tempi_ex= a0*  sin(-sq*x)* exp(-sq*x)
     log_tempr_ex= log10( abs(tempr_ex)+ small )
     tempa_ex= a0* exp(-sq*x)
     log_tempa_ex= log10( abs(tempa_ex)+ small )
equations
 dx[-k*dx(tempr)]+ dy[-k*dy(tempr)]- heat- omega* rcp* tempi= 0
 dx[-k*dx(tempi)]+ dy[-k*dy(tempi)]+ omega* rcp* tempr= 0
boundaries
                  region 1
 start (0,0)
 natural(tempr)= 0  natural(tempi)= 0     line to (Lx,0) to (Lx,Ly) to (0,Ly)
 value(tempr)= a0  value(tempi)= 0   line to finish
     { Input temperature, real and imaginary part }
monitors
 elevation( log_tempr) from (0,Ly/2) to (Lx,Ly/2)
plots
 elevation( tempr, tempi) from (0,Ly/2) to (Lx,Ly/2)   as 'tempr,   tempi'
 elevation( phase) from (0,Ly/2) to (Lx,Ly/2)
 elevation( log_tempr, log_tempi, log_tempa)  from (0,Ly/2) to (Lx,Ly/2)
      as 'log_tempr, log_tempi, log_tempa'
 elevation( log_tempr, log_tempr_ex)  from (0,Ly/2) to (Lx,Ly/2)
 elevation( log_tempa, log_tempa_ex)  from (0,Ly/2) to (Lx,Ly/2)
 surface( log_tempr) interactive
end
```

The figure below is a *surface* plot of the absolute value of the real part of the solution produced by PDEase. As the linear plots indicate, the real part oscillates between positive and negative values, hence passing through zero. If it were not for the normal errors of the FEA calculations, the values would periodically drop to 10^{-12}, which is our token for an infinitely small value. In other words, we should interpret the minima in the surface plot as being effectively zeros. The figure shows absolute values; hence the interval from one zero to the next is one half wavelength. The *surface* plot reproduces the first two full wavelengths rather well, but the third one is strongly corrupted by computational noise.

The comparison with the exact solution shows excellent agreement between the absolute values, down to a level of 10^{-6} of the input amplitude. The absolute value is the amplitude of the temperature oscillation: at every point of the object there is an oscillation at the same frequency as that used to excite it.

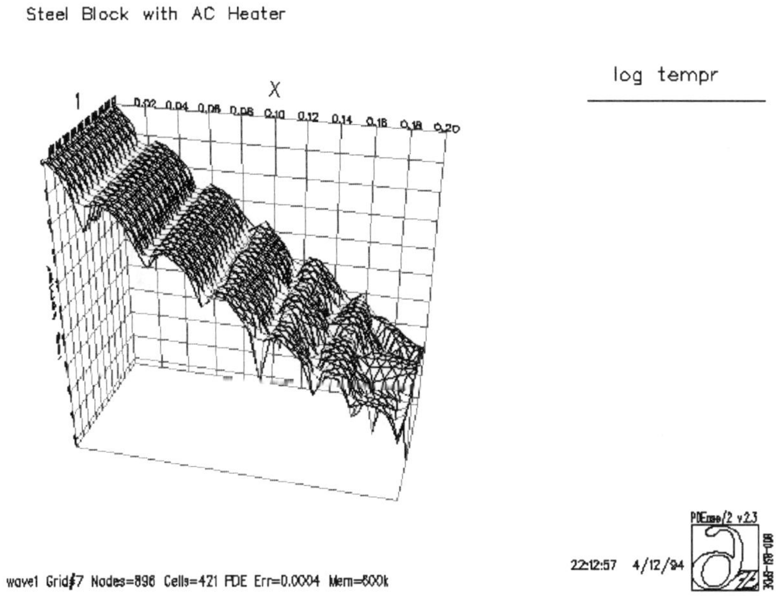

The interpretation of the above plot is the following. According to the rules of the "$j\omega$-method" the expression for the actual temperature increment is

$$\mathrm{Re}\left[(T_r(\mathbf{r}) + iT_i(\mathbf{r}))\exp(i\omega t)\right]$$

For particular values of the time, corresponding to $\omega t = 2\pi n$ the exponential factor becomes equal to 1, and for these times the temperature increment is simply given by

$$\mathrm{Re}\left[(T_r(\mathbf{r}) + iT_i(\mathbf{r}))\exp(i\omega t)\right] = T_r(\mathbf{r})$$

Thus the real part $T_r(\mathbf{r})$ shown in the above figure is a "snapshot" of the temperature distribution at a particular time. Similarly, the imaginary part yields the temperature distribution at another point of time.

The agreement with respect to wavelengths, shown by the plots of the real part, seems satisfactory. The real or imaginary part of a temperature oscillation may be measured by lock-in techniques, using the heater voltage as the reference signal.

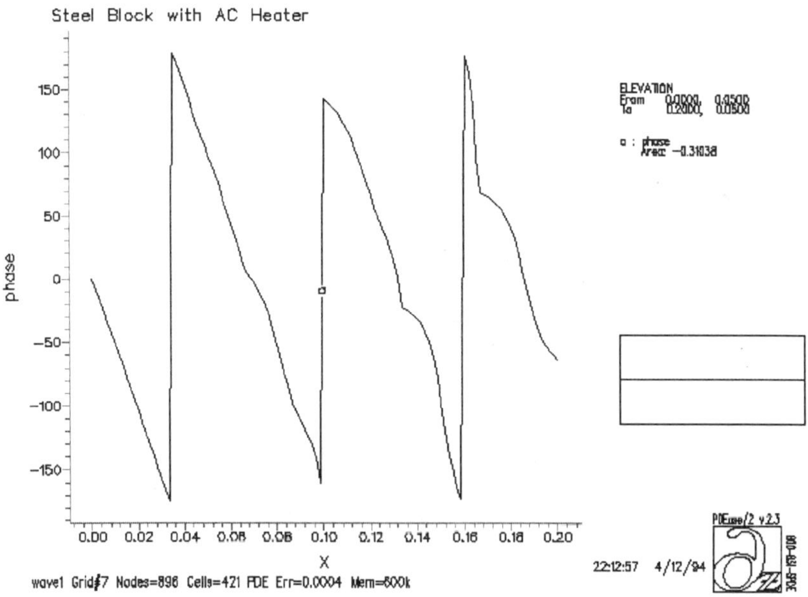

The preceding figure shows curves of the phase angle of the solution versus x. Evidently the angle continuously decreases as we move to the right, until the value reaches -180^0, where the angle is shifted by 360^0 to display the next period. If we wish, we could ignore the vertical jumps and consider the phase angle to be presented by a continuous curve, forever decreasing. This plot shows in even more detail how far the calculations reproduce the wave: after two wavelengths the distortion becomes appreciable.

Temperature waves are commonly observed natural phenomena. In summertime the sun heats the surface of the earth much more intensely than during winters, and the result is a temperature wave, which of course is most noticeable on the ground. These oscillations may, however, also be measured deep down in a mine shaft - attenuated and retarded as suggested by our calculations.

Steel Sheet Under Sinusoidal Volume Heating

The next case for study is a steel sheet, clamped between copper jaws, such that the left and right edges are forced to have a constant temperature of 300K. The jaws are also used to conduct an electric current through the sheet, shaped in time to produce a sinusoidally varying power. Most of the preceding descriptor still applies, and the following list contains the modifications. Previous comparisons with an exact solution are no longer relevant and should be erased.

```
...
heat= 1e0                                { Amplitude of volume power }
...
            region 1
start (0,0)
natural(tempr)= 0  natural(tempi)= 0  line to (Lx,0)
value(tempr)= 0   value(tempi)= 0   line to (Lx,Ly)
natural(tempr)= 0  natural(tempi)= 0  line to (0,Ly)
value(tempr)= 0   value(tempi)= 0   line to finish
...
```

The plot below shows that the real and imaginary parts both are zero at the left and right edges, as required. In the middle of the sheet the imaginary part dominates, i.e. the temperature oscillations are retarded by about 90^0 with respect to the power applied. The logarithmic plots of the real part indicate the wavelength.

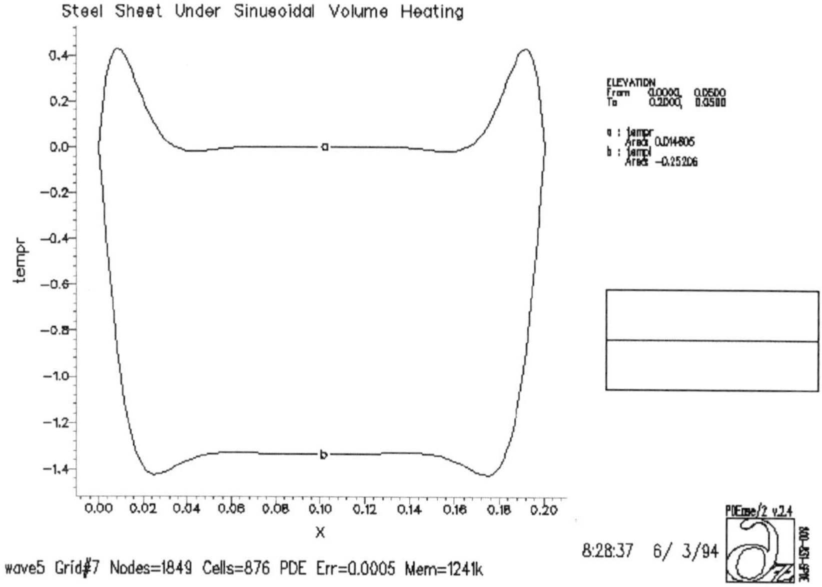

The two examples above were effectively one-dimensional, although we never declared this in the descriptor. PFEase successfully deduced from the PDE and the boundary conditions that the solution does not vary with *y*. If the thermal properties had varied in space, however, we would have had a genuinely two-dimensional situation, and the program would have solved the problem all the same.

Sinusoidal Heating of a Block by a Strip

We shall now consider an oscillating solution in 2D. A heater strip, extending in the *z* direction, provides heat flux over the middle part of the upper face of a block. The lower face is at 300 K, and the others are insulated.

```
...
errlim= 5e-4                    paint= off
...
Lx= 0.2               Ly= 0.2              d= 0.03
...
heat= 0                                         { No volume heating }
f 0= 1e4                                     { Amplitude of strip flux }
...
         region 1
start (0,0)
natural(tempr)= 0  natural(tempi)= 0  line to (Lx,0)
value(tempr)= 0  value(tempi)= 0  line to (Lx,Ly)
natural(tempr)= 0  natural(tempi)= 0  line to (0,Ly) to (0,Ly/2+d/2)
natural(tempr)= -f 0  natural(tempi)= 0  line to (0,Ly/2-d/2)
                                              { Outward flux density }
natural(tempr)= 0  natural(tempi)= 0  line to finish
...
```

The following *surface* plot shows a snapshot of the temperature wave created by the heater strip. We can clearly distinguish five zero crossings, but at large distance from the source the errors become conspicuous.

A simple resistive heater can only heat, not cool, the specimen. Hence there will also be a certain amount of average heating, which we have not taken into account. We could well solve for steady conduction separately, however, just adding the two solutions to obtain the total temperature field.

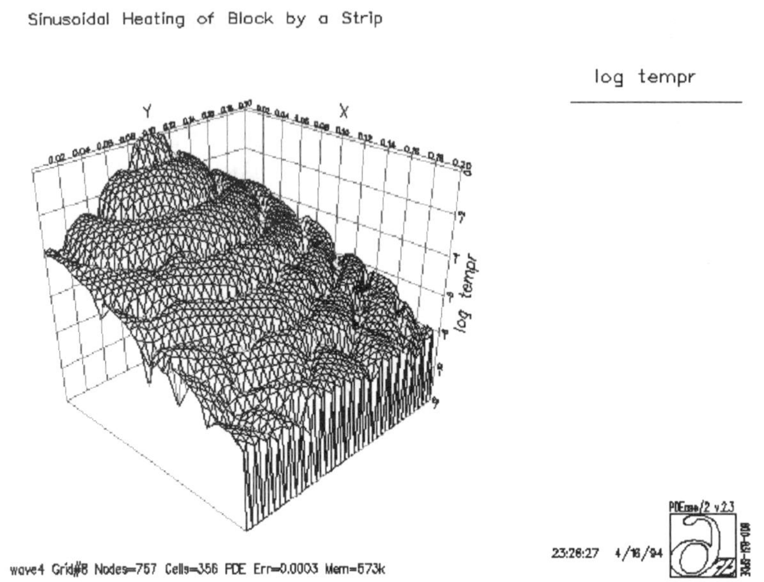

Exercises

❑ In the first example the curve of absolute values diverges gradually from the exact one as we approach the right edge. Test if this is due to the influence of the right boundary, by moving it further to the right. Could other factors also contribute to this divergence?

❑ Change the first descriptor slightly, to let the program "know" that the problem is one-dimensional. Does this improve the accuracy?

❑ Modify the first example by inputting an oscillating flux density on the left surface, instead of an oscillating temperature. What is the major change you observe in the solution? Modify the exact solution, by allowing for a complex amplitude (*a*0), and compare to the FEA solution.

❑ Starting from the last example, write a descriptor for a block of pine wood with the fibers in the *y* direction. Choose a suitable frequency to bring out the temperature wave.

Transient Problems in One Space Dimension

In the preceding chapter we treated cases of oscillating conditions, giving rise to sinusoidal solutions. We shall now consider more general problems, explicitly containing the time variable. In order to introduce difficulties gradually we first confine ourselves to two independent variables, t and x.

Iron Bar With a Temperature Step

The descriptor below specifies a rectangular iron bar, initially at 300 K. The keyword *initial values* now has a completely different sense than on earlier occasions. It no longer signifies a guess value, intended to help the program on its way, but it states that at time zero the object has exactly the temperature given in this segment. The initial value need not be a constant as in the present case, but could be any function of space.

The other temperature given is that at the left boundary of the object. Here we have entered a constant value, 400 K, which means that the left face will keep this temperature for $t>0$.

The equation is the time-dependent PDE for heat conduction already used (p.130). We have assumed equal conductivities (k) in the x and y directions, but the program would handle anisotropic situations as well.

There is an exact solution to a similar problem[4p63], which we may use for comparison to conquer our natural skepticism. That solution applies to a bar with infinite extension along the positive x-axis, but for sufficiently short times, when the temperature at the right end of our bar has not increased noticeably, it should be an excellent approximation. It is easy to verify, by direct substitution, that the following function satisfies the PDE.

$$T = (T_i - T_0)\mathrm{erf}\left(\frac{x}{2\sqrt{\lambda t / \rho c_p}}\right) + T_0 \text{ , where } \mathrm{erf}(y) = \frac{2}{\sqrt{\pi}}\int_0^y \exp(-\eta^2)d\eta$$

The new keyword *time* in the descriptor applies when the problem contains the time explicitly. The plot statement involving *for...by...to* obviously specifies at what times we wish to see a snapshot of the solution. Since we are allowed to set up several successive loops, it is easy to cover a large range of time without filling the disk with hundreds of figures.

```
select                                                    { trans1.pde }
  errlim= 1e-3
variables
  temp
definitions
  Lx= 1                   Ly= 0.1
  { conductivity,     heat capacity  of Fe }
  k= 82                   rcp= 7.87e3* 449
  tempi= 300              temp0= 400
  temp_ex= (tempi-temp0)* erf[ x/(2* sqrt(k*t/rcp))]+ temp0
  rel_error= (temp-temp_ex)/ (temp0-tempi)
initial values
  temp= tempi
equations
  dx[-k*dx(temp)]+ dy[-k*dy(temp)]+ rcp* dt(temp)= 0
boundaries
        region 1
  start (0,0)
  natural(temp)= 0  line to (Lx,0) to (Lx,Ly) to (0,Ly)
  value(temp)= temp0   line to finish           { Temperature for time> 0 }
TIME
  0 TO 10000
monitors
plots
  FOR t= 500  BY  500  TO  2000
    elevation( temp, temp_ex) from (0,Ly/2) to (Lx,Ly/2)
    contour( rel_error) zoom(1e-3*Lx,0,Lx,Ly)
  for t= 3000 by 1000 to 10000
    elevation( temp, temp_ex) from (0,Ly/2) to (Lx,Ly/2)
end
```

The first few snapshots show excellent agreement with the exact solution, as demonstrated by the *contour* plot of the relative error (below). The deviation from 1D behavior is evidently also negligible.

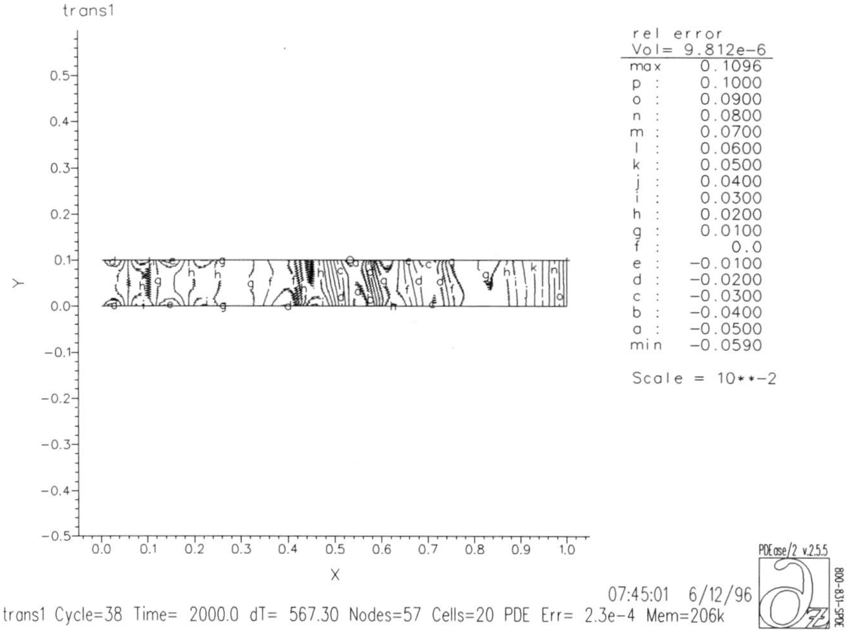

At large times, however, the *elevation* curves begin to diverge as one approaches the right end. The fault is not with the numerical solution: the difference is a natural consequence of the boundary conditions at the right face. The exact solution applies to the case of infinite extension of the bar, whereas we have specified a finite length. We only need to move the right face further along the *x*-axis to restore the agreement.

Insulated Bar With a Temperature Step Half-Way

We shall now modify the preceding descriptor to solve an essentially different problem. Let us insulate all boundaries of the iron bar and introduce a temperature step in the middle. In practice, we could have two equal parts, kept at different temperatures, and join them at time zero. We need to introduce this temperature step in the *initial values* segment, by means of a discontinuous function as shown below.

```
...                                                    { trans7.pde }
initial values
  temp= 200+ 200* ustep(x-Lx/2)          { Unit step function, ustep( ) }
  ...
  region 1
  start(0,0)
  natural(temp)= 0  line to (Lx,0) to (Lx,Ly) to (0,Ly) to finish
  ...
time
  0 to 30000
monitors
plots
  for t= 1, 3, 10, 30, 100, 300, 1000, 3000, 10000, 30000
    elevation(temp) from (0,Ly/2) to (Lx,Ly/2)
    contour(temp)
end
```

In the plot segment we use a variant of the previous snapshot formula, which is expedient when you need time intervals of non-uniform lengths. The *elevation* plot below shows the solution after *t*=100, which is late enough to round off the initial step function. The series of *elevation* and *contour* plots demonstrate that the integral of temperature (*Area*, *Vol*) is independent of time, a result which we readily understand in terms of energy conservation.

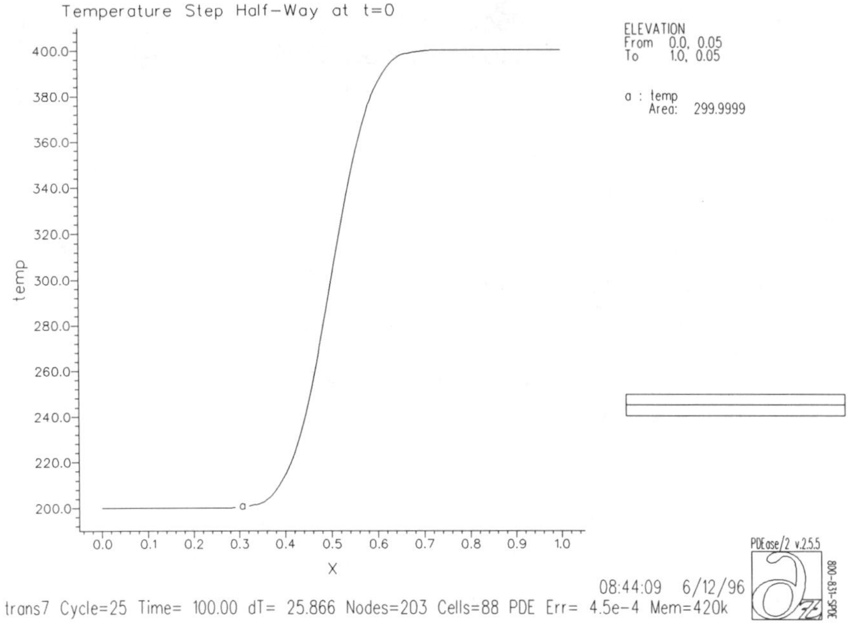

Iron Bar Soldered to a Copper Bar

Even in transient problems the materials properties may vary in space. In the following example we let the first half of the specimen be of Cu, the second half of Fe. We again impose *T*=400 on the left face, as in the first example (p.140).

This descriptor contains the new keywords *history* and *histories*, which permit us to draw curves of the time dependence at various points in space.

```
...                                                          { trans3.pde }
fluxd_x= -k*dx(temp)
...
            region 1
k= 82           rcp= 7.87e3* 449                              { For Fe }
start (0,0)  natural(temp)= 0  line to (Lx,0) to (Lx,Ly) to (0,Ly)
value(temp)= temp0   line to finish
            region 2
```

```
  k= 400                 rcp= 8.96e3* 385                                    { For Cu }
  start (0,0)  line to (0.2*Lx,0) to (0.2*Lx,Ly)  to (0,Ly) to finish
time
  0 to 10000
plots
  for t= 500 by 500 to 2000
    elevation( temp) from (0,Ly/2) to (Lx,Ly/2)
  for t= 3000 by 1000 to 10000
    elevation( temp) from (0,Ly/2) to (Lx,Ly/2)
    elevation( fluxd_x) from (0,Ly/2) to (Lx,Ly/2)
HISTORIES
  HISTORY( temp) at (0.1*Lx,Ly/2) at (0.2*Lx,Ly/2) at (0.3*Lx,Ly/2)
           at (0.4*Lx,Ly/2) at (0.5*Lx,Ly/2)
end
```

The plot below shows the temperature variation of the composite bar at five different points. The insert to the right of the plot indicates the positions of the points where these recordings were taken.

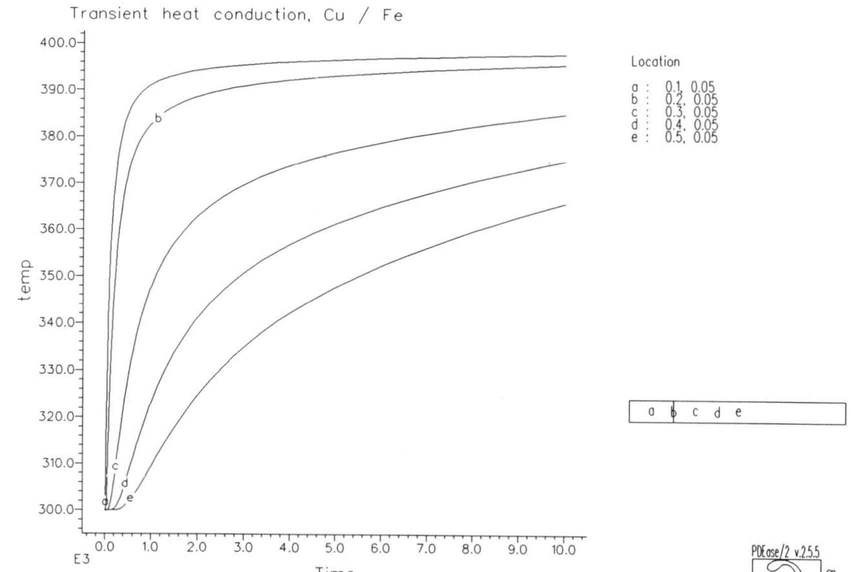

The *elevation* plots illustrate what happens at the Cu/Fe junction. Here the discontinuity in thermal conductivity causes a change of slope of the temperature curve. The *x*-component of the flux density varies continuously with *x*, however, as it should.

Ramp Function at Boundary

Boundary values may be any reasonable function of space and time. Let us now pass from a constant value to the simplest extension: a ramp function. The descriptor is the same as the preceding one, with the small differences listed below.

```
 ...                                              { trans5.pde }
variables
  temp( RANGE= 300, 500)
  ...
  tempi= 300          temp0= 300+ 0.01*t
  ...
```

We have now imposed a time-dependent temperature on the left face, starting at T=300, which happens to be equal to the initial temperature. We have added a *range* declaration for the temperature variable, giving the program a hint about tolerable errors.

Even if we omit the *range* statement, the calculations might eventually be successful, but the program would work long and hard in the initial phase. It would correctly conclude that the temperature increase is very slight in the beginning and try to produce a solution with an error less than 0.001 of that small increment.

The following plot displays the temperature distribution after t=3000. It shows the location of the step in thermal conductivity at the junction between copper and iron.

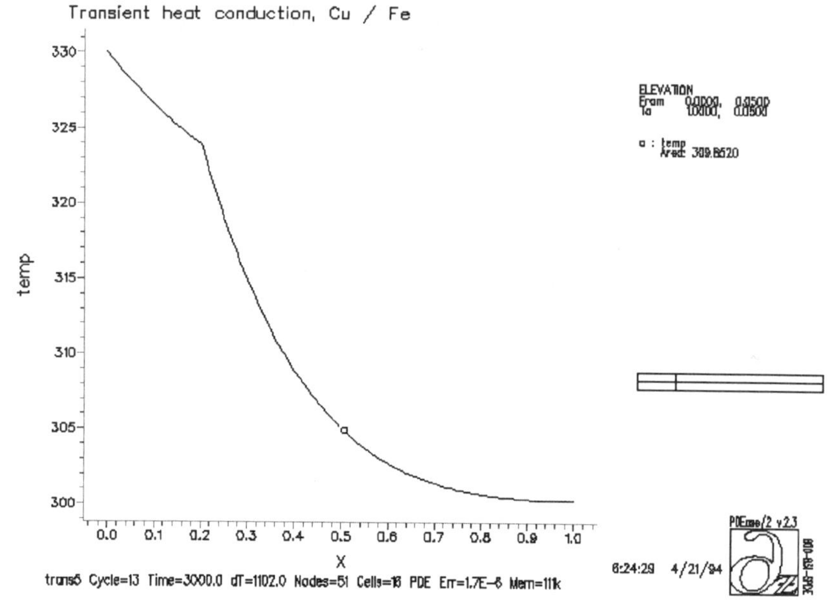

Exercises

❑ Modify the first descriptor by inputting a suitable flux density through the left face.

❑ Starting with the second descriptor, involving the unit step function, introduce a different material (Cu) in the right half of the bar. Will the integral still be independent of time? Could you find a function other than the temperature which would yield an invariant integral?

❑ Modify the example with the Cu/Fe bar (p.143) by keeping the right face at 300 K.

❑ Try another function of time at the left face than the linear slope proposed in the final example. Choose, for instance, the function *temp*0= 300+ 100* *ustep*(1000-*t*).

Transient Problems in (x,y) and (ρ,z) Space

In the preceding chapter all examples were effectively in one space dimension only, but nevertheless PDEase rendered solutions in 2D. Hence, there is nothing more to learn about descriptors in order to treat genuine problems in two space dimensions and the time. We only need to supply the appropriate boundary conditions. The calculations in this chapter will generally take a longer time, however, than previous examples.

Atomic Diffusion of Nickel into Copper

The random diffusion of atoms in solids is somewhat analogous to electric or thermal conduction. If there is a gradient of concentration, Fick's law (analogous to the law of heat conduction[4p7]) tells us that

$$\nabla(-D\nabla c)+\frac{\partial c}{\partial t}=0 \qquad \bullet$$

where D is the diffusion coefficient and c the concentration. Since the equation is homogeneous we may use any unit for concentration, e.g. mass per unit volume, which we shall use in this example. The flux density is $\varphi=-D\nabla c$.

In the following descriptor we have a triangular plate of Ni in the center of a circular plate of Cu, the two materials being put into contact with each other at zero time.

Transient problems generally take long to solve, especially if discontinuous functions are involved. In order to save time, we use the new command *nodelimit* under select. The program tries to reduce the RMS error of the solution below *errlim*, but if this should require more than 1000 nodes, a limit is put at approximately that value, at the expense of reduced accuracy.

Ni atoms diffuse into Cu, but at the same time Cu atoms diffuse into Ni, although we do not solve for that process. Ni atoms also diffuse among other Ni atoms, however, and we assume the diffusion coefficient for Ni atoms to be the same in both environments. We choose a value of D that would apply at some elevated temperature.

```
select                                                          { trans11.pde }
  errlim= 1e-3          nodelimit= 1000
  contours= 10          surfacegrid= 30          paint= off
variables
  conc(range= 0, 10e3)
definitions
  mm= 1e-3              r0= 100*mm
  s= 30*mm              hyp= s/ cos(pi/6)                  { Size of triangle }
  D= 1e-8                                              { Diffusion coefficient }
  factor                                                     { Domain factor }
  j_x= -D*dx(conc)      j_y= -D*dy(conc)
  jabs= sqrt( j_x**2+ j_y**2 )
initial values
  conc= 8.9e3* factor                                         { Density of NI }
equations
  dx(-D*dx(conc))+ dy(-D*dy(conc))+ dt(conc) = 0
boundaries
          region 1                                                 { Circular }
  factor= 0                                        { Zero concentration of Ni }
  start (r0,0)
  natural(conc)= 0
  arc to (0,r0) to (-r0,0) to (0,-r0) to finish
          region 2                                               { Triangular }
  factor= 1                                                          { All Ni }
  start (-s,-hyp/2)  line to (s,-hyp/2)  to (0,hyp)  to finish
time
  0 to 1e4
monitors
plots
  for t= 1e1, 3e1, 1e2, 3e2, 1e3, 3e3, 1e4
    contour( conc)        surface(conc)
    elevation( jabs) from (0,0) to (0,r0)
histories
  history( conc) at (0,0) at (0,r0)
end
```

The following figure shows the concentration of Ni after 10,000 seconds. The distribution was initially flat inside a triangular region, but after this time it has become rounded off, and nickel is visibly dispersing into the copper region.

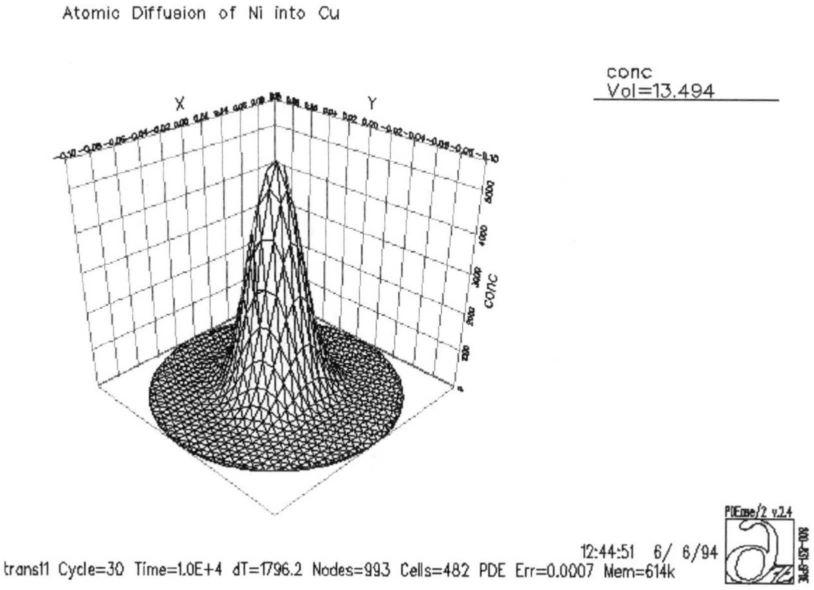

Internally Heated Copper Bar With Conductive Loss

In the following example a copper bar with square cross section is heated by electromagnetic induction, such that the heat dissipation is uniform and constant. Wads of glass fiber insulate the bar from the cylindrical enclosure, which is kept at 300 K by external cooling.

This descriptor will take 10-60 minutes to run, depending on the speed of your computer. In order to reduce the number of figures, while still reproducing the main features of the heating process, we use an approximately logarithmic scale of times in the plot commands.

```
  ...                                                        { trans21.pde }
variables
  temp
definitions
  mm= 1e-3   Lx= 30*mm      Ly= 30*mm     r0= 100*mm
  k          rcp         heat
  tempi= 300
  fluxd_x= -k*dx(temp)
initial values
  temp= tempi
equations
  dx(-k*dx(temp))+ dy(-k*dy(temp))- heat+ rcp* dt(temp) = 0
boundaries
          region 1
  k= 0.5            rcp= 2e6        heat= 0
  { conductivity, heat capacity  of glass fiber }
  start (r0,0)
  value(temp)= tempi
  arc to (0,r0) to (-r0,0) to (0,-r0) to finish
          region 2
  k= 45           rcp= 7.87e3* 449         heat= 1e5                { steel }
  start (-Lx,-Ly) line to (Lx,-Ly) to (Lx,Ly) to (-Lx,Ly) to finish
time
  0 to 1e5
monitors
plots
  for t= 3e2, 1e3, 3e3, 1e4, 3e4, 1e5
    contour( temp)       surface( temp)
    elevation( temp) from (-r0,0) to (r0,0)
    elevation( fluxd_x) from (-r0,0) to (r0,0)
histories
  history( temp) at (0,0) at (0.1*r0,0) at (0.2*r0,0) at (0.5*r0,0)
end
```

The *surface* plot below shows the temperature distribution in the bar at 1000 s. The full series of figures show that the distribution inside the bar is clearly convex for small times, while it becomes nearly constant at the end of the process. This may perhaps surprise you at first, but the explanation of this behavior lies in the amount of heat leaving the bar at various times.

The *elevation* plots of the flux density demonstrate that this function is continuous, as it should be. Three of the *history* curves fall accurately on top of each other, which confirms the space constancy of temperature inside the bar. In particular, these curves show that a steady state is reached at about half of the maximum time.

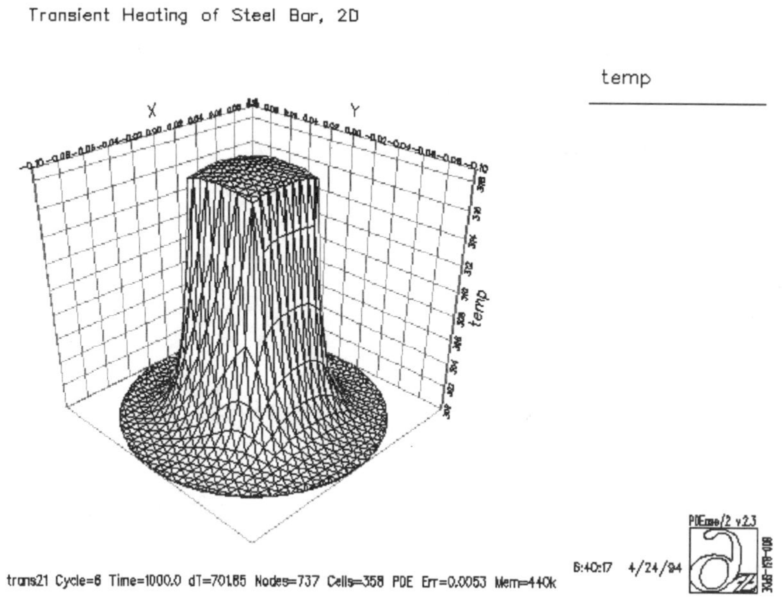

Capsule and Sample in a Scanning Calorimeter

When using a differential scanning calorimeter (DSC), one encloses a sample in a metal capsule, which is heated by contact with a base plate of linearly varying temperature. The instrument measures the power necessary to make the sample follow this temperature ramp. The question arises to what extent the sample retains a uniform temperature during this process.

The following descriptor (listed in part) defines the geometry and the thermal properties of an aluminum capsule, filled with a polymeric sample. The capsule is axially symmetric, and we thus use the PDE pertinent to cylindrical coordinates. We assume that the temperature at the bottom of the capsule varies linearly at a rate of 10 K/s.

In the plot section we use the keyword *viewpoint* for the first time. The *viewpoint* command replaces *interactive* in cases where it is not convenient to rotate the figure by means of the keyboard in order to obtain the best view. The first two arguments of *viewpoint* are the x and y coordinates of the eye of the viewer, and the third argument is its angular elevation above the xy plane.

```
...                                                        { trans31.pde }
variables
  temp(range= 300, 1000)
definitions
  mm= 1e-3       D= 0.1*mm       r0= 3.3*mm          Ly= 1.6*mm
  k              rcp             rate= 10
  tempi= 300         heat= 0
initial values
  temp= tempi
equations
  (1/x)*dx(-k*x*dx(temp))+ dy(-k*dy(temp))- heat+ rcp* dt(temp)= 0
boundaries
          region 1
  k= 0.3                 rcp= 3e6                   { Polymer specimen }
  start (0,0)
  value(temp)= tempi+ rate*t        line to (r0,0)
  natural(temp)= 0       line to (r0,Ly) to (0,Ly) to finish
          region 2
  k= 238                 rcp= 3e6                   { Aluminum capsule }
  start (0,0)  line to (r0,0) to (r0,Ly) to (0,Ly)
     to (0,Ly-D) to (r0-D,Ly-D) to (r0-D,D) to (0,D) to finish
time
  0 to 100
monitors
plots
  for t= 3, 10, 30, 100
    elevation( temp) from (0,0) to (0,Ly)
    contour( temp)
    surface( temp)  VIEWPOINT(-r0,Ly/2,0)
end
```

The figure below shows the *contour* plot of the temperature distribution after 10 seconds. The heating is from below, and we see that the aluminum transmits the heat very well. In the middle of the specimen, however, the temperature has only just started to rise above the initial value.

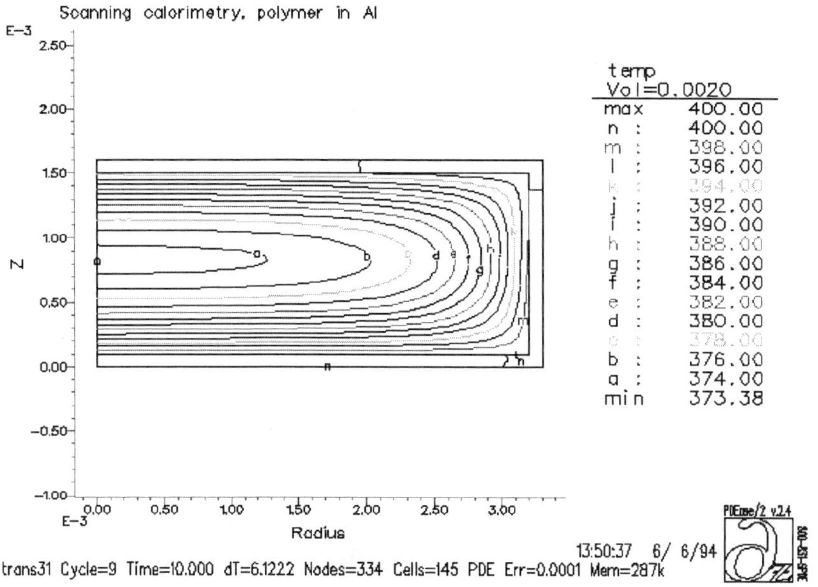

The next figure shows a temperature *surface* at 30 s, seen from the inside of the capsule. Here the heat enters from the right, and it is clear that the top of the capsule has not reached the same temperature as the bottom. The interior of the specimen lags behind even more. It is interesting to observe how these figures change as time goes on.

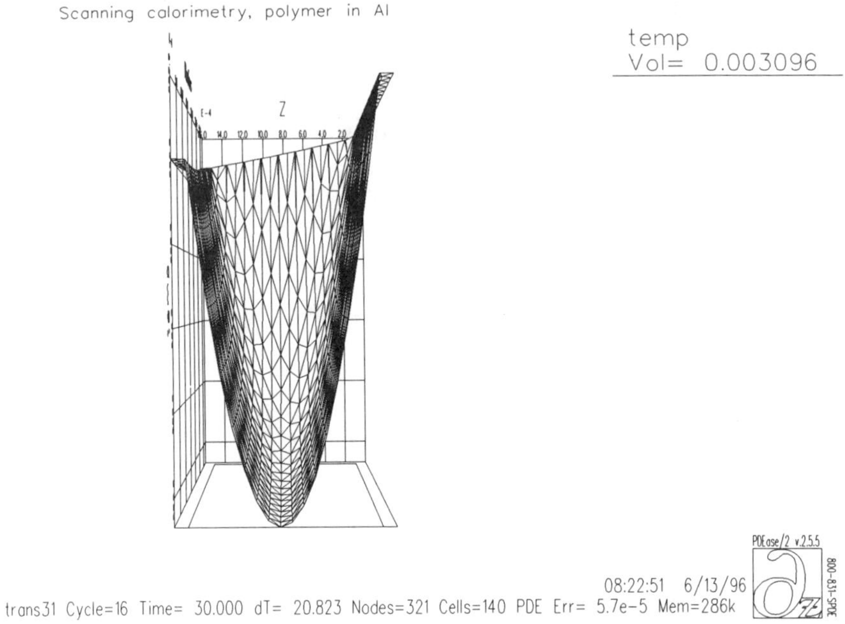

Transient Cooling of an MgO Plate

Finally we shall venture to solve a non-linear, transient problem in two spatial dimensions. We borrow the general configuration from an earlier problem (p.117), which we already solved in the steady regime. The temperature is initially 1000 K, when cooling is abruptly applied to the right edge of the plate, keeping the temperature there at 300 K.

The following descriptor is required. We choose a *viewpoint*, which permits us to watch the temperature surface from the front.

```
variables                                              { trans41.pde }
  temp(range= 300, 1000)
definitions
  Lx= 0.6         Ly= 0.4          r0= 0.1
  k= 15000/temp                    { Thermal conductivity versus T }
  rcp= 3e6                                         { Heat capacity }
```

```
initial values
  temp= 1000
equations
  dx[-k*dx(temp)]+ dy[-k*dy(temp)]+ rcp*dt(temp)= 0
boundaries
                    region 1
  start (0,0)
    natural(temp)= 0            line to (Lx,0)
    value(temp)= 300            line to (Lx,Ly)
    natural(temp)= 0            line to (0,Ly)
    value(temp)= 1000           line to finish
  start (Lx/2-r0,Ly/2)    natural(temp)= 0
    arc to (Lx/2,Ly/2+r0) to (Lx/2+r0,Ly/2) to (Lx/2,Ly/2-r0) to finish
time
   0 to 1e4
monitors
plots
  for t= 3e1, 1e2, 3e2, 1e3, 3e3, 1e4
    surface( temp)  viewpoint(1,-1,45)
end
```

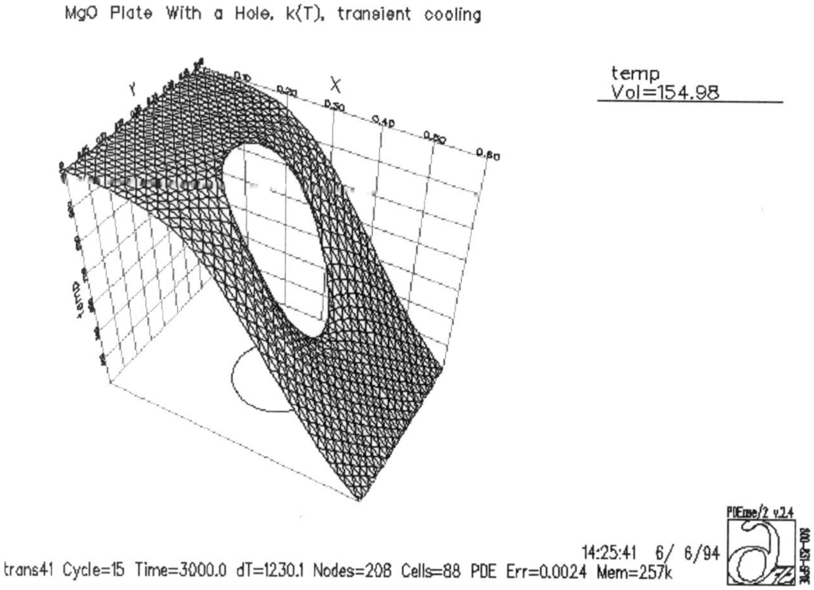

The above figure shows the temperature *surface* at 3000 s. The *surface* plot at the shortest time demonstrates that the program has difficulties handling abrupt changes in the dependent variable.

Exercises

❑ Study the diffusion of Cu into Ni using the same plate geometry and assuming the diffusion coefficients to be the same.

❑ Modify the problem concerning the square steel bar to display the flux leaving the bar. Compare that value to the heat dissipated by induction.

❑ How does the scanning calorimeter perform at lower rates of heating, say at 0.1 K/s, with respect to temperature uniformity in the specimen?

❑ Run the calorimeter descriptor again, this time including a thermal resistance between the base plate and the bottom of the capsule. Simulate the thermal resistance by an extra plate of suitable thermal conductivity and negligible heat capacity.

❑ Let the MgO plate initially be at 300 K and apply sudden heating to 1000 K at the left edge. Display the magnitude of the heat flux density as well.

Elementary Elasticity in (x,y) Space

The study of elastic deformation begins with the equations of mechanical equilibrium for a volume element[5p288]:

$$\begin{cases} \dfrac{\partial \sigma_x}{\partial x} + \dfrac{\partial \tau_{xy}}{\partial y} + \dfrac{\partial \tau_{xz}}{\partial z} + F_x = 0 \\ \dfrac{\partial \tau_{xy}}{\partial x} + \dfrac{\partial \sigma_y}{\partial y} + \dfrac{\partial \tau_{yz}}{\partial z} + F_y = 0 \\ \dfrac{\partial \tau_{xz}}{\partial x} + \dfrac{\partial \tau_{yz}}{\partial y} + \dfrac{\partial \sigma_z}{\partial z} + F_z = 0 \end{cases}$$

where σ are the normal and τ the shear components of stress associated with the three axes. The components of the external force **F** refer to unit volume, and may be gravitational, centrifugal or even magnetic.

We shall now assume that all stresses are independent of *z*, and that there is complete symmetry with respect to the *xy* plane. This reduces the equilibrium equations to

$$\begin{cases} \dfrac{\partial \sigma_x}{\partial x} + \dfrac{\partial \tau_{xy}}{\partial y} + F_x = 0 \\ \dfrac{\partial \tau_{xy}}{\partial x} + \dfrac{\partial \sigma_y}{\partial y} + F_y = 0 \end{cases} \qquad \bullet$$

The most general form of Hooke's law may be written[5p289]

$$\begin{cases} \varepsilon_x = \dfrac{1}{E}\left[\sigma_x - \mu\left(\sigma_y + \sigma_z\right)\right] + \alpha\Delta T \\ \varepsilon_y = \dfrac{1}{E}\left[\sigma_y - \mu\left(\sigma_x + \sigma_z\right)\right] + \alpha\Delta T \\ \varepsilon_z = \dfrac{1}{E}\left[\sigma_z - \mu\left(\sigma_x + \sigma_y\right)\right] + \alpha\Delta T \end{cases} \qquad \bullet$$

and

$$\gamma_{xy} = \frac{1}{G}\tau_{xy} \text{ , } \gamma_{xz} = \frac{1}{G}\tau_{xz} \text{ , } \gamma_{yz} = \frac{1}{G}\tau_{yz} \quad \text{with } \frac{1}{G} = \frac{2(1+\mu)}{E}$$

where ε are the normal strain components, γ the shear strain components, μ Poisson's ratio, α the linear thermal expansivity and ΔT a temperature increment.

We now wish to replace the stress components in the equilibrium equations by strain components, using Hooke's law. In the case of shear this is simple, but the manipulation of normal strains is slightly more complicated. Let us restate these equations in concise matrix form, after subtracting the term $\alpha\Delta T$ on both sides.

$$\begin{Bmatrix} \varepsilon_x - \alpha\Delta T \\ \varepsilon_y - \alpha\Delta T \\ \varepsilon_z - \alpha\Delta T \end{Bmatrix} = \frac{1}{E}\begin{bmatrix} 1 & -\mu & -\mu \\ -\mu & 1 & -\mu \\ -\mu & -\mu & 1 \end{bmatrix}\begin{Bmatrix} \sigma_x \\ \sigma_y \\ \sigma_z \end{Bmatrix} \qquad \bullet$$

Before we invert the matrix to obtain stresses in terms of strains, we must note that the third dimension may not be decoupled simply by "insulation" or by considering the cross-section of a long bar. According to Hooke's law, stresses in the *xy* plane generally produces strain in the *z* direction as well.

In elasticity we may choose one of two simple assumptions. The first alternative is vanishing *z*-stress, which means that no external forces oppose the deformations along the third axis. The other one is vanishing strain in the *z* direction. The latter would correspond to an object being constrained without friction between two absolutely rigid and immobile plates, which is a rare situation. Although we could treat either of those cases without difficulty, we choose the more commonly occurring condition of vanishing σ_z, known as plain stress, for the following examples.

With $\sigma_z = 0$ in the third row of the matrix equation, the two first rows become

$$\begin{Bmatrix} \varepsilon_x - \alpha\Delta T \\ \varepsilon_y - \alpha\Delta T \end{Bmatrix} = \frac{1}{E}\begin{bmatrix} 1 & -\mu \\ -\mu & 1 \end{bmatrix}\begin{Bmatrix} \sigma_x \\ \sigma_y \end{Bmatrix}$$

and after inverting the matrix (by hand, or using Derive®)

$$\begin{Bmatrix} \sigma_x \\ \sigma_y \end{Bmatrix} = \frac{E}{1-\mu^2}\begin{bmatrix} 1 & \mu \\ \mu & 1 \end{bmatrix}\begin{Bmatrix} \varepsilon_x - \alpha\Delta T \\ \varepsilon_y - \alpha\Delta T \end{Bmatrix} =$$

$$\frac{E}{1-\mu^2}\begin{Bmatrix} \varepsilon_x + \mu\varepsilon_y - (1+\mu)\alpha\Delta T \\ \mu\varepsilon_x + \varepsilon_y - (1+\mu)\alpha\Delta T \end{Bmatrix}$$

Returning now to the equations of equilibrium, we substitute expressions for the stresses and obtain the PDEs

$$\begin{cases} \dfrac{\partial}{\partial x}\left(\dfrac{E}{1-\mu^2}\left[\varepsilon_x + \mu\varepsilon_y - (1+\mu)\alpha\Delta T\right]\right) + \\ \qquad\qquad \dfrac{\partial}{\partial y}\left(\dfrac{E}{2(1+\mu)}\gamma_{xy}\right) + F_x = 0 \\ \dfrac{\partial}{\partial x}\left(\dfrac{E}{2(1+\mu)}\gamma_{xy}\right) + \\ \dfrac{\partial}{\partial y}\left(\dfrac{E}{1-\mu^2}\left[\varepsilon_y + \mu\varepsilon_x - (1+\mu)\alpha\Delta T\right]\right) + F_y = 0 \end{cases}$$

We now only need to introduce the differential expressions for the strains in terms of displacements in the x and y directions

$$\varepsilon_x = \frac{\partial u}{\partial x}, \quad \varepsilon_y = \frac{\partial v}{\partial y}, \quad \text{and} \quad \gamma_{xy} = \frac{\partial v}{\partial x} + \frac{\partial u}{\partial y}$$

in order to have the final equations for a PDEase descriptor

$$\begin{cases} \dfrac{\partial}{\partial x}\left(\dfrac{E}{1-\mu^2}\left[\dfrac{\partial u}{\partial x} + \mu\dfrac{\partial v}{\partial y} - (1+\mu)\,\alpha\Delta T\right]\right) + \\ \qquad\qquad \dfrac{\partial}{\partial y}\left(\dfrac{E}{2(1+\mu)}\left[\dfrac{\partial v}{\partial x} + \dfrac{\partial u}{\partial y}\right]\right) + F_x = 0 \\ \dfrac{\partial}{\partial x}\left(\dfrac{E}{2(1+\mu)}\left[\dfrac{\partial v}{\partial x} + \dfrac{\partial u}{\partial y}\right]\right) + \\ \qquad \dfrac{\partial}{\partial y}\left(\dfrac{E}{1-\mu^2}\left[\dfrac{\partial v}{\partial y} + \mu\dfrac{\partial u}{\partial x} - (1+\mu)\,\alpha\Delta T\right]\right) + F_y = 0 \end{cases}$$

The general procedure will be first to solve for the displacements, then compute strains by differentiation, and finally extract stresses by Hooke's law.

The boundary conditions to apply are rather obvious in the case of given displacements at external surfaces. Then we only need to specify the function describing the displacements in the x and y directions by *value*( ) commands. This procedure will become clear as we make use of it.

The situation where surface forces act on the object is more subtle. If a surface is perpendicular to the x-axis, for instance, we may consider the equilibrium of a small volume element limited on one side by the boundary. This immediately gives us the boundary conditions

$$\sigma_x = \overline{X} \text{ and } \tau_{xy} = \overline{Y},$$

where $\overline{X}$ is the external force component per unit area in the x direction, etc..

It now remains to introduce the external loads into boundary conditions. Here we must remember that PDEase makes no distinction between different kinds of problems. The program does not "know" that we are attempting to solve a problem of elasticity, rather than the heating of an object by an electric current. In a heat conduction problem (p.106) we used *natural*(*temp*) to specify the heat flux density imposed by the outer world. The equation we then used was

$$\frac{\partial}{\partial x}\left(-\lambda\frac{\partial T}{\partial x}\right)+\frac{\partial}{\partial y}\left(-\lambda\frac{\partial T}{\partial y}\right)-h= \quad , \quad \text{or } \frac{\partial f_x}{\partial x}+\frac{\partial f_y}{\partial y}-h=0,$$

if we substitute the components of flux density, **f**. In previous descriptors we entered the boundary condition *natural*(*temp*)= *flux_density_along_outward_normal*, or in terms of the Cartesian components

natural(*temp*)= $f_x \cos\theta + f_y \sin\theta$,

θ being the angle from the x-axis to the surface normal

The PDEs for elasticity, on the other hand, are based on the relations of mechanical equilibrium

$$
\begin{cases}
\dfrac{\partial \sigma_x}{\partial x} + \dfrac{\partial \tau_{xy}}{\partial y} + F_x = 0 \\
\dfrac{\partial \tau_{xy}}{\partial x} + \dfrac{\partial \sigma_y}{\partial y} + F_y = 0
\end{cases}
$$

where the first one refers to forces directed along the x (or u) direction, and so forth. By analogy with the case of heat conduction we should specify, for the first equation,

natural(u)= $\sigma_x \cos\theta + \tau_{xy} \sin\theta$.

By considering the equilibrium of a wedge-shaped volume element[6p28] we find that

$\overline{X} = \sigma_x \cos\theta + \tau_{xy} \sin\theta$,

where $\overline{X}$ is the x-component of the surface force per unit area. This expression is evidently identical to the one we obtained by analogy. Hence, the boundary condition for u should be

natural(u)= $\overline{X}$

and correspondingly for the other component of displacement. PDEase also lets us use the equivalent statement

load(u)= $\overline{X}$

and so on, in order to guide our intuition.

Steel Plate Under Compressive Strain

We shall now convince ourselves that the above system of PDEs really yields correct answers when solved by PDEase. For this purpose we choose the simplest possible situation, so that we may easily check the solution by pencil and paper. Let us take a thick plate of steel and shorten it by applying compressive strain only. Practically, we could press it between two rigid, smooth and well lubricated anvils, such that the plate is free to slide sidewise, except at the middle point of one of the faces.

In the descriptor below we have declared two dependent variables, *u* and *v*, which are the displacements along the axes. There are also two PDEs, as must be, and we have simplified them by assuming constant temperature and neglecting volume forces, such as gravitation.

In the boundaries segment we use the command *load* for the first time. It is completely equivalent to *natural*, and we use it only because this word is more suggestive in cases where we are applying forces on boundary surfaces. On the bottom surface we thus apply no load, neither horizontally (*u*) nor vertically (*v*). The point value serves to fix the plate in the vertical direction, to avoid drift during calculations.

The *grid* plot, which is also new, displays the boundaries of the steel plate, including displacements, such that the deformation of the plate becomes enlarged by the factor 200. Most of the other plots are of the *contour* type, but you could try *surface* plots to explore if they are easier to read.

```
select
  errlim= 1e-6            paint= off
variables
  u  v
definitions
  mm= 1e-3        Lx= 40*mm          Ly= 30*mm
  mu = 0.3        E  = 200e9                                      {Steel}
  c= E/(1-mu**2)        G = E/[2*(1+mu)]
  dabs= sqrt(u**2+ v**2)
  ex= dx(u)         ey= dy(v)          exy= dx(v)+ dy(u)           { Strains }
                        { Stresses }
  sx= c*(ex+ mu*ey)   sy= c*(mu*ex+ ey)      sxy= G*exy
initial values
equations                  { Constant temperature, no volume forces }
  dx( c*(dx(u) + mu*dy(v)) ) + dy( G*(dx(v)+ dy(u)) )= 0
  dx( G*( dx(v)+ dy(u)) )+  dy( c*(dy(v) + mu*dx(u)) )= 0
boundaries
                  region 1
  start (0,0)       { No surface forces }
  LOAD(u)= 0     load(v)= 0          line to (Lx,0)
  value(u)= -1e-3*Lx          load(v)= 0          line to (Lx,Ly)
```

```
  load(u)= 0        load(v)= 0             line to (0,Ly)
  value(u)= 0       load(v)= 0             line to (0,Ly/2)
  point value(v)= 0
  value(u)= 0    load(v)= 0                line to finish
monitors
plots
  GRID( x+200*u, y+200*v)
  contour( u)               contour( v)
  vector( u/dabs, v/dabs)
  contour(ex)               contour(ey)               contour(exy)
  contour(sx)               contour(sy)               contour(sxy)
end
```

The following figure shows contours of constant vertical displacement. The values are negative in the lower half of the plate and positive in the upper half, in quantitative agreement with Hooke's law. The contours are also straight lines, as expected.

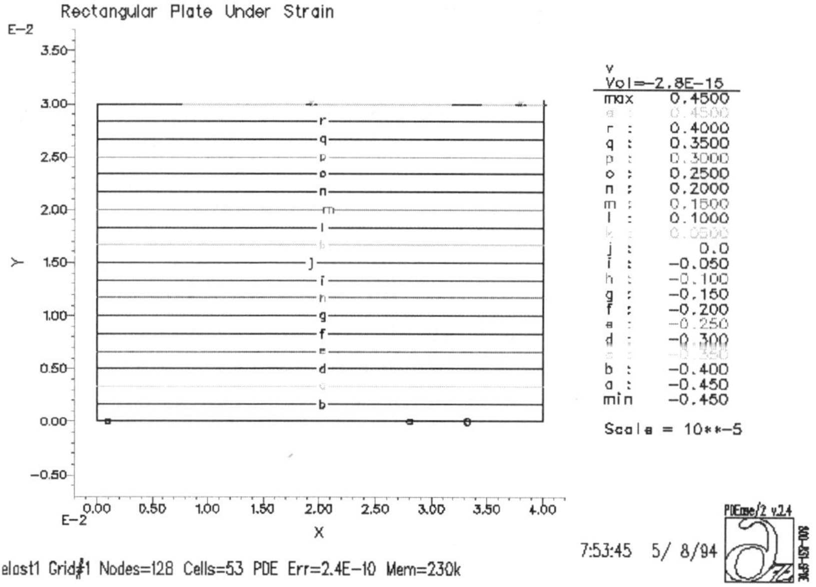

The next plot shows how the atoms move as the plate is compressed. It is instructive to compare this figure with the *grid* plot of the deformation. The vertical displacement is independent of x, but the existing slope of the vector displacement in the upper-right corner is hardly visible. At the left face of the plate, however, the horizontal displacement vanishes, and at that end the vertical motion thus dominates.

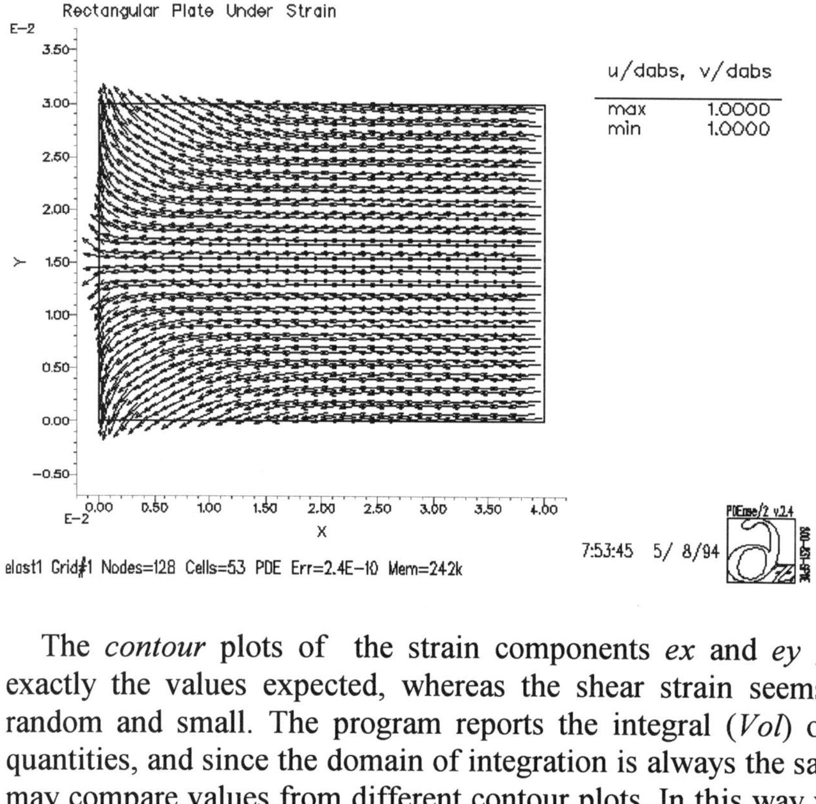

The *contour* plots of the strain components *ex* and *ey* give us exactly the values expected, whereas the shear strain seems to be random and small. The program reports the integral (*Vol*) of these quantities, and since the domain of integration is always the same, we may compare values from different contour plots. In this way we find that the average value of the shear strain *exy* is smaller than the vertical strain by the factor 10^6. The plots of the stress components also agree beautifully with elementary calculations.

The above example is the simplest possible, but the PDEs are far from trivial, and the same is true of the numerical methods exploited. The fact that we obtain such excellent agreement with elementary calculations is thus an impressive demonstration of the accuracy and efficiency of PDEase.

Plate, Constrained Vertically at Ends

This is a variation of the previous problem, the difference being that we assume that the anvils surfaces are sufficiently rough to prevent sliding.

The descriptor modifications are listed below. Since both ends of the plate are fixed by *value* statements, we no longer need the point value.

```
...
errlim= 5e-6
...
            region 1
start (0,0)
load(u)= 0             load(v)= 0        line to (Lx,0)
value(u)= -1e-3*Lx     value(v)= 0       line to (Lx,Ly)
load(u)= 0             load(v)= 0        line to (0,Ly)
value(u)= 0            value(v)= 0       line to finish
```

The following figure shows the resulting *grid* plot of the deformation. The free surfaces have now become clearly convex. Because of the constraints on the vertical faces the solution has large curvature at the corners, and these regions require denser grids for the calculations.

The second plot below shows that the constrained sides produce shear stress, especially near the corners. The contour for zero shear stress is forked, and the variation is small in the region of those curves, as indicated by the distance to the next contour. This is one of the reasons why these contours appear somewhat irregular. The other reason is, of course, that they are based on derivatives of the solution for the displacements.

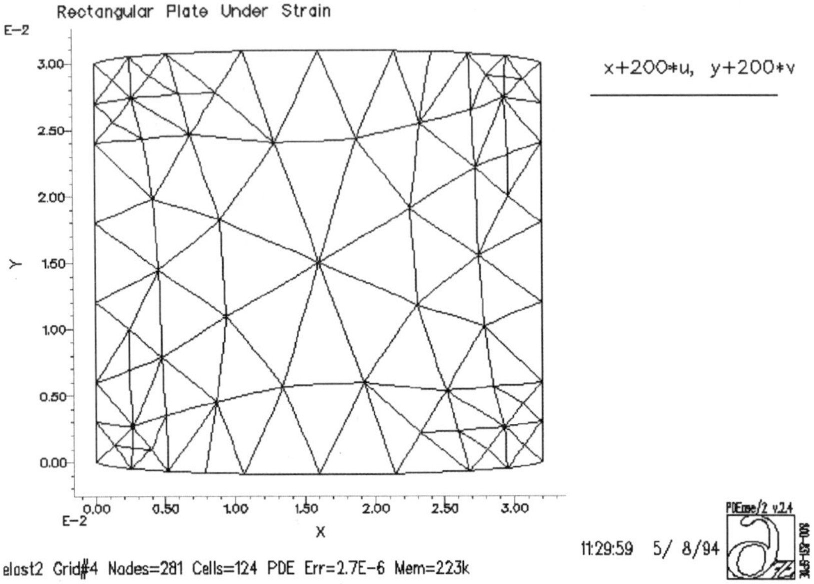
Rectangular Plate Under Strain
x+200*u, y+200*v
Y
X
E-2
11:29:59 5/ 8/94
PDEase/2 v2.4
elast2 Grid#4 Nodes=281 Cells=124 PDE Err=2.7E-6 Mem=223k

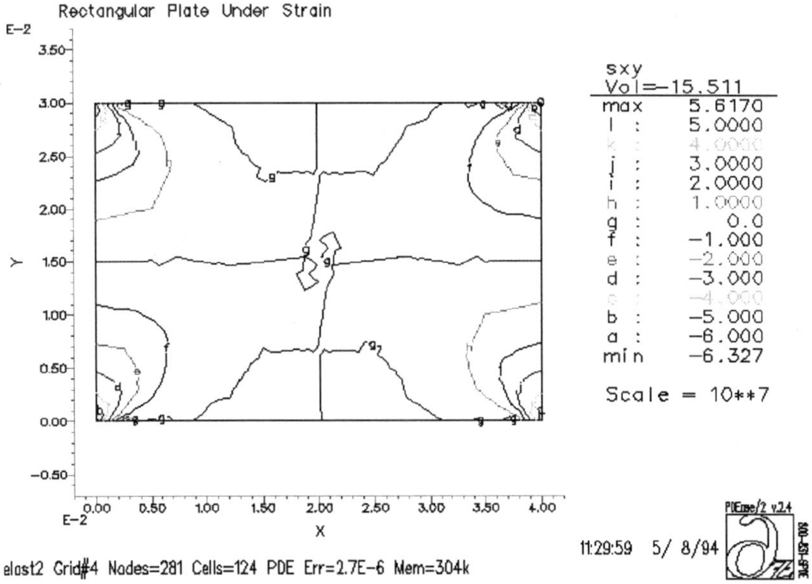
Rectangular Plate Under Strain
sxy
Vol=-15.511
max 5.6170
l : 5.0000
k : 4.0000
j : 3.0000
i : 2.0000
h : 1.0000
g : 0.0
f : -1.000
e : -2.000
d : -3.000
c : -4.000
b : -5.000
a : -6.000
min -6.327
Scale = 10**7
Y
X
E-2
11:29:59 5/ 8/94
PDEase/2 v2.4
elast2 Grid#4 Nodes=281 Cells=124 PDE Err=2.7E-6 Mem=304k

Plate Subjected to Pressure at One End

The next problem is asymmetrical. We keep the left end of the plate fixed and apply pressure at the right end, according to the rules of derivative boundary conditions.

```
errlim= 1e-6
...
            region 1
start (0,0)
load(u)= 0              load(v)= 0      line to (Lx,0)
load(u)= -3e8           load(v)= 0      line to (Lx,Ly)   { Pressure }
load(u)= 0       load(v)= 0          line to (0,Ly)
value(u)= 0      value(v)= 0         line to finish
```

The next figure shows the distribution of σ_x. As you can see, the right half of the plate has the expected stress, which means that we succeeded in imposing our derivative boundary conditions.

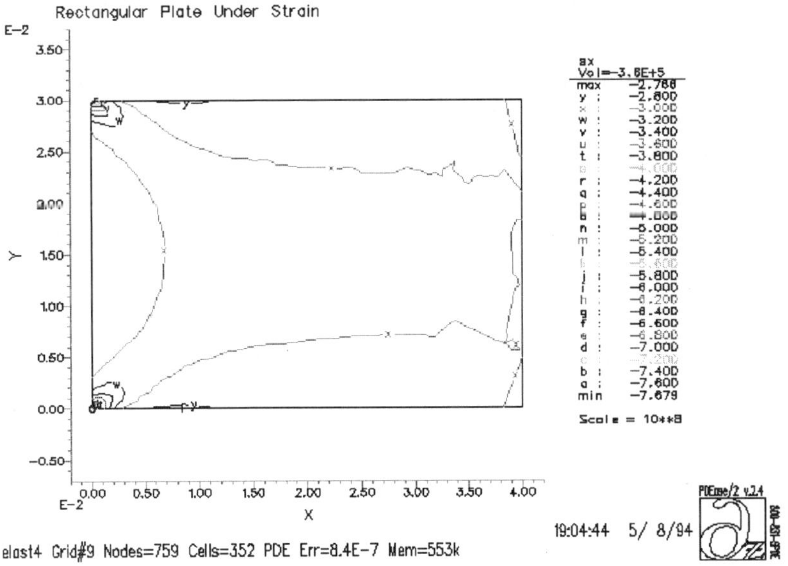

Plate Subjected to Shear Forces

As a final example in our round of the elementary phenomena of elasticity we shall now study the same plate as before, but with tangential load on all faces, as specified in the following descriptor. The force per unit area is the same on all four faces, except for sign.

```
...
errlim= 1e-9
...
c= E/(1-mu**2)          G = E/[2*(1+mu)]          shear= 2e8
...
          region 1
start (0,0)             point value(u)= 0
load(u)= -shear         value(v)= 0               line to (Lx,0)
load(u)= 0              load(v)= shear            line to (Lx,Ly)
load(u)= shear          load(v)= 0                line to (0,Ly)
load(u)= 0              load(v)= -shear           line to finish
...
```

The figure below shows the shape of the plate after the forces have been applied, the deformation being multiplied by 200 for clarity. It is interesting to interpret the other plots in detail. The average vertical displacement (v) is smaller than the horizontal one by a factor of 10^9, as you may see from the integral values. In fact, theory leads us to expect that v should vanish completely.

The shear strain comes out as exactly 0.2600E-2, whereas the mean vertical strain is much smaller, also by the factor 10^9. In short, the numerical results are in marvellous agreement with simple theory.

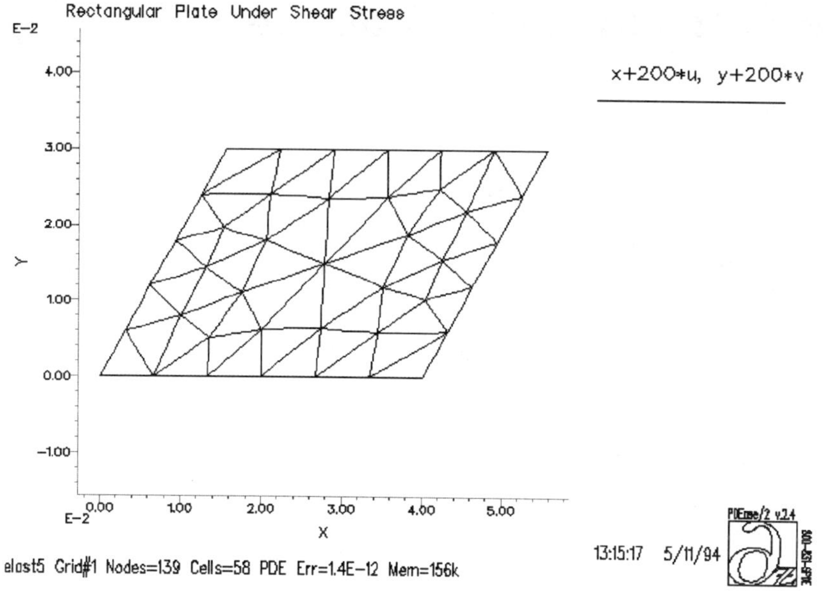

Exercises

❑ Assume the plate in the second example to be welded to the anvils and reverse the sign of the displacement at the right face. Study the deformation and compare.

❑ Shear the rectangular plate again, using *value* statements only. Fix the bottom face and constrain the upper face to move by a certain distance to the right, keeping the height constant. Impose x-deformations proportional to y on the vertical faces, such that the boundary becomes continuous.

More Elasticity in (x,y) Space

In this chapter we shall approach problems of some technological importance and also elaborate on the graphical representation of the results. Although technical applications usually are quite complex, involving many parts, we shall insist on the simplest possible example of any given kind. It is rather easy to build a complex model from elementary constituents, but complexity is not very instructive.

Tube With Internal Pressure

We shall now consider a case of slightly more complicated loads than before. In this first example the surface forces are zero on the outer surface of a tube, and liquid pressure acts on the inside. We can separate this pressure (force per unit area) into its x and y components and use those in the derivative (*natural* or *load*) boundary conditions. As discussed in the preceding chapter (p.160), the components of internal pressure must balance the stresses in the solid material along the two Cartesian directions. The following descriptor, which is based on what we used before, shows the simple expressions required for the loads.

This problem strictly involves only *load* boundary conditions. Since the PDEase solver performs better if *value* boundary conditions are present, we prefer to treat one-quarter of the cylinder, which also reduces the number of nodes. On the radial surfaces there will then be zero displacement along one axis and zero *load* (normal derivative) along the other one, both for symmetry reasons.

PDEase versions 2.5 and later permit us to write PDEs in terms of expressions given in the *definitions* segment. Here we exploit this convenient feature to bring out the inherent simplicity of elastic theory. In the *equations* segment we just specify the conditions of equilibrium of the forces acting on a volume element.

```
select                                                        { elast11.pde }
  errlim= 1e-4
variables
  u     v
definitions
  mm = 1e-3             r1 = 2*mm              r2 = 10*mm
  mu = 0.3              E = 200e9                                 {Steel}
  c = E/(1-mu**2)       G = E/2/(1+mu)
  dabs= sqrt(u**2+ v**2)
  ex= dx(u)             ey= dy(v)              exy= dx(v)+ dy(u)
  sx= c*(ex+ mu*ey)     sy= c*(mu*ex+ ey)      sxy= G*exy
  p1= 1e8               p1x= p1*x/r1           p1y= p1*y/r1
equations
  dx( sx) + dy( sxy)= 0                                  { Simplified PDEs }
  dx( sxy)+  dy( sy)= 0                                  { No volume forces }
boundaries
          region 1
  start (r1,0)
  load(u)= 0        value(v)= 0     line to (r2,0)
  load(u)= 0        load(v)= 0      arc(center= 0,0) to (0,r2)
  value(u)= 0       load(v)= 0      line to (0,r1)
  load(u)= p1x      load(v)= p1y                            { Pressure }
  arc(center= 0,0)  to finish
monitors
  contour(dabs)
plots
  grid(x+200*u, y+200*v)
  contour(u)        contour(v)          contour(dabs)
  vector(u,v)       vector(u/dabs, v/dabs)
  contour(ex)       contour(ey)         contour(exy)
  contour(sx)       contour(sy)         contour(sxy)
end
```

The first figure on the next page shows contours of constant absolute value of the displacement. As we expect from the symmetry of pressure load, the contours are circular, although the expressions for the loads on the boundaries do not explicitly suggest this simple behavior.

The next *contour* plot below shows the distribution of shear stress. We see that the vertical contour (u) corresponds to τ_{xy}=0, and so does the horizontal contour. This means that all stresses on these lines have their principal directions along the coordinate axes. This is of course true of any choice we might have made for the direction of our coordinate system, i.e. the principal axes are generally along (and perpendicular to) the radius.

The *vector* plot (not shown here) of the displacements (u,v) illustrates the axial symmetry of this problem, and it also demonstrates that the radial displacement is largest close to the inner surface. The *contour* plot below of the magnitude of the displacement, *dabs*, of course also demonstrates this fact. The other plots illustrate how complex the results may appear when we employ Cartesian coordinates in a case of axial symmetry.

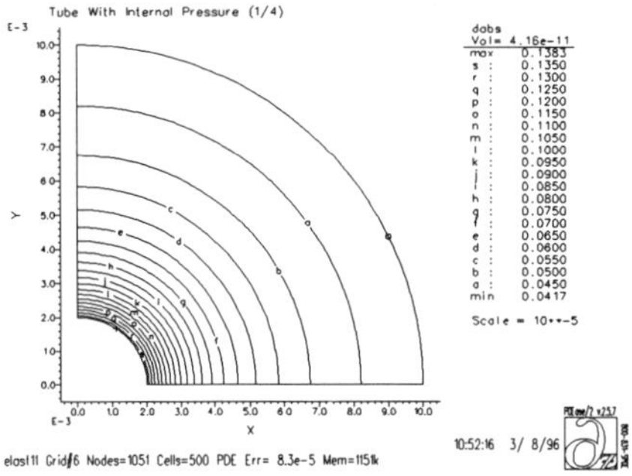

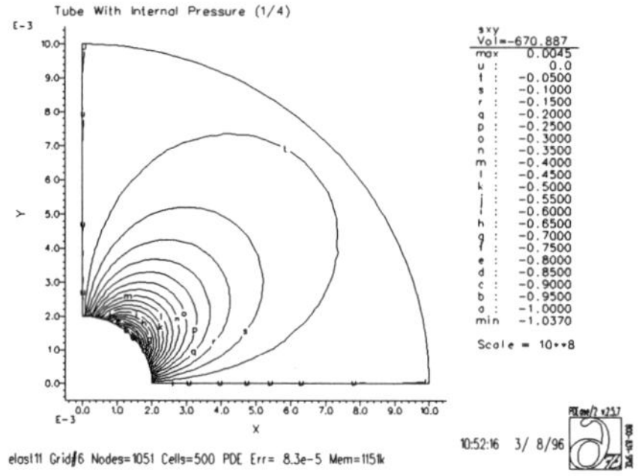

Comparison With Exact Solution

The deformation of a tube subjected to inner and outer pressure is one of the few elastic problems that may be solved exactly in terms of elementary functions[5p295]. We now modify the descriptor by adding the exact expression *ur_ex* for the radial displacement via the radial and tangential stresses. Although the outer pressure is zero in this first example, we enter the full expression including *p2*.

For comparison we also need an expression for the radial displacement in terms of the finite-element solution. We obtain the radial displacement, *ur*, as the vector sum of the Cartesian values u and v. This will have the same magnitude as *dabs* but may take the opposite sign when an outer pressure is applied.

```
                                                    { elast13.pde }
definitions
  ...
  rad= sqrt( x^2+ y^2)
  ur= u*x/rad+ v*y/rad                  { Vector sum of displacements }
  p2= 0                                              { Outer pressure }
  q21= r2/r1
  sr_ex= [-p1*((r2/rad)**2- 1)- p2*(q21**2- (r2/rad)**2)]
                / (q21**2- 1)
  st_ex= -[-p1*((r2/rad)**2+ 1)+ p2*(q21**2+ (r2/rad)**2)]
                / (q21**2- 1)
  ur_ex= rad/E*( st_ex- mu*sr_ex)                          { Hooke }
  ...
plots
  contour( [ur-ur_ex]/ ur_ex)
  ...
```

We choose to plot the contours of *relative* error, as seen at the end of the list.

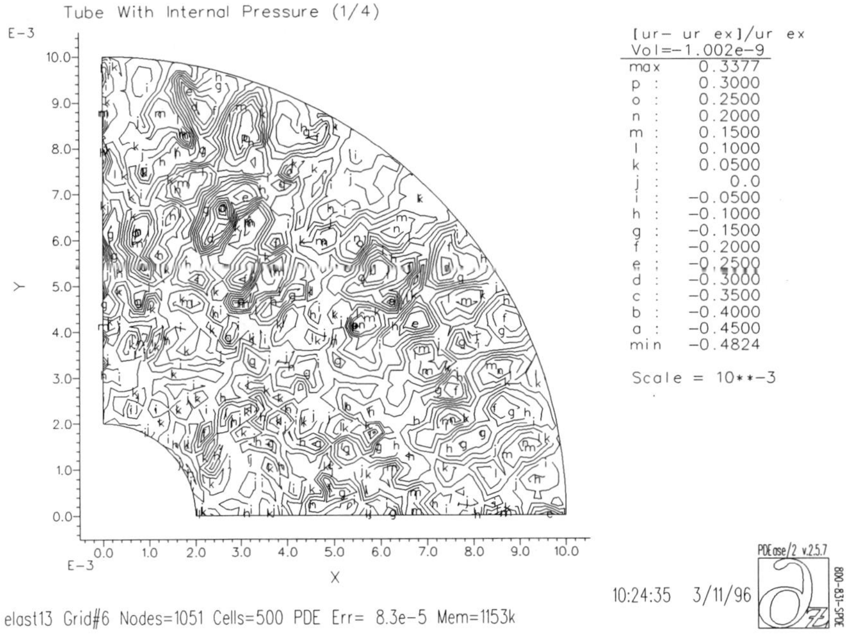

We find several contours of zero error, distributed randomly over the cross-section of the tube. The maximum error is about 0.05%, from which we estimate a standard deviation of 0.02%.

Further Analysis of Stresses

The normal and shear stresses transform under a rotation (θ) of the coordinate system according to the following equations

$$\begin{cases} \sigma_{x'} = \dfrac{\sigma_x + \sigma_y}{2} + \dfrac{\sigma_x - \sigma_y}{2}\cos 2\theta + \tau_{xy}\sin 2\theta \\ \tau_{x'y'} = -\dfrac{\sigma_x - \sigma_y}{2}\sin 2\theta + \tau_{xy}\cos 2\theta \\ \sigma_{y'} = \dfrac{\sigma_x + \sigma_y}{2} - \dfrac{\sigma_x - \sigma_y}{2}\cos 2\theta - \tau_{xy}\sin 2\theta \end{cases}$$ •

The primed stresses are what we obtain in the new coordinate system. A central fact of elasticity is that the shear stress vanishes for certain angles of rotation, corresponding to the principal system of coordinates. From the middle equation we immediately obtain the angle from the *x*-axis to one of the two principal axes

$$\theta = 0.5\arctan\frac{2\tau_{xy}}{\sigma_x - \sigma_y}.$$ •

We know that the other principal axis is at right angles to this direction, but since we can only show one of the two directions on a plot we need some criterion for selecting it . We choose to show the axis corresponding to the highest algebraic value of principal stress. This means tensile stress, except in ambiguous cases where both principal stresses are compressive.

The additional descriptor lines below indicate how to select the main principal axis (x'). Using arctan(x) we first calculate the angle to a principal axis. Then we calculate both principal stresses and test whether the actual *x*-axis corresponds to the highest value. If so, we keep the answer, otherwise we add *pi*/2. Once we have chosen the principal system we calculate its principal stresses: *sxp* for the "tensile" axis and *syp* for the perpendicular direction.

Since we have expressions for all the stress components, we may also calculate an equivalent stress[5p316] (von Mises), which permits us to decide if the object is likely to flow plastically.

```
  ...                                                    { elast14.pde }
  p_ang= 0.5* arctan( 2*sxy/(sx-sy) )                        { Radians }
  sxp0= 0.5*(sx+sy)+ 0.5*(sx-sy)*cos(2*p_ang)+ sxy*sin(2*p_ang)
  syp0= 0.5*(sx+sy)+ 0.5*(sx-sy)*cos(2*(p_ang+pi/2))
                                       + sxy*sin(2*(p_ang+pi/2))
                                            { Test for highest value }
  p_angl=  if sxp0>syp0 then p_ang   else p_ang+ pi/2
  sxp= 0.5*(sx+sy)+ 0.5*(sx-sy)*cos(2*p_angl)+ sxy*sin(2*p_angl)
  syp= 0.5*(sx+sy)- 0.5*(sx-sy)*cos(2*p_angl)- sxy*sin(2*p_angl)
  p_angle= p_angl * 180/pi                                   { Degrees }
  mises= sqrt( 0.5*( (sx-sy)**2+ sx**2+ sy**2)+ 3*sxy**2)
initial values
  ...
plots
  ...
  contour( p_angle)
  vector( cos(p_angl), sin(p_angl))  as 'Direction of "tension" '
  contour( sxp)           contour( syp)
  contour( Mises)
  ...
```

The next figure shows directions of the principal stress $\sigma_{x'}$ in the tube. In order to indicate directions in a plot, we would have preferred line segments without arrow-heads, since opposite directions are equivalent. We have to make do with the symbols included in PDEase, however, and should just disregard the way arrows point. In fact, as we see from the *vector* plot, half of the arrows circulate in the opposite direction.

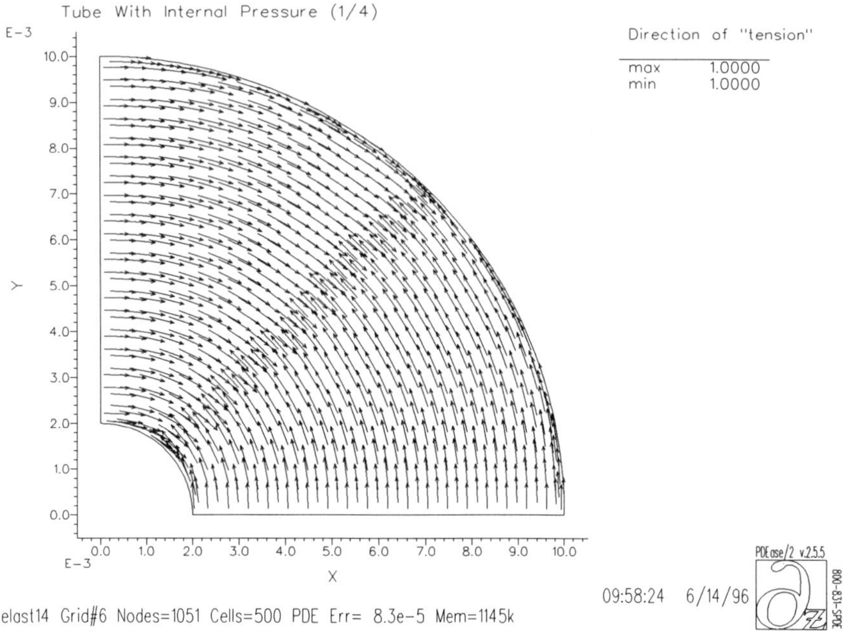

The following figures show the principal stresses in the system finally chosen. The left one presents the stress component in the directions shown in the vector plot, and we see that all contour values are in fact positive. The figure to the right indicates the stress in the perpendicular (radial) direction.

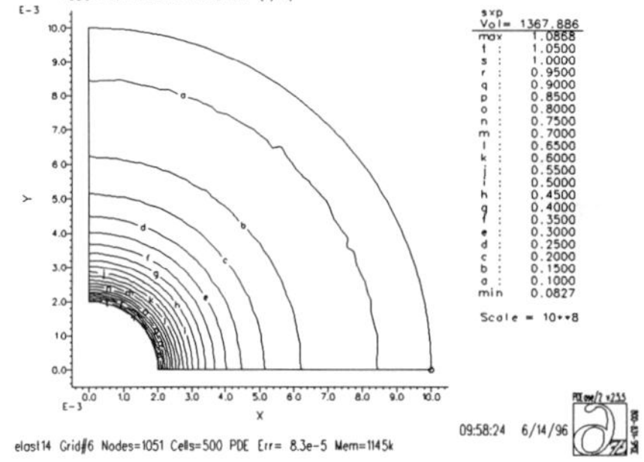

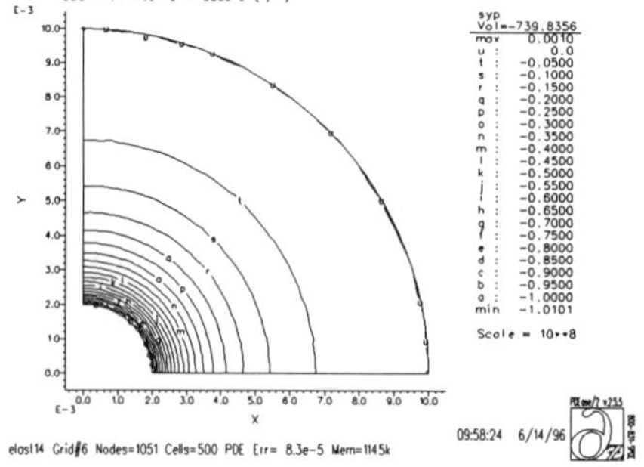

In summary, the figures show that radial stress is compressive, as it certainly must be close to the bore, and that the tangential stresses are tensile.

Tube With Internal Tangential Stress

An interesting case to explore is a tube deformed by tangential stress, caused by friction against a rotating rod inside it, while the outer surface remains fixed. Friction could hardly exist without a normal force, but let us disregard this component for the moment. We may superimpose the additional stresses later, if we wish.

In the descriptor that follows we have specified a friction force f_1 of 1e8 per unit area. The x and y components of that tangential force are what we need for the *load* statements referring to the inner surface.

```
...
f1= 1e8                                                     { Friction }
...
            region 1
start (r2,0)
value(u)= 0               value(v)= 0
    arc to (0,r2) to (-r2,0) to (0,-r2) to finish
start (r1,0)                                                { Cut-out }
load(u)= - f1*y/r1        load(v)= f1*x/r1
    arc to (0,-r1) to (-r1,0) to (0,r1) to finish
...
```

The first of the following plots shows the vector displacement, which takes its maximum value at the inner surface and vanishes at the outer surface, as expected.

The shear forces do not change the inner radius, as we glean from the contour plots of u and v, keeping in mind the evident axial symmetry of *dabs*.

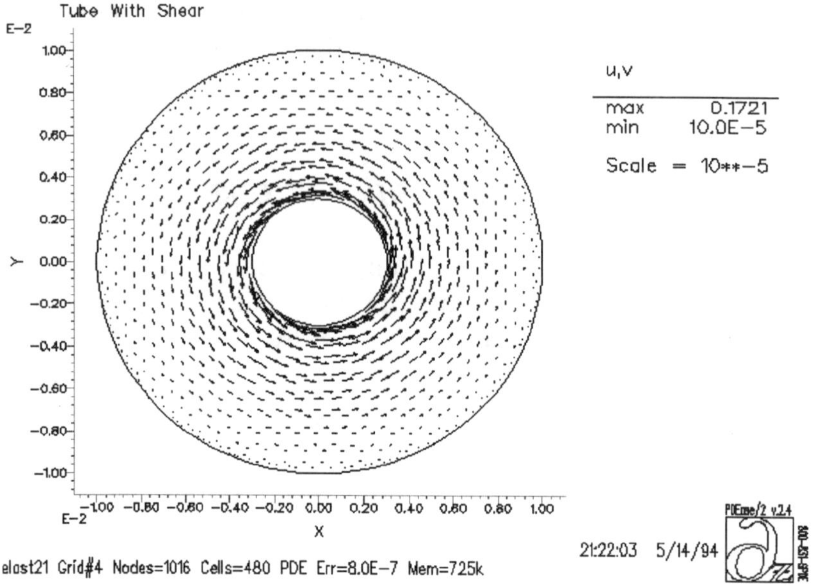
Tube With Shear
E−2
1.00
0.80
0.60
0.40
0.20
Y 0.00
−0.20
−0.40
−0.60
−0.80
−1.00
−1.00 −0.80 −0.60 −0.40 −0.20 0.00 0.20 0.40 0.60 0.80 1.00
E−2
X
u,v
max 0.1721
min 10.0E−5
Scale = 10**−5
21:22:03 5/14/94
PDEase/2 v2.4
elast21 Grid#4 Nodes=1016 Cells=480 PDE Err=8.0E−7 Mem=725k

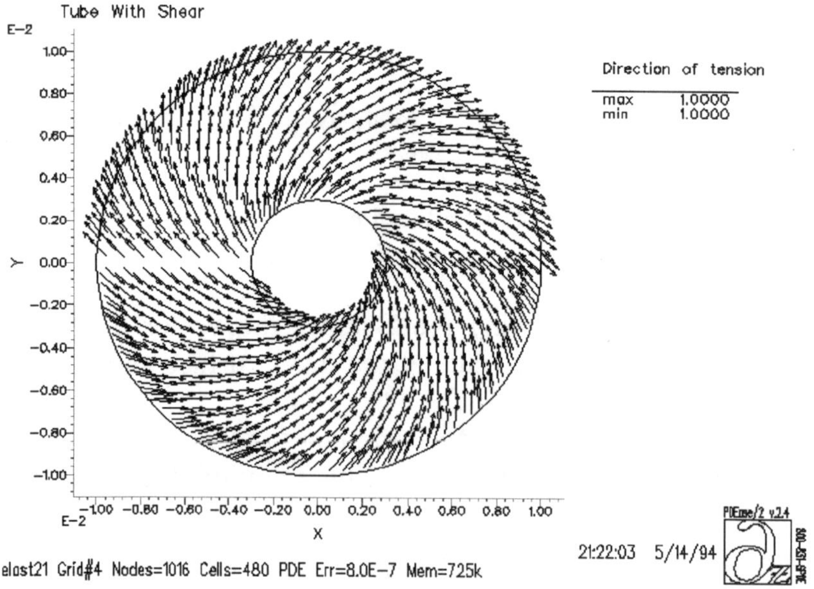
Tube With Shear
E−2
1.00
0.80
0.60
0.40
0.20
Y 0.00
−0.20
−0.40
−0.60
−0.80
−1.00
−1.00 −0.80 −0.60 −0.40 −0.20 0.00 0.20 0.40 0.60 0.80 1.00
E−2
X
Direction of tension
max 1.0000
min 1.0000
21:22:03 5/14/94
PDEase/2 v2.4
elast21 Grid#4 Nodes=1016 Cells=480 PDE Err=8.0E−7 Mem=725k

The contours of constant principal angle look like spokes, i.e. the inclination of a principal stress is constant along any radius. The spiral pattern in the preceding figure, formed by the directions of tensile stress, confirms this feature. It might seem as if tensile radial stress acted on the outer rim, but at any point on that boundary there is also compressive stress of the same magnitude. The *contour* plots of *sx* and *sy* clearly show that the radial stress vanishes at the ends of radial and horizontal radii.

Ring Strained Between Two Edges

This example illustrates a case of more complex stress. Using the same geometry as before we impose down displacement at the top and an equally large up displacement at the bottom, such that the center position remains unchanged. This corresponds to unspecified vertical loads on two lines, perpendicular to the figure. Otherwise there is no load on the cylindrical boundaries.

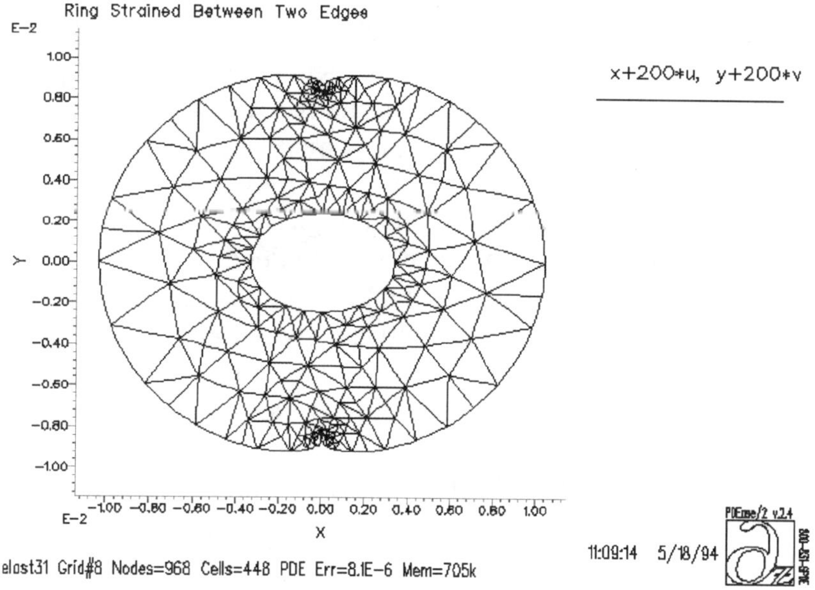

Most of the preceding descriptor still applies, but we need to add or modify the following lines.

```
errlim= 3e-3                                                    { elast31.pde }
...
r1 = 3*mm               r2 = 10*mm              delr= 0.01*mm
...
        region 1
start (r2,0)
load(u)= 0    load(v)= 0    arc(center= 0,0) to (0,r2)
point value(u)= 0    point value(v)= -delr
  arc to (-r2,0) to (0,-r2)
point value(u)= 0    point value(v)= delr
  arc(center= 0,0)  to finish
start (r1,0)                                                    { Cut-out }
load(u)= 0    load(v)= 0
  arc to (0,-r1) to (-r1,0) to (0,r1) to finish
...
contour( sxp) zoom(-r2,-0.8*r2, 2*r2,1.6*r2)
contour( syp) zoom(-r2,-0.8*r2, 2*r2,1.6*r2)
contour( Mises)               contour( Mises)  zoom(0,0, 1.2*r1,1.2*r1)
end
```

The stresses will be very high in the vicinity of the edges, and as a result most of the contour curves will be concentrated in that region. In order to make the plots more informative we use the *zoom* command to exclude these singularities, which are not of interest here.

We obtain the *contour* plot of principal angles shown below. This figure is simply related to the pattern obtained in an optical experiment with a model of the ring, made of photo-elastic material. In such a test we place the model between two foils of Polaroid, having perpendicular directions of polarization, and illuminate one side by monochromatic light. When the ring is viewed from the other side, dark lines will appear wherever a principal axis coincides with the polarization plane. By rotating the crossed Polaroids with respect to the strained ring we may thus trace all the curves shown in the contour plot.

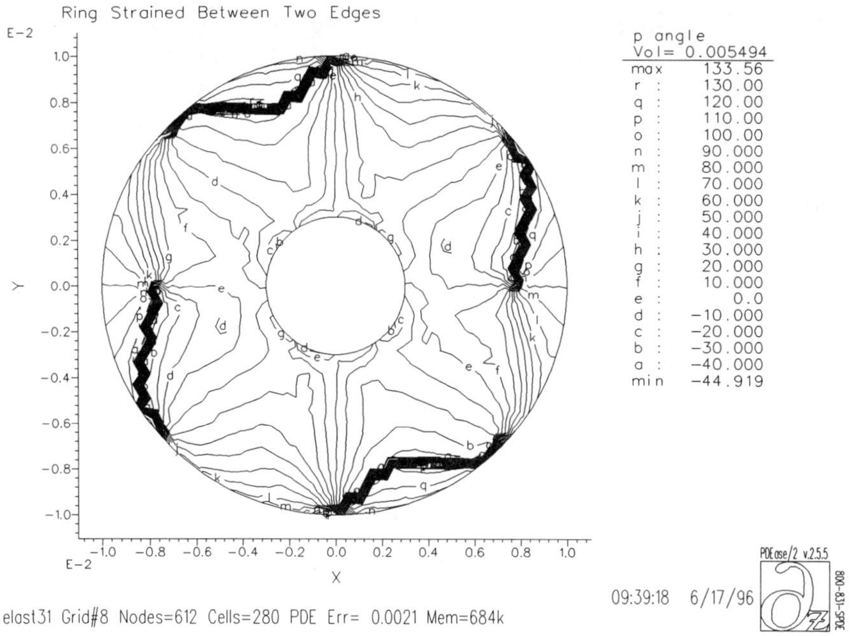
Ring Strained Between Two Edges
E-2
1.0
0.8
0.6
0.4
0.2
0.0
-0.2
-0.4
-0.6
-0.8
-1.0
Y
-1.0 -0.8 -0.6 -0.4 -0.2 0.0 0.2 0.4 0.6 0.8 1.0
E-2
X
p angle
Vol= 0.005494
max 133.56
r : 130.00
q : 120.00
p : 110.00
o : 100.00
n : 90.000
m : 80.000
l : 70.000
k : 60.000
j : 50.000
i : 40.000
h : 30.000
g : 20.000
f : 10.000
e : 0.0
d : -10.000
c : -20.000
b : -30.000
a : -40.000
min -44.919
elast31 Grid#8 Nodes=612 Cells=280 PDE Err= 0.0021 Mem=684k
09:39:18 6/17/96
PDEase/2 v.2.5.5
800-831-SPDE

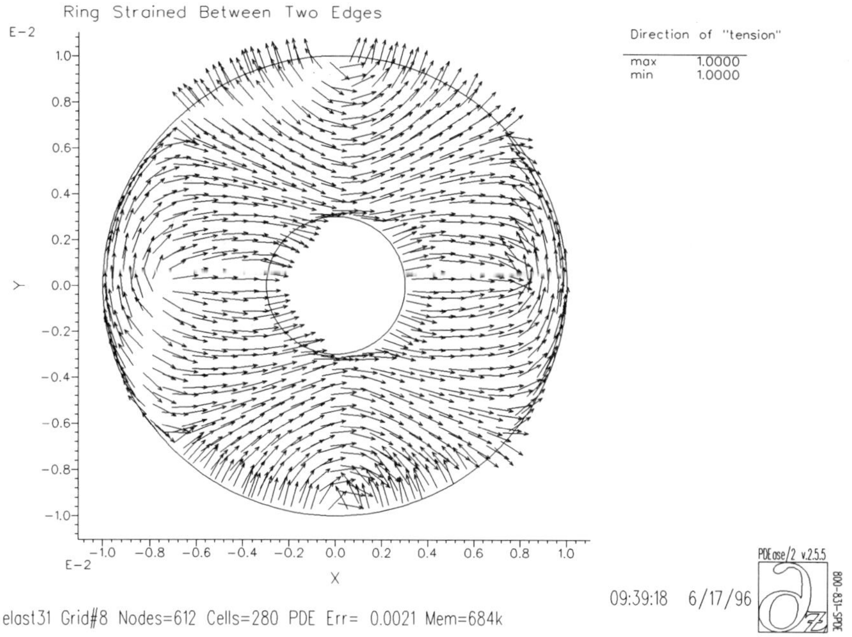
Ring Strained Between Two Edges
E-2
1.0
0.8
0.6
0.4
0.2
0.0
-0.2
-0.4
-0.6
-0.8
-1.0
Y
-1.0 -0.8 -0.6 -0.4 -0.2 0.0 0.2 0.4 0.6 0.8 1.0
E-2
X
Direction of "tension"
max 1.0000
min 1.0000
elast31 Grid#8 Nodes=612 Cells=280 PDE Err= 0.0021 Mem=684k
09:39:18 6/17/96
PDEase/2 v.2.5.5
800-831-SPDE

The preceding *vector* plot indicates the direction of the principal axis of highest stress. Compressive stress (perpendicular to the tensile axis) appears to be fanning out from the top edge to meet at the bottom edge. In regions on either side of the bore, both principal stresses are compressive, but the vertical stress dominates. The contour plots of the principal stresses σ_x' (*sxp*) and σ_y' (*syp*) show the state of stress in more detail.

The *von Mises* (equivalent) stress obviously tends to infinity at the edge supports. High values also occur close to the inner boundary, however, so this region is explored by a separate plot.

Although the local stresses at the edges are infinite, the forces applied must remain finite. We specified the displacements rather than the force, but from the stress distribution we can now compute the load at an edge by considering the equilibrium of a chosen part of the object, say, the upper half. It is sufficient to integrate σ_y over a horizontal cross-section by means of an elevation plot. The value of *Area* then yields the force per unit thickness in the *z* direction.

Exercises

❑ Change the first example (p.170) from *load* to *value* boundary conditions. Keep the outer boundary fixed and apply a radial strain of 1e-3 to the inner surface.

❑ Modify the second example (with the exact solution) by applying pressure on the outer surface, but not on the inner one.

❑ Try pressure inside and outside the tube at the same time. What happens when the pressures become equal?

❑ Calculate the force applied to the ring (p.180) by means of *elevation* plots along lines at $y = 0$, $y = r1$, $y = 0.5*r2$ and $y = 0.9*r2$.

❑ Change the ring descriptor (p.180) into a solid cylinder. Find the regions of maximum *von Mises* stress by suitably zoomed *contour* or *surface* plots.

Bending of Beams

Transverse loads on rectangular beams are of great importance in engineering. In these cases a number of simple, approximate solutions exist which may be used for comparison.

Beam With Moments at Ends

We first consider a problem of the utmost simplicity, in which opposite bending moments act on the ends of a beam. In fact there will also be a gravity force, but we shall neglect that for the moment. We may assume, for instance, that our x axis is oriented along the field of gravity, in which case the latter may safely be neglected in our example.

Starting with the last descriptor (p.180) from the previous chapter we add the following lines. The bending moment is *Mb* and the second moment of cross-sectional area is *Izz* ("moment of inertia"). In terms of these quantities we may express the y displacement *v_ex*, using beam theory and first-order geometry of deflection[5p531].

```
...                                                   { beam1.pde }
errlim= 1e-10
...
L= 1.0          h= 0.2
Mb= 1.0e5       Izz= h**3/12          v_ex= Mb/( E*Izz)* x**2/ 2
...
           region 1                              { Right half of beam }
start (0,-h/2)    point value(v)= 0
load(u)= 0        load(v)= 0    line to (L,-h/2)
load(u)= -(Mb/Izz)*y          load(v)= 0          line to (L,h/2)
load(u)= 0        load(v)= 0    line to (0,h/2)
value(u)= 0       load(v)= 0    line to finish
...
contour( v-v_ex)  as 'Error in v'
...
```

In order to simplify the descriptor we solve for only the right half of the beam, as indicated by the figure below. For reasons of symmetry the left face must then remain plane and vertical, i.e. the displacement u must be zero. Since there is no *value* for v in this problem, we also specify a *point value* at the left face.

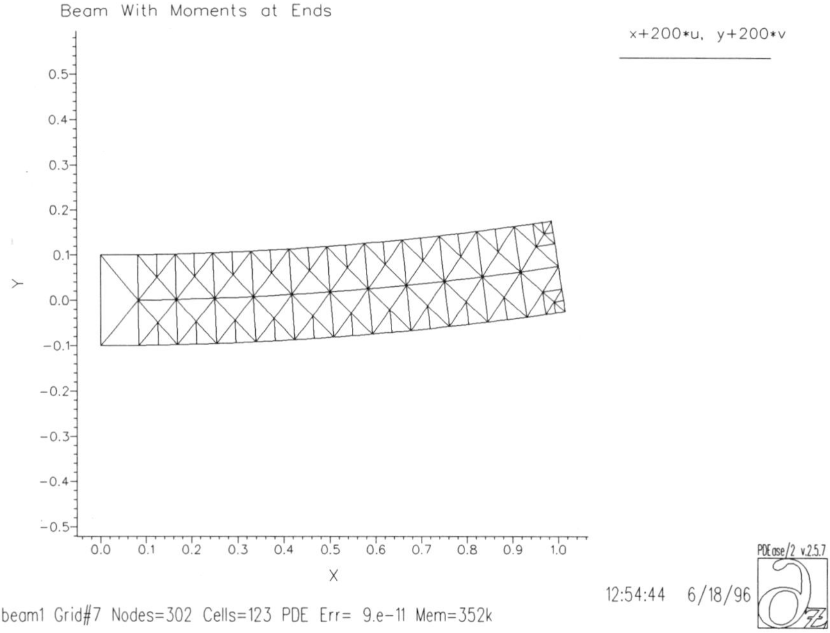

The bending moment we introduce by a distribution of normal forces at the right face, such that the force changes sign in the middle of the cross-section. The axial force thus sums to zero, while the first moment becomes *Mb* as required.

The plots show that the magnitude of the y displacement v is about 3×10^{-4}, while the error is of the order of 10^{-6} in the same unit.

The following *contour* plot shows that the x-component of stress is independent of x, as expected from elementary theory. Comparing the plots of the stress components we find that σ_y and τ_{xy} are smaller than 10^{-6} of σ_x, which also agrees with theoretical predictions.

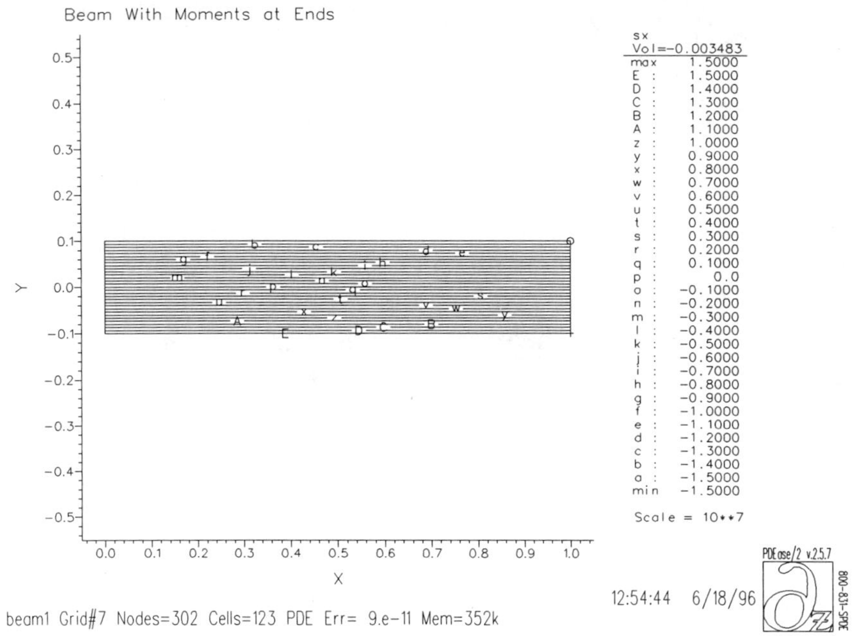

Gravity Bending

Let us now consider a cantilever beam loaded by its own weight and welded to a vertical support at one end. Gravity acts on each volume element, so we have to introduce the term *Fy* in the second PDE (p.157). Starting from the previous descriptor we add or replace the following lines.

In addition we eliminate all loads on boundaries but specify *value*(*u*)= 0 and *value*(*v*)= 0 on the left face.

```
...                                                  { beam11.pde }
errlim= 1e-4
...
Fy= -7.8e3*9.8                        w0= Fy* h
v_approx= w0* x**2/(24*E*Izz)* (x**2+ 6*L**2- 4*L*x)
sxy_approx= w0*(L-x)/(2*Izz)*[h**2/4-y**2]
...
```

```
dx( sxy)+  dy( sy)+ Fy= 0
...
grid( x+3e3*u, y+3e3*v)
contour( v-v_approx)                    contour( sxy- sxy_approx)
```

There are approximate theoretical expressions for the vertical displacement[5p531] as well as for shear stress[5p436], both of which we use for comparison. The following plot shows the distribution of shear stress and the next one displays the difference with respect to the approximate formula. There are several contours (*m*) of zero error over the domain, which indicates reasonable agreement.

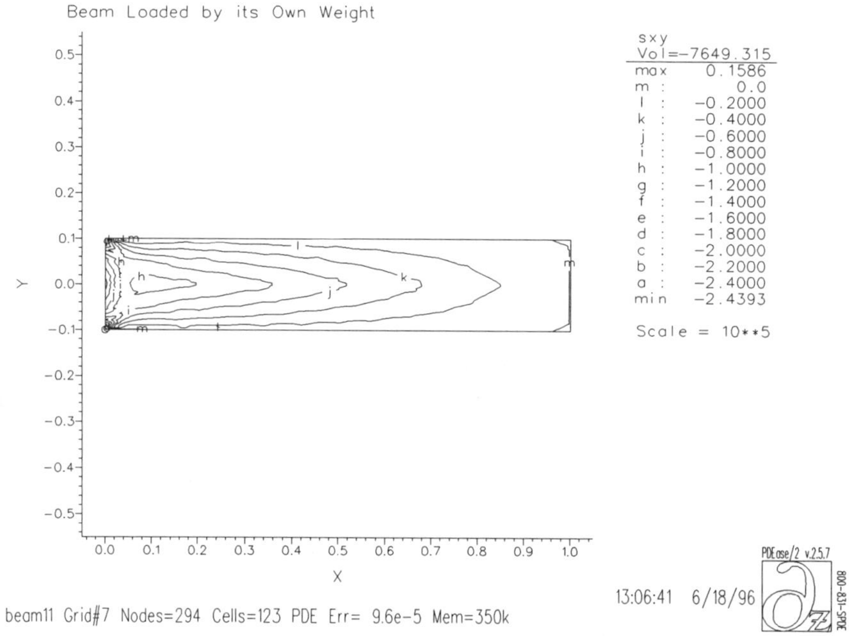

The agreement with the approximate expression for the vertical displacement is equally good. The maximum deflection is about 10^{-5}, while the discrepancy is roughly 5×10^{-7} at the same point. The plot of von Mises stress shows that the maximum risk of failure exists at the left corners.

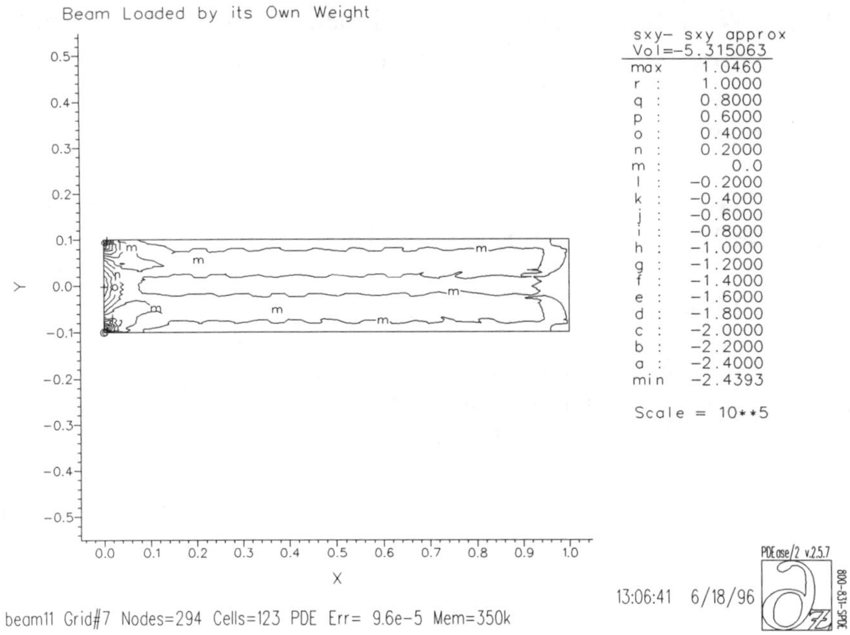

Forces and Moments

Having obtained a solution in terms of displacements, we may compute the force (per unit extension in the z direction) acting on a surface. In the case of the beam in the preceding example we may isolate the major part of it by a vertical cut just to the right of the welded end. The integral (*Area*) of τ_{xy} should yield the shear force at the intersection, which in turn should balance the force of gravity on the beam. In order to make this calculation we need to add the following lines (and perhaps eliminate some plots that we have already seen).

In the *select* segment we now introduce two new commands, *converge* and *subconverge*, in order to enhance precision in the solution process. These additional error statements impose strong conditions on the iteration procedure used when PDEase solves the system of linear or non-linear algebraic equations resulting from the analysis. With these statements included, a new grid cannot be created before the solution attains the specified accuracy. We shall keep this line in all of the following descriptors. It is of particular value when analyzing slender beams.

```
...                                                    { beam12.pde }
CONVERGE= 1e-6          SUBCONVERGE= 1e-6
...
small= 1e-6*L
...
elevation(sxy) from (small,-h/2) to (small,h/2)
elevation(sx) from (small,-h/2) to (small,h/2)
elevation(sx* y) from (small,-h/2) to (small,h/2)
```

The next figure shows the pertinent *elevation* plot, and we note that the integral agrees well with the weight of the beam.

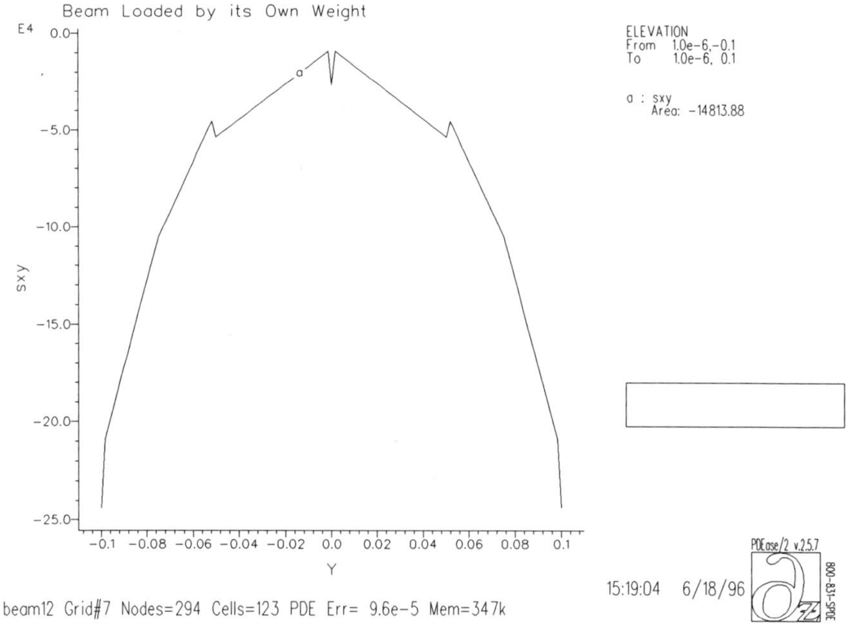

The second *elevation* plot gives us the integrated force in the x direction on the face of the isolated part of the beam. It shows that this force is in fact small compared to the weight of the beam.

Another interesting verification is to compute the bending moment generated by the distribution of σ_x over the cross-section of the cut. This bending moment must balance that of gravity. The plot of *sx*y* confirms that the two moments are closely the same.

Semi-Circular Beam

A beam need not be straight, nor does it have be of the same material throughout. We shall now study a steel hook, welded to a support at one end and uniformly loaded by a weight at the other one. We neglect the effect of gravity ($F_y = 0$).

There is no simple engineering approximation to this problem, and hence the descriptor will not contain any functions for comparison. The other modifications should be evident from the following list.

```
...                                                    { beam21.pde }
r1= 0.06          r2= 0.1          F= 1e4
...                    { Load }
Fy= 0
...
start (-r2,0)
load(u)= 0        load(v)= 0
   arc to (0,-r2) to (r2,0)
load(u)= 0        load(v)= -F/(r2-r1)    line to (r1,0)
load(u)= 0        load(v)= 0
   arc to (0,-r1) to (-r1,0)
value(u)= 0       value(v)= 0    line to finish
...
grid( x+1e3*u, y+1e3*v)
```

The following plot of vertical displacement should make the object geometry clear. The zero contour suggests that v changes sign on a nearly vertical line below the welded face.

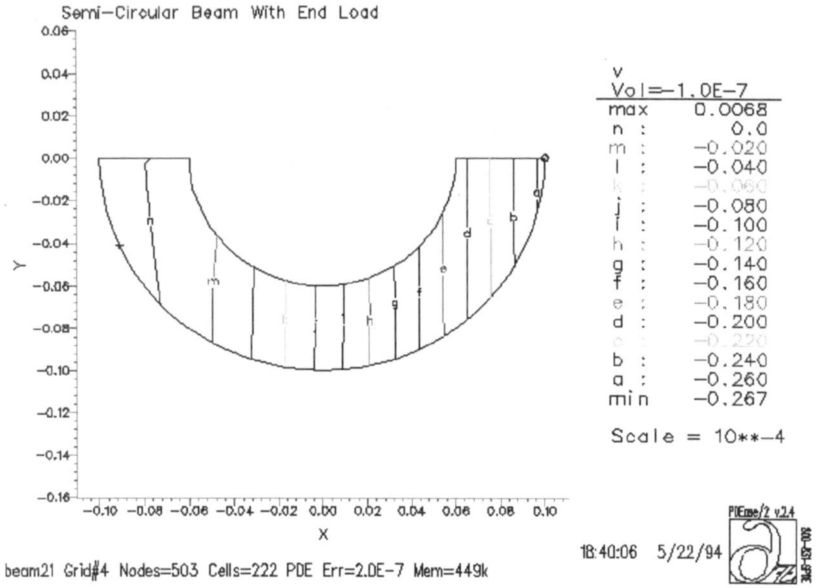

Semi-Circular Beam With End Load
v
Vol=-1.0E-7
max 0.0068
n : 0.0
m : -0.020
l : -0.040
k : -0.060
j : -0.080
i : -0.100
h : -0.120
g : -0.140
f : -0.160
e : -0.180
d : -0.200
c : -0.220
b : -0.240
a : -0.260
min -0.267
Scale = 10**-4
X
Y
18:40:06 5/22/94
PDEase/2 v.2.4
beam21 Grid#4 Nodes=503 Cells=222 PDE Err=2.0E-7 Mem=449k

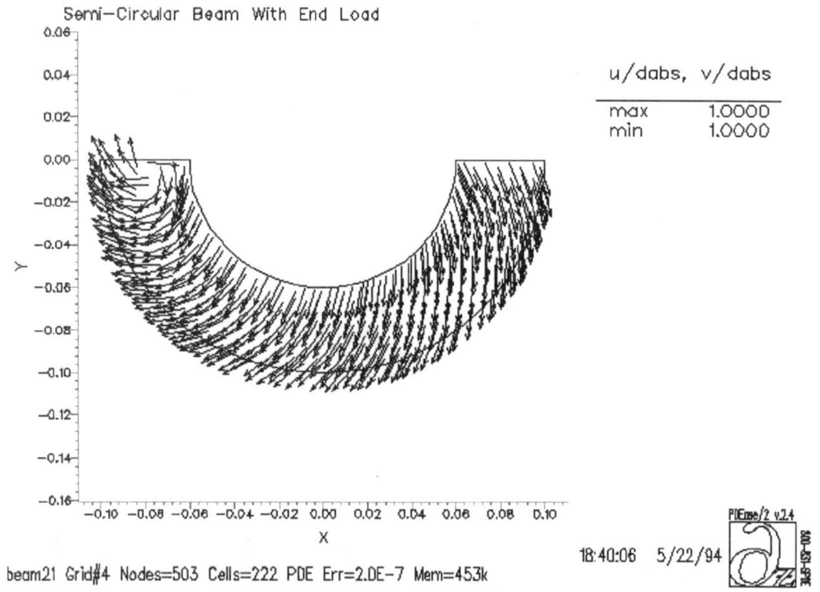

Semi-Circular Beam With End Load
u/dabs, v/dabs
max 1.0000
min 1.0000
X
Y
18:40:06 5/22/94
PDEase/2 v.2.4
beam21 Grid#4 Nodes=503 Cells=222 PDE Err=2.0E-7 Mem=453k

The *vector* plot above provides more details about how matter moves as load is applied to the right face of the beam. The motion to the right, as well as down, at the loaded end may seem surprising at first. Since the vectors are normalized to unit length we can even see the directions of displacement close to the fixed end, where the vectors are very short.

The plots related to the principal system of stress reveals an interesting region on the left where both principal stresses are tensile. By comparing the *contour* plots of *sxp* and *syp* with the *vector* plot of the principal angle we may verify that the normal tension goes to zero at the boundaries.

The plot of von *Mises* stress is difficult to read in the present version. The contours suggest, however, that the highest equivalent stress might be on the left part of the inside cylindrical surface, but we would need a *surface* plot or perhaps a zoomed plot to confirm this conclusion.

Composite Semi-Circular Beam

As a final example we shall study an aluminum beam with a steel reinforcement along the upper boundary. The shape and size are the same as in the preceding case, and we only need to modify the boundaries segment by adding a lower region consisting of aluminum.

```
...
r1= 0.06          r2= 0.1          rm= 0.07
...
        region 2          E= 70e9                    { Aluminum }
start (-r2,0)
arc to (0,-r2) to (r2,0) line to (rm,0)
arc to (0,-rm) to (-rm,0) line to finish
...
```

PDEase solves this problem as easily as for the uniform hook, but the results are of course more complex. The tensile stress is largely confined to the steel cladding over the left part of the object.

The *surface* plot below shows the distribution of equivalent stress. It evidently takes its largest values on the inside surface near the support. It is about half as large at the opposite side of the welded face, however, which means that the aluminum will probably flow before the steel ruptures.

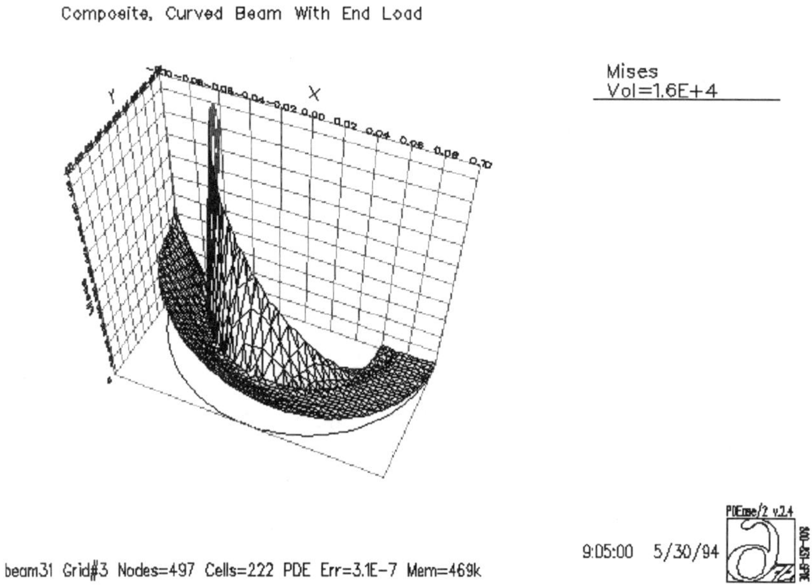

Exercises

❑ Replace the distributed forces at the end of the beam with two opposite *u* displacements at the right corners, choosing the values resulting from the first example (p.183).

❑ Remove one of the two point values in the preceding exercise.

❑ Check overall mechanical equilibrium for the hook by an *elevation* plot of *sy* just below the loaded face, and also of *sxy* along the vertical symmetry line.

❑ Study the stress distribution in a wooden ski with a steel sole. Represent the weight of the skier by pressure over a reasonable length near the middle.

Simple Torsion

In this chapter we shall be concerned with objects extending in the z direction and having a cross-section of constant shape and size. A twisting moment acts on and distorts this object. At first sight this seems to be a three-dimensional elastic problem, since the displacements u and v definitely vary with the axial (z) coordinate. Due to the subtle symmetry of torsion in case of cylindrical objects, however, we may treat the deformation adequately while staying in the xy plane. In fact, we shall only need one PDE.

Rigorous theory of elasticity[6p291] shows that a bar twisted by appropriate distributions of shear stress at the ends will retain its original cross-sectional shape, as projected on the xy plane. On the other hand it may warp, or bend out of that plane. The resulting displacement w along the z-axis may then be formally written

$$w = \theta \psi(x, y), \qquad \bullet$$

where θ is the angle of twist per unit axial length.

The crucial device for solving the problem of torsion is an auxiliary stress function ϕ, from which we obtain the shear stress components as follows[6p295]

$$\tau_{zx} = \frac{\partial \phi}{\partial y}, \qquad \tau_{zy} = -\frac{\partial \phi}{\partial x}. \qquad \bullet$$

This function has been shown to obey the PDE

$$\frac{\partial^2 \phi}{\partial x^2} + \frac{\partial^2 \phi}{\partial y^2} + 2G\theta = 0. \qquad \bullet$$

The boundary conditions are particularly simple if the cross-section does not have a cut-out area. In that case only the outer surface is relevant, and the boundary condition simply becomes $\varphi = 0$. We shall restrict ourselves here to this simple case and to solid cylinders of uniform composition.

Twisted Circular Shaft

The solid, circular shaft is an attractive first project, since an exact solution exist in very simple form. In the descriptor below we input the geometrical dimensions and the shear modulus G to calculate the stress function ϕ. Assuming that there is no distortion of the cross-section we specify a value of θ, the angular rotation per unit axial length (*twist*), and eventually obtain the shear stress distribution required to achieve this geometrical condition. This is the reverse of the previous procedure, where we found the deformation caused by a given distribution of forces.

The displacements u and v in the xy plane are simple geometric consequences of our assumption: no distortion of the cross-section.

The shear stress components we obtain as derivatives of *phi*, and we use these to calculate the twisting moment per unit area of the cross-section. In a *contour* plot of this quantity PDEase automatically computes the integral (*Vol*) over the solution domain, which gives us the total twisting moment.

```
  ...
  errlim= 1e-5
variables
  phi                                                    { Stress function }
definitions
  rad= 1e-2          L= 1.0
  twist= 1e-3                                                 { Radians }
  G= 80e9                                                       { Steel }
  tau_zx= dy(phi)              tau_zy= -dx(phi)           { Shear stress }
  tau_abs= sqrt(tau_zx**2+ tau_zy**2)
  moment= x*tau_zy- y*tau_zx                     { Moment per unit area }
  u= -L*twist*y                v= L*twist*x             { Rigid rotation }
  dabs= sqrt(u**2+ v**2)                    { u,v are x,y displacements }
equations
  dxx(phi)+ dyy(phi)+ 2*G*twist = 0
boundaries
            region 1
  value(phi)=0
  start(rad,0)  arc to (0,rad) to (-rad,0) to (0,-rad) to finish
monitors
```

```
plots
  contour( tau_abs)          contour( moment)
  vector( u, v)              surface( dabs)
  surface( phi) interactive
end
```

The following plot shows the moment per unit area. The total moment agrees very well with the exact expression[5p373], $M_t = G\theta\pi r^4 / 2$, where r is the radius of the shaft.

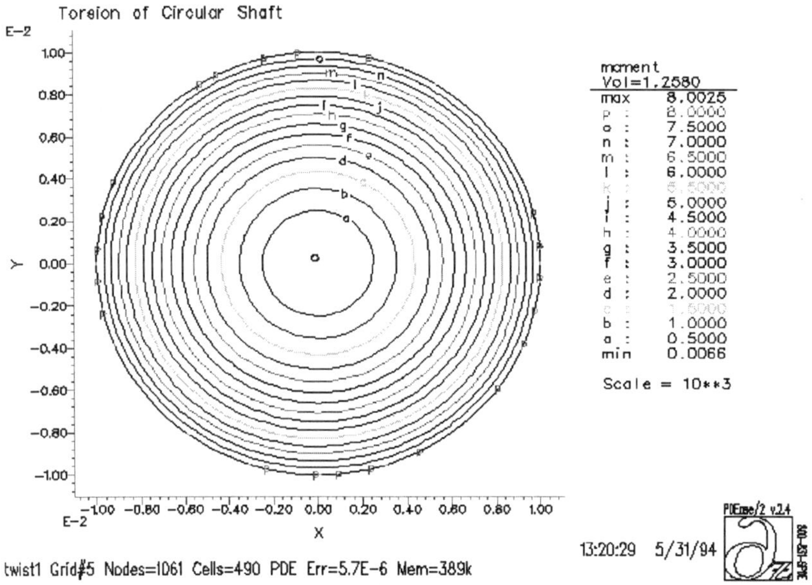

Twisted Square Bar

Although the problem of a twisted rectangular bar does not have an elementary solution, it may be treated by means of Fourier series. In the special case of a square cross-section the solution yields a simple expression for the maximum shear stress and the integrated moment. Thus, even in this more complicated case we may check our results against theory.

Only very few modifications of the preceding descriptor are necessary, as explained in the following list.

```
...
errlim= 1e-5          nodelimit= 2000
...
Lx= 1e-2          Ly= 1e-2          L= 1.0
...
                    region 1
value(phi)=0
start(-Lx,-Ly) line to (Lx,-Ly) to (Lx,Ly) to (-Lx,Ly) to finish
...
```

The following *vector* plot shows the motion of matter as twist is applied. The cross-section rotates like a solid disk according to our assumption.

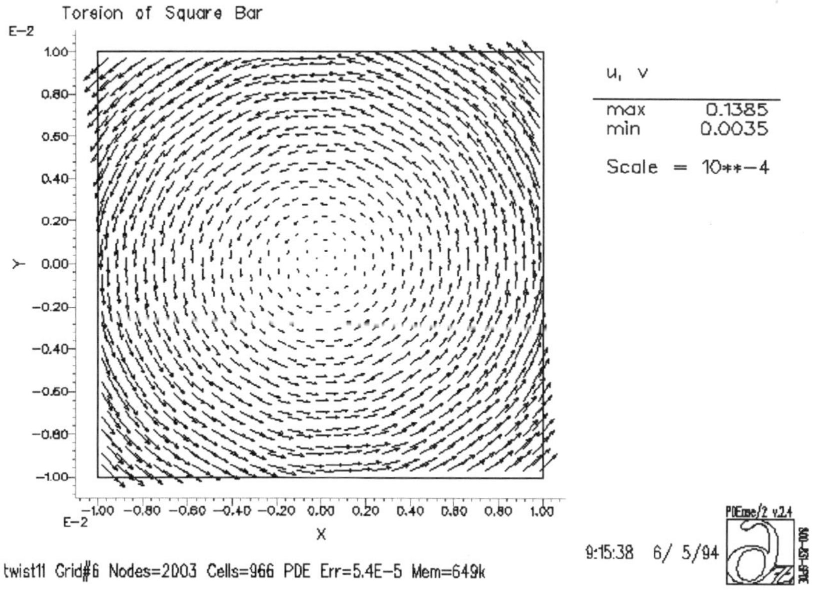

The next figure shows the distribution of shear moment per unit area. If we compare the integrated value with the theoretical expression[6p313], $M_t = 0.1406G\theta(2a)^4$ with a=Lx, we find the same value within 0.1%. The maximum shear stress, taken from the plot of *tau_abs*, agrees equally well[6p313].

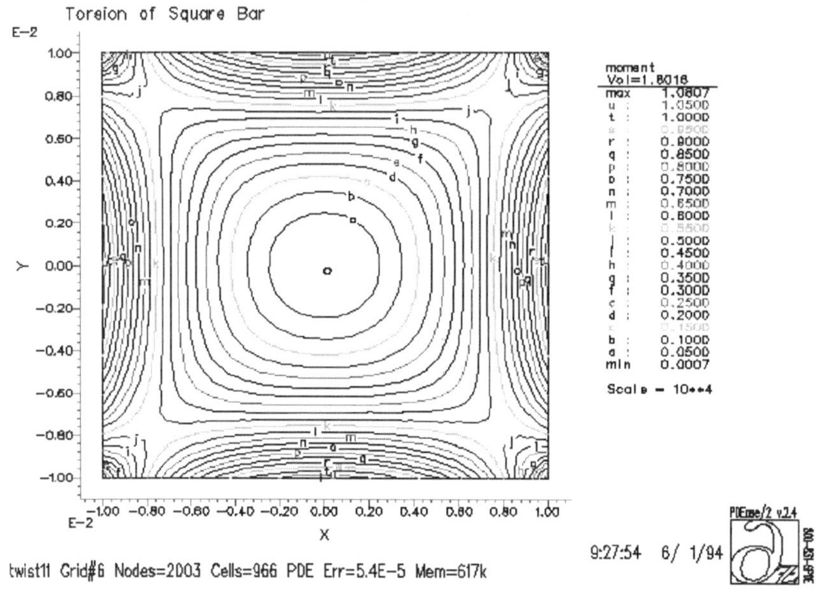

We may pursue the analysis one step further by exploring how a plane, square cross-section warps under torsion. The following differential relations apply[6p293]

$$\gamma_{zx} = \theta\left(\frac{\partial\psi}{\partial x} - y\right), \quad \gamma_{zy} = \theta\left(\frac{\partial\psi}{\partial y} + x\right)$$

from which we obtain the z-component of displacement (w) by line integrals

$$w = \theta\,\psi(x,y) = \int_{x_0,y_0}^{x,y_0} \left(\gamma_{zx}(\xi,\eta) + \theta\eta\right)d\xi + \int_{x,y_0}^{x,y} \left(\gamma_{zy}(\xi,\eta) - \theta\xi\right)d\eta$$

This formula yields the warping at any point (x,y) with respect to the value at a reference point (x_0,y_0). Using *elevation* plots we may readily calculate these line integrals, which yield the warping displacements at points (x,y_0) and (x,y).

In order to incorporate the above line integrals we need to add a few *elevation* plots to the descriptor, and at the same time we include a *contour* plot of *von Mises* equivalent stress.

```
mises= sqrt[ (3/2)* (tau_zx**2+ tau_zy**2)]
...
elevation( gamma_zx+ twist*y) from (0,0) to (Lx,0) as 'Warping'
elevation( gamma_zy- twist*x) from (Lx,0) to (Lx,Ly) as 'Warping'
elevation( gamma_zy- twist*x) from (Lx,-Ly) to (Lx,Ly) as 'Warping'
contour( mises)
```

The integrals of the *elevation* plots yield w= 5e-13 (from a very noisy integrand) at (Lx,0) and w= -2e-10 at (Lx,Ly). These values of w should be compared to the maximum displacement in the xy plane (1.4e-5). Evidently, the warping at the midpoint of a side is negligible, but it may reach measurable values at other points.

If we inspect the last *elevation* plot, which is taken vertically along the right side, we find that the derivative of w (the integrand) is largest at the midpoint and crosses zero at about $\pm 0.6 * Ly$. This means that the highest magnitude of warping should be found at these points.

The *contour* plot reveals that the highest values of von Mises equivalent stress are encountered at the midpoints of the sides.

Twisted L-Beam

In the example of the square bar we noticed that the shear stress was small close to the corners. We shall now investigate the stress distribution in an L-shaped beam, which has a reentrant corner as shown in the figure below. The twist evidently does not distort the cross-section, but only rotates it.

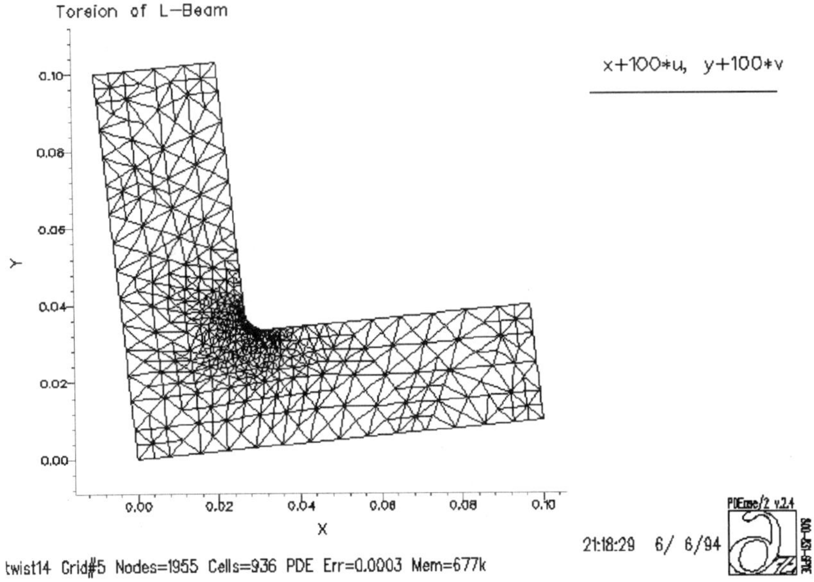

The new features required in the descriptor are listed below. We choose a small but finite inner radius of curvature. In this case we refrain from calculating the warping and concentrate on possible stress enhancements.

```
select
  errlim= 1e-5        nodelimit= 2000
  contours= 10        surfacegrid= 30        paint= off
variables
  phi                              { Stress function }
definitions
  Lxy= 0.1     d= 0.03      r0= 5e-3       L= 1.0
  twist= 1e-3         G= 80e9
  u= -L*twist*y                     v= L*twist*x
  dabs= sqrt(u**2+ v**2)
  tau_xz= dy(phi)                   tau_yz= -dx(phi)
  tau_abs= sqrt(tau_xz**2+ tau_yz**2)
  moment= x*tau_yz- y*tau_xz
  gamma_xz= (1/G)* tau_xz             gamma_yz= (1/G)* tau_yz
  mises= sqrt[ (3/2)* (tau_xz**2+ tau_yz**2)]
equations
```

```
  dxx(phi)+ dyy(phi)+ 2*G*twist = 0
boundaries
        region 1
  value(phi)=0
  start(0,0) line to (Lxy,0) to (Lxy,d) to (d+r0,d)
    arc(center= d+r0, d+r0) to (d,d+r0)
    line to (d,Lxy) to (0,Lxy) to finish
monitors
plots
  grid( x+100*u, y+100*v)
  contour( tau_abs)              contour( moment)
  vector( u, v)                  surface( dabs)
  surface( mises) interactive
end
```

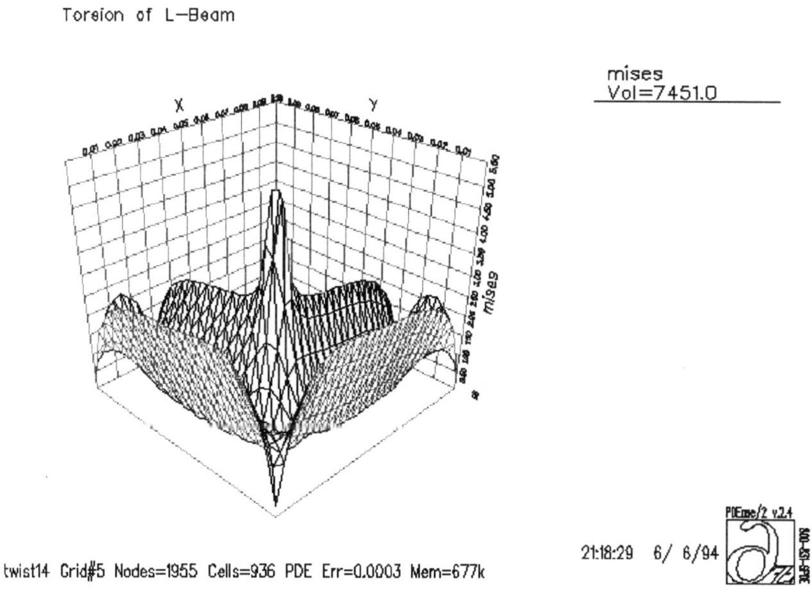

The *surface* plot above illustrates that the equivalent (von Mises) stress generally takes high values along the outside edges of the "L", but twice as high values occur in the region of the reentrant corner.

Exercises

❑ Solve the problem of a twisted strip with *Lx*=1e-3 and *Ly*=1e-2. The exact solution in the limit of vanishing *Lx* yields τ_{max}=2**Lx***G***twist* and M_t=16**Ly***Lx***3**G***twist*/3.
❑ Calculate the warping of the cross-section in the case of the circular shaft and compare with the maximum displacement in the *xy* plane.
❑ Find the maximum value of warping in the case of the square bar.
❑ Change the origin of the coordinate system to a corner of the square bar. How does this influence the twisting moment, the maximum shear stress and the warping? Explain the results.
❑ Modify the descriptor for the L-shaped beam by putting the radius of the reentrant corner equal to 0.03. How does this change the *von Mises* equivalent stress?

Elasticity in (ρ,z) Space

The derivation of the PDEs in cylindrical coordinates (ρ,θ,z) closely follows that for Cartesian coordinates (p.157). We again start with the equations of mechanical equilibrium[5p255]

$$\begin{cases} \dfrac{\partial\sigma_\rho}{\partial\rho}+\dfrac{1}{\rho}\dfrac{\partial\tau_{\rho\theta}}{\partial\theta}+\dfrac{\partial\tau_{\rho z}}{\partial z}+\dfrac{\sigma_\rho-\sigma_\theta}{\rho}=0 \\ \dfrac{\partial\tau_{\rho\theta}}{\partial\rho}+\dfrac{1}{\rho}\dfrac{\partial\sigma_\theta}{\partial\theta}+\dfrac{\partial\tau_{\theta z}}{\partial z}+2\dfrac{\tau_{\rho\theta}}{\rho}=0 \\ \dfrac{\partial\tau_{\rho z}}{\partial\rho}+\dfrac{1}{\rho}\dfrac{\partial\tau_{\theta z}}{\partial\theta}+\dfrac{\partial\sigma_z}{\partial z}+\dfrac{\tau_{\rho z}}{\rho}=0 \end{cases} \bullet$$

Assuming that the object as well as the loads acting on it are axially symmetric we may neglect all derivatives with respect to θ. Furthermore there are no shear forces in the $\rho\theta$ and θz planes, i.e. the stress components $\tau_{\rho\theta}$ and $\tau_{\theta z}$ must vanish. Hence the second equation vanishes identically and we are left with

$$\begin{cases} \dfrac{\partial\sigma_\rho}{\partial\rho}+\dfrac{\partial\tau_{\rho z}}{\partial z}+\dfrac{\sigma_\rho-\sigma_\theta}{\rho}=0 \\ \dfrac{\partial\tau_{\rho z}}{\partial\rho}+\dfrac{\partial\sigma_z}{\partial z}+\dfrac{\tau_{\rho z}}{\rho}=0 \end{cases} \bullet$$

PDE for a Tube With Radial Load

In order to illustrate the method of analysis as simply as possible we first assume that all loads act in a radial direction (pressure), which also means that there is no axial load (plane stress). Finally we take the radius, the elastic properties and the pressure to be independent of *z*, which implies such symmetry that the last equation vanishes identically and only the first one remains (with $\tau_{\rho z}=0$)

$$\frac{\partial \sigma_\rho}{\partial \rho} + \frac{\sigma_\rho - \sigma_\theta}{\rho} = 0$$

It may seem strange that we have two unknown stress components, σ_ρ and σ_z, but only one equation. After we have introduced the displacements, however, only one dependent variable will remain.

Using the same strategy as in (x,y) space (p.157) we first express the stresses in terms of strains and these finally in terms of displacements. Hooke's law becomes, assuming plain stress (σ_z=0),

$$\begin{Bmatrix} \varepsilon_\rho - \alpha\Delta T \\ \varepsilon_\theta - \alpha\Delta T \\ \varepsilon_z - \alpha\Delta T \end{Bmatrix} = \frac{1}{E}\begin{bmatrix} 1 & -\mu & -\mu \\ -\mu & 1 & -\mu \\ -\mu & -\mu & 1 \end{bmatrix}\begin{Bmatrix} \sigma_\rho \\ \sigma_\theta \\ 0 \end{Bmatrix} = \frac{1}{E}\begin{Bmatrix} \sigma_\rho - \mu\sigma_\theta \\ -\mu\sigma_\rho + \sigma_\theta \\ -\mu\sigma_\rho - \mu\sigma_\theta \end{Bmatrix}$$

where we only need the first two rows for the present

$$\begin{Bmatrix} \varepsilon_\rho - \alpha\Delta T \\ \varepsilon_\theta - \alpha\Delta T \end{Bmatrix} = \frac{1}{E}\begin{Bmatrix} \sigma_\rho - \mu\sigma_\theta \\ -\mu\sigma_\rho + \sigma_\theta \end{Bmatrix} = \frac{1}{E}\begin{bmatrix} 1 & -\mu \\ -\mu & 1 \end{bmatrix}\begin{Bmatrix} \sigma_\rho \\ \sigma_\theta \end{Bmatrix}$$

After inverting the matrix we obtain the stresses

$$\begin{Bmatrix} \sigma_\rho \\ \sigma_\theta \end{Bmatrix} = \frac{E}{1-\mu^2}\begin{bmatrix} 1 & \mu \\ \mu & 1 \end{bmatrix}\begin{Bmatrix} \varepsilon_\rho - \alpha\Delta T \\ \varepsilon_\theta - \alpha\Delta T \end{Bmatrix} = \frac{E}{1-\mu^2}\begin{Bmatrix} \varepsilon_\rho + \mu\varepsilon_\theta - (1+\mu)\alpha\Delta T \\ \mu\varepsilon_\rho + \varepsilon_\theta - (1+\mu)\alpha\Delta \end{Bmatrix}$$

The strain components are geometrically related to the radial displacement u as follows

$$\varepsilon_\rho = \frac{\partial u}{\partial \rho} \qquad \varepsilon_\theta = \frac{u}{\rho} \ ,$$

and substituting these we obtain

$$\begin{Bmatrix} \sigma_\rho \\ \sigma_\theta \end{Bmatrix} = \frac{E}{1-\mu^2}\begin{Bmatrix} \dfrac{\partial u}{\partial \rho} + \mu\dfrac{u}{\rho} - (1+\mu)\alpha\Delta T \\ \mu\dfrac{\partial u}{\partial \rho} + \dfrac{u}{\rho} - (1+\mu)\alpha\Delta \end{Bmatrix}$$

With these expressions for the stresses the PDE of equilibrium degenerates into an *ordinary* differential equation

$$\frac{d}{d\rho}\left\{\frac{E}{1-\mu^2}\left[\frac{du}{d\rho}+\mu\frac{u}{\rho}-(1+\mu)\alpha\Delta T\right]\right\}+\frac{1-\mu}{\rho}\frac{E}{1-\mu^2}\left\{\frac{du}{d\rho}-\frac{u}{\rho}\right\}=0 \quad \bullet$$

which is in a form accepted by PDEase. We now assume the elastic properties and the temperature to be uniform and constant. Under these particular conditions the above equation simplifies to $d^2u/d\rho^2+(1/\rho)du/d\rho-u/\rho^2=0$. In PDEase we still use the preceding form, however, since it must be compatible with derivative (*natural*) boundary conditions (p.160).

Tube Subjected to Internal Pressure

Our first object of axial symmetry is a steel tube of uniform temperature under internal pressure. This problem has an elementary solution in terms of a rational function, which we can use to check the accuracy of the numerical algorithms. Although we solve in (ρ,z) space we are able to calculate the strain in the tangential (θ) direction as well.

```
select
  errlim = 1e-10                nodelimit= 500
  converge= 1e-6                subconverge= 1e-6
  contours= 10                  vectorgrid= 30          paint= off
  alias(x) = 'Radius'           alias(y) = 'Z'
variables
  u
definitions
  mm= 1e-3
  r1 = 10*mm     r2 = 20*mm     q21= r2/r1      L = 10*mm
  E= 200e9       mu= 0.30                                   { Steel }
  c= E/(1-mu**2)        G= E/[2*(1+mu)]
  p1= 1e6               p2= 0                   { Inner and outer pressure }
    sr= c*( dx(u)+ mu*u/x)
    st= c*( mu*dx(u)+ u/x)
    sz= 0
  epsz= (1/E)*(-mu)*( sr+st)          w= L*epsz
  mises= sqrt( 0.5*[ (sr-st)**2+ sr**2+ st**2] )          { Exact solution ... }
```

```
  sr_ex= (-p1*[(r2/x)**2- 1]+ p2*[q21**2- (r2/x)**2] )/ (q21**2- 1)
  st_ex= -(-p1*[(r2/x)**2+ 1]- p2*[q21**2+ (r2/x)**2)]/ (q21**2- 1)
  u_ex= (x/E)*(st_ex- mu*sr_ex)
  epsz_ex= (1/E)*(-mu)*( sr_ex+st_ex)          w_ex= y*epsz_ex
equations
  dx( c*[dx(u)+mu*u/x])+ (1-mu)/x* c*[dx(u)-u/x]= 0
boundaries
  region 1
  start (r1,0)
  load(u)= 0       line to (r2,0)
  load(u)= p2      line to (r2,L)
  load(u)= 0       line to (r1,L)
  load(u)= p1      line to finish
monitors
plots
  contour( u)              contour( w)
  elevation( u) from (r1,L/2) to (r2,L/2)
  elevation( (u- u_ex)/ u_ex) from (r1,L/2) to (r2,L/2)
  elevation( sr, st, sr+st) from (r1,L/2) to (r2,L/2)
  elevation( sr- sr_ex)  from (r1,L/2) to (r2,L/2)
  elevation( st- st_ex)  from (r1,L/2) to (r2,L/2)
  elevation( w, w_ex)  from (r1,L/2) to (r2,L/2)
  elevation( mises)  from (r1,L/2) to (r2,L/2)
end
```

The following figure displays the relative deviation of the solution *u* from the exact expression. Evidently the difference is mostly less than 0.01 %. Stresses are derivatives of strain and hence more sensitive to round-off errors, as is evident from the other comparative plots. All the errors may of course be further reduced at the expense of longer calculation times.

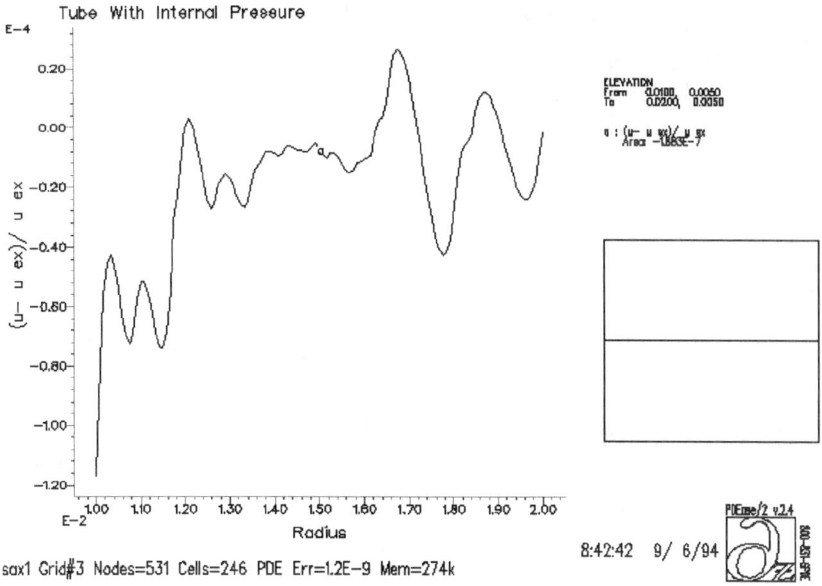

The General Case of Axial Symmetry

We are now ready to eliminate the special assumptions. The equations of mechanical equilibrium still are

$$\begin{cases} \dfrac{\partial \sigma_\rho}{\partial \rho} + \dfrac{\partial \tau_{\rho z}}{\partial z} + \dfrac{\sigma_\rho - \sigma_\theta}{\rho} = 0 \\[2ex] \dfrac{\partial \tau_{\rho z}}{\partial \rho} + \dfrac{\partial \sigma_z}{\partial z} + \dfrac{\tau_{\rho z}}{\rho} = 0 \end{cases}$$

Since we no longer have plane stress, however, Hooke's law must be written in its general form

$$\begin{Bmatrix} \varepsilon_\rho - \alpha \Delta T \\ \varepsilon_\theta - \alpha \Delta T \\ \varepsilon_z - \alpha \Delta T \end{Bmatrix} = \frac{1}{E} \begin{bmatrix} 1 & -\mu & -\mu \\ -\mu & 1 & -\mu \\ -\mu & -\mu & 1 \end{bmatrix} \begin{Bmatrix} \sigma_\rho \\ \sigma_\theta \\ \sigma_z \end{Bmatrix}$$

which we invert into

$$\begin{Bmatrix} \sigma_\rho \\ \sigma_\theta \\ \sigma_z \end{Bmatrix} = \frac{E}{(1-2\mu)(1+\mu)} \begin{bmatrix} 1-\mu & \mu & \mu \\ \mu & 1-\mu & \mu \\ \mu & \mu & 1-\mu \end{bmatrix} \begin{Bmatrix} \varepsilon_\rho - \alpha\Delta T \\ \varepsilon_\theta - \alpha\Delta T \\ \varepsilon_z - \alpha\Delta T \end{Bmatrix}$$

Multiplying the column vector by the matrix we obtain

$$\begin{Bmatrix} \sigma_\rho \\ \sigma_\theta \\ \sigma_z \end{Bmatrix} = Q \begin{Bmatrix} (1-\mu)\varepsilon_\rho + \mu\varepsilon_\theta + \mu\varepsilon_z - (1+\mu)\alpha\Delta T \\ \mu\varepsilon_\rho + (1-\mu)\varepsilon_\theta + \mu\varepsilon_z - (1+\mu)\alpha\Delta T \\ \mu\varepsilon_\rho + \mu\varepsilon_\theta + (1-\mu)\varepsilon_z - (1+\mu)\alpha\Delta T \end{Bmatrix} ,$$

and if we express the strains in terms of displacements

$$\begin{Bmatrix} \sigma_\rho \\ \sigma_\theta \\ \sigma_z \end{Bmatrix} = Q \begin{Bmatrix} (1-\mu)\dfrac{\partial u}{\partial \rho} + \mu\dfrac{u}{\rho} + \mu\dfrac{\partial w}{\partial z} - (1+\mu)\alpha\Delta T \\ \mu\dfrac{\partial u}{\partial \rho} + (1-\mu)\dfrac{u}{\rho} + \mu\dfrac{\partial w}{\partial z} - (1+\mu)\alpha\Delta T \\ \mu\dfrac{\partial u}{\partial \rho} + \mu\dfrac{u}{\rho} + (1-\mu)\dfrac{\partial w}{\partial z} - (1+\mu)\alpha\Delta T \end{Bmatrix} ,$$

where

$$Q = \frac{E}{(1-2\mu)(1+\mu)}$$

The shear stress takes much the same form as before

$$\tau_{\rho z} = G\gamma_{\rho z} = G\left(\frac{\partial u}{\partial z} + \frac{\partial w}{\partial \rho}\right), \quad \text{where} \quad G = \frac{E}{2(1+\mu)} .$$

When we substitute these expressions for the stresses the equations of equilibrium become

$$\begin{cases}\dfrac{\partial}{\partial\rho}\left\{Q\left[(1-\mu)\dfrac{\partial u}{\partial\rho}+\mu\dfrac{u}{\rho}+\mu\dfrac{\partial w}{\partial z}-(1+\mu)\alpha\Delta T\right]\right\}+\dfrac{\partial}{\partial z}\left\{G\left(\dfrac{\partial u}{\partial z}+\dfrac{\partial w}{\partial\rho}\right)\right\}+\\ \qquad\dfrac{Q}{\rho}\left[(1-\mu)\dfrac{\partial u}{\partial\rho}+\mu\dfrac{u}{\rho}+\mu\dfrac{\partial w}{\partial z}-(1+\mu)\alpha\Delta T\right]-\\ \qquad\dfrac{Q}{\rho}\left[\mu\dfrac{\partial u}{\partial\rho}+(1-\mu)\dfrac{u}{\rho}+\mu\dfrac{\partial w}{\partial z}-(1+\mu)\alpha\Delta T\right]=0\\ \dfrac{\partial}{\partial\rho}\left\{G\left(\dfrac{\partial u}{\partial z}+\dfrac{\partial w}{\partial\rho}\right)\right\}+\dfrac{\partial}{\partial z}\left\{Q\left[\mu\dfrac{\partial u}{\partial\rho}+\mu\dfrac{u}{\rho}+(1-\mu)\dfrac{\partial w}{\partial z}-(1+\mu)\alpha\Delta T\right]\right\}+\\ \qquad\dfrac{G}{\rho}\left\{\dfrac{\partial u}{\partial z}+\dfrac{\partial w}{\partial\rho}\right\}=0\end{cases}$$

After collecting pairs of terms in the second and third rows we arrive at the final system of PDEs, ready for use with PDEase,

$$\begin{cases}\dfrac{\partial}{\partial\rho}\left\{Q\left[(1-\mu)\dfrac{\partial u}{\partial\rho}+\mu\dfrac{u}{\rho}+\mu\dfrac{\partial w}{\partial z}-(1+\mu)\alpha\Delta T\right]\right\}+\\ \qquad\dfrac{\partial}{\partial z}\left\{G\left(\dfrac{\partial u}{\partial z}+\dfrac{\partial w}{\partial\rho}\right)\right\}+\ \dfrac{Q(1-2\mu)}{\rho}\left\{\dfrac{\partial u}{\partial\rho}-\dfrac{u}{\rho}\right\}=0\\ \dfrac{\partial}{\partial\rho}\left\{G\left(\dfrac{\partial u}{\partial z}+\dfrac{\partial w}{\partial\rho}\right)\right\}+\\ \qquad\dfrac{\partial}{\partial z}\left\{Q\left[\mu\dfrac{\partial u}{\partial\rho}+\mu\dfrac{u}{\rho}+(1-\mu)\dfrac{\partial w}{\partial z}-(1+\mu)\alpha\Delta T\right]\right\}+\\ \qquad\dfrac{G}{\rho}\left\{\dfrac{\partial u}{\partial z}+\dfrac{\partial w}{\partial\rho}\right\}=0\end{cases}$$

We now have the equations required for a problem of axial symmetry. Before typing the descriptor, however, we have to consider a practical point concerning finite element analysis. As will be further explained at the end of this volume, the program solves PDEs by minimizing an integral expression containing the equations. The problem is the factor $1/\rho$ appearing in some of the terms. Unless the object has a central bore (as in the first example), this factor will over-emphasize the region around the axis of symmetry.

In the following descriptors we hence multiply all the terms in the equations by ρ, which does not change their validity. As we shall discuss in a later chapter, however, we also must multiply the expression entering the *natural* boundary conditions by the same factor. No modifications of the *value* boundary conditions are necessary.

Test of the Full System of PDEs

The PDEs just developed for the general axially symmetric case appear to be quite complicated, and you may have some doubts about their validity. In order to put your mind at ease we first apply the general formalism to the pressurized tube of the previous example.

A few changes in the first descriptor are required, as shown below. We no longer calculate w from the stress components, since w now is a dependent variable. The *definitions* and *equations* should be obvious from the preceding pages.

In this modification we need a more complete expression for the *von Mises* equivalent stress, since the third component of normal stress (*sz*) no longer vanishes.

```
variables
  u  w
  ...
  Q= E/[(1-2*mu)*(1+mu)]       G= E/[2*(1+mu)]
  p1= 1e6           p2= 0
    sr= Q*[ (1-mu)*dx(u)+ mu*u/x+ mu*dy(w) ]
    st= Q*[ mu*dx(u)+ (1-mu)*u/x+ mu*dy(w) ]
    sz= Q*[ mu*dx(u)+ mu*u/x+ (1-mu)*dy(w) ]
    srz= G*[ dy(u)+ dx(w)]
  mises= sqrt( 0.5*[ (sr-st)**2+ (sr-sz)**2+ (st-sz)**2]+ 3*srz**2)
  ...
equations                                       { Uniform temperature }
  x* dx( Q*[ (1-mu)*dx(u)+ mu*u/x+ mu*dy(w) ) ]+
        x* dy( G*[dy(u)+ dx(w) ] )+ Q*(1-2*mu)*( dx(u)- u/x)= 0
  x* dx( G*[ dy(u)+ dx(w) ])+
```

```
        x* dy( Q*[ mu*dx(u)+ mu*u/x+ (1-mu)* dy(w) ] )+
        G*[ dy(u)+ dx(w) ]= 0
...
               region 1
start (r1,0)          point value(w)= 0
load(u)= 0            load(w)= 0          line to (r2,0)
load(u)= x*p2                line to (r2,L)
load(u)= 0            line to (r1,L)
load(u)= x*p1                line to finish
...
```

Although we now have two impressive-looking equations the problem does not take much longer time to run. The following *elevation* plot displays the radial stress error, to be compared to an applied pressure of 1e6. We note that the maximum error is somewhat larger than in the preceding one-dimensional calculation, but only moderately so.

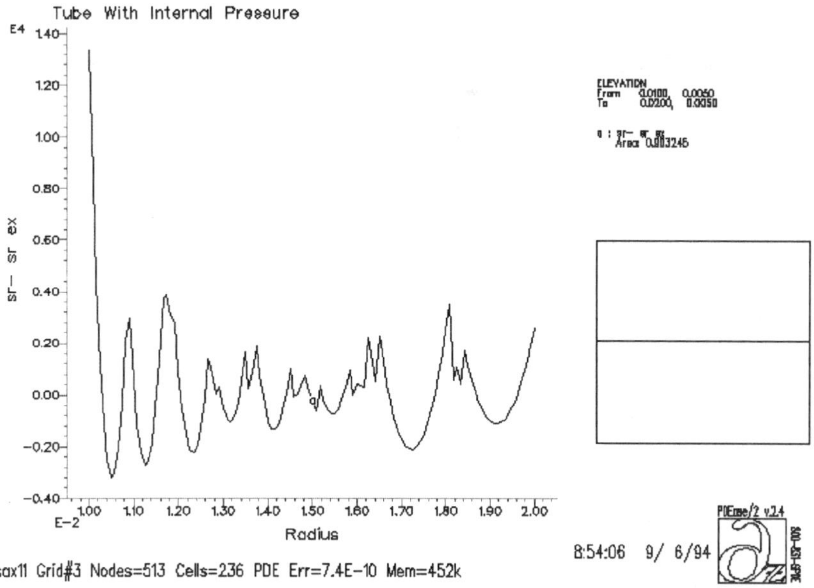

Tube Pressurized Over Part of Its Length

After having convinced ourselves that the system of PDEs does yield correct results in a simple case, we proceed to a problem which has no elementary solution. The geometry remains the same, but we modify the *boundaries* segment to exert pressure only over the central third of the bore.

Since we expect the stress distribution to be more complex in this case, we also introduce expressions for the principal stresses, similar to those used before in Cartesian coordinates (p.174). Rotating the coordinate system in the ρz plane does not modify the normal stress in the tangential (θ) direction. The shear stress $\tau_{\theta z}$ vanishes, since no twisting moment acts on the tube, and for reasons of axial symmetry $\tau_{\rho\theta}$ must also be zero. Hence the tangential axis is already a principal one, and we may use the same equations as before (p.174) for rotating the axes in the ρz plane. Apart from the usual figures indicating the direction of a principal axis we request a *grid* plot of the deformation.

```
 ...
 p_ang= 0.5* arctan( 2*srz/(sr-sz) )                    { Radians }
 sxp0= 0.5*(sr+sz)+ 0.5*(sr-sz)*cos(2*p_ang)+ srz*sin(2*p_ang)
 p_angl=  p_ang+ ustep(-sxp0)*pi/2
 sxp= 0.5*(sr+sz)+ 0.5*(sr-sz)*cos(2*p_angl)+ srz*sin(2*p_angl)
 syp= 0.5*(sr+sz)- 0.5*(sr-sz)*cos(2*p_angl)- srz*sin(2*p_angl)
 p_angle= p_angl * 180/pi                               { Degrees }
 ...
            region 1
 start (r1,0)       point value(w)= 0
 load(u)= 0         load(w)= 0        line to (r2,0)
 load(u)= x*p2             line to (r2,L)
 load(u)= 0         line to (r1,L) to (r1,2*L/3)
 load(u)= x*p1             line to (r1,L/3)
 load(u)= 0         line to finish
 ...
 grid( x+3e4*u, y+3e4*w)          contour( p_angle)
 vector[ cos(p_angl), sin(p_angl)]  as 'Direction of  tension'
 contour( sxp)       contour( syp)
end
```

The first figure below shows the deformation of the object, exaggerated by a factor of 3e4. The hole is distorted over more than one third of its length, and even the circular end surfaces have become slightly curved by the load.

The second figure displays principal directions. It should be studied together with the *contour* plots of the principal stresses, which reveal tension in the radial direction close to the flat faces. In the vicinity of the pressurized region of the hole both principal stresses are compressive.

Comparing these results to those of the first example we also notice that the maximum *von Mises* stress is far smaller when the tube is only pressurized over part of its length.

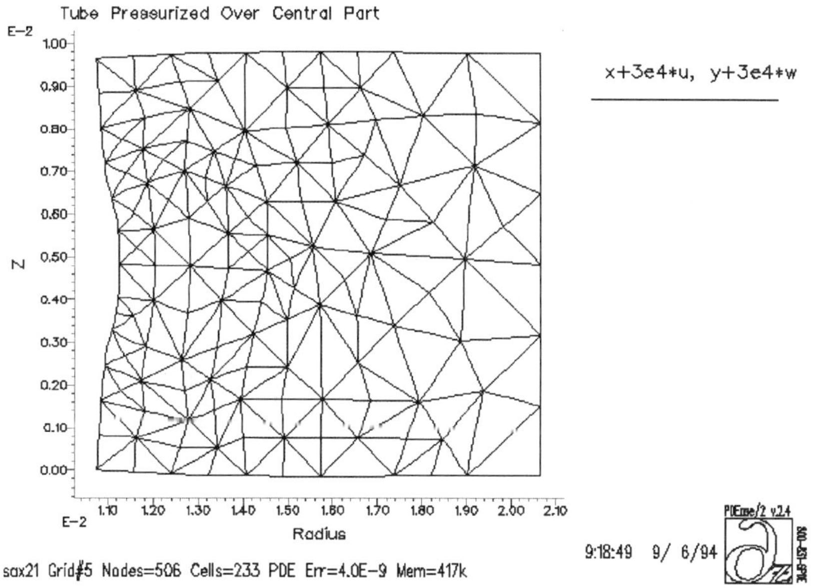

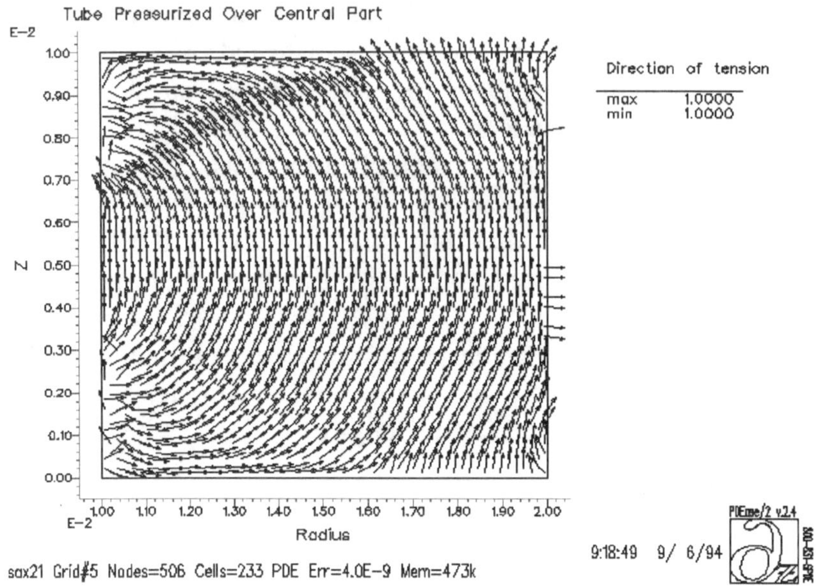

Stress Concentration in a Shaft

We shall now consider another case of practical importance: the stress concentration in the vicinity of a groove on an otherwise smooth surface. Our example concerns a shaft of circular cross-section with a circumferential groove, as shown in the grid plot below.

Curves of stress concentration factors for various geometries are available in the literature[5p302]. Such curves are generally based on a combination of theoretical results for simple situations and photo-elastic tests. It is customary to present the results in terms of the parameters $k_1 = r/d$ (r= the radius of the groove, d= minimum shaft diameter) and $k_2 = h/r$ (h= depth of groove), and we adapt to this notation in order to compare results. For the same reason we calculate the von *Mises* stress in units of nominal stress (*snom*), which is defined as the total tensile force divided by the smallest cross-sectional area of the shaft.

The following descriptor defines the problem of an axially symmetric shaft, on which we impose tensile stress (p) at the ends.

```
...
mm= 1e-3              d2= 10*mm              L= 5*mm
k1= 0.1               k2= 1.0                { Input parameters }
r0= k1*d2/(1+2*k1*k2)          h= k2*r0      { Radius and depth }
L0= sqrt(r0**2- [r0-h]**2)     p= 2e8
snom= p* (d2/2)**2/ (d2/2-h)**2              { Nominal stress }
...
              region 1
start (0,-L)
load(u)= 0     load(w)= -x*p         line to (d2/2,-L)
load(u)= 0     load(w)= 0       line to (d2/2,-L0)
arc(center= d2/2-h+r0, 0)  to (d2/2-h,0)  to (d2/2,L0)
line to (d2/2,L)
load(u)= 0     load(w)= x*p     line to (0,L)
value(u)= 0    load(w)= 0       line to finish
...
grid( x+1e2*u, y+1e2*w)
contour( u)           contour( w)
contour( sr)      contour( st)       contour( sz)         contour( srz)
contour( p_angle)
vector( cos(p_angl), sin(p_angl))  as 'Direction of tension'
surface( sxp)         surface( syp)
contour( mises/snom)                         { Divided by nominal stress }
contour( mises/snom) zoom(d2/2-1.5*h,-r0,  1.5*h,2*r0)
end
```

The first figure below shows the deformation of the shaft under the conditions of load imposed. The distortion of the groove generates visible effects even at the ends of the shaft. The presence of the groove also induces tensile radial stress in the central part of the object, by mechanisms that may be analyzed by elementary theory.

The second figure is a *contour* plot of equivalent stress, displaying the distribution in the region close to the groove. As may be expected the highest value occurs where the shaft radius is smallest.

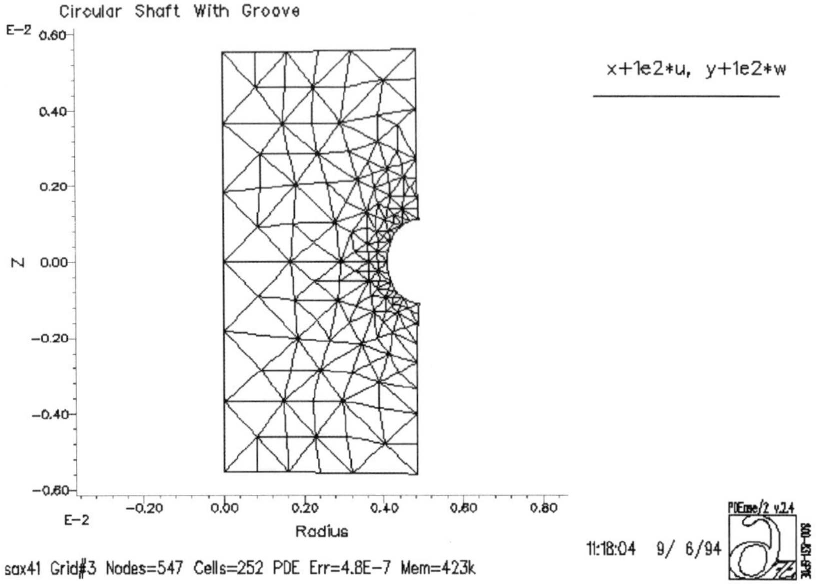
Circular Shaft With Groove
x+1e2*u, y+1e2*w
Z
Radius
E-2
sax41 Grid#3 Nodes=547 Cells=252 PDE Err=4.8E-7 Mem=423k
11:18:04 9/ 6/94
PDEase/2 v2.4

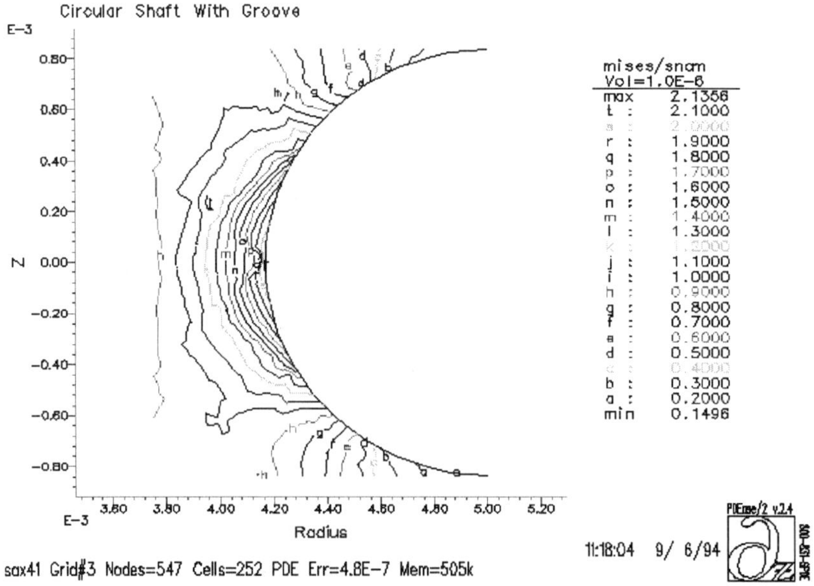
Circular Shaft With Groove
mises/snom
Vol=1.0E-6
max 2.1356
t : 2.1000
s : 2.0000
r : 1.9000
q : 1.8000
p : 1.7000
o : 1.6000
n : 1.5000
m : 1.4000
l : 1.3000
k : 1.2000
j : 1.1000
i : 1.0000
h : 0.9000
g : 0.8000
f : 0.7000
e : 0.6000
d : 0.5000
c : 0.4000
b : 0.3000
a : 0.2000
min 0.1496
Z
Radius
E-3
sax41 Grid#3 Nodes=547 Cells=252 PDE Err=4.8E-7 Mem=505k
11:18:04 9/ 6/94
PDEase/2 v2.4

Exercises

❑ Apply pressure only on the outside of the tube in the first example (p.203). Compare again to the exact solution.
❑ Modify the first example (p.203) by imposing a radial displacement of 0.1 % of *r1*, using no load on the other surfaces.
❑ Using an *elevation* plot, integrate the tangential (θ) stress over a plane going through the axis of a tube (p.203) and compare with the total force in that direction due to the internal pressure.
❑ Apply suitable axial displacements to the ends of the shaft (p.215) and compare the maximum equivalent stress to that obtained in the last example. Same for a shaft twice as long.

Thermo-Elasticity in (x,y) and (ρ,z) Space

In previous chapters we assumed uniform temperature throughout the object under study. We shall now investigate the effect of a space-varying temperature on the stress field. The derivation of the PDEs (p.157) already included a term involving the temperature increment ΔT, and we need only introduce this new term in the *definitions* segment.

Steel Block Strained by a Temperature Gradient

We begin with a somewhat hypothetical example: a steel block fixed to a completely rigid base at 300 K, the opposite face being kept at 600 K. Much of the following descriptor is similar to that for the tube with internal pressure (p.175).

Since the problem of heat conduction has an elementary solution we may introduce that under *definitions* and solve only for the displacements.

```
...                                                        { thel1.pde }
mm= 1e-3
Lx= 100*mm              Ly = 100*mm
mu= 0.3                 E = 200e9                               {Steel}
c= E/(1-mu**2)     G = E/[2*(1+mu)]
alpha= 1e-5                                     { Thermal expansivity }
dtemp= (y+Ly)/(2*Ly)* 300               { Temperature above ambient }
dabs= sqrt(u**2+ v**2)
ex= dx(u)          ey= dy(v)             exy= dx(v)+ dy(u)
sx= c*[ex+ mu*ey-(1+mu)*alpha*dtemp]
sy= c*[mu*ex+ ey-(1+mu)*alpha*dtemp]         sxy= G*exy

p_ang= 0.5* arctan( 2*sxy/(sx-sy) )                         { Radians }
sxp0= 0.5*(sx+sy)+ 0.5*(sx-sy)*cos(2*p_ang)+ sxy*sin(2*p_ang)
syp0= 0.5*(sx+sy)+ 0.5*(sx-sy)*cos(2*(p_ang+pi/2))
```

```
                                    + sxy*sin(2*(p_ang+pi/2))
                                                  { Test for highest value }
  p_angl=   if sxp0>syp0 then p_ang    else p_ang+ pi/2
  sxp= 0.5*(sx+sy)+ 0.5*(sx-sy)*cos(2*p_angl)+ sxy*sin(2*p_angl)
  syp= 0.5*(sx+sy)- 0.5*(sx-sy)*cos(2*p_angl)- sxy*sin(2*p_angl)
  p_angle= p_angl * 180/pi                                    { Degrees }
  mises= sqrt( 0.5*( (sx-sy)**2+ sx**2+ sy**2)+ 3*sxy**2)
initial values
equations
  dx( sx) + dy( sxy)= 0
  dx( sxy)+  dy( sy)= 0
boundaries
          region 1
  start(-Lx,-Ly)
  value(u)= 0            value(v)= 0        line to (Lx,-Ly)
  load(u)= 0             load(v)= 0
  line to (Lx,Ly) to (-Lx,Ly) to finish
monitors
plots
  grid(x+100*u, y+100*v)
  contour(dtemp)
  contour(u)             contour(v)         contour(dabs)
  vector(u,v)            vector(u/dabs, v/dabs)
  contour(sx)            contour(sy)         contour(sxy)
  contour(p_angle)
  vector(cos(p_angl), sin(p_angl))  as 'Direction of "tension" '
  contour(sxp)           contour(syp)        contour(mises)
end
```

The *grid* plot below depicts the distortion of the cross-section due to the thermal expansion of the upper part of the block.

In the *contour* plot of σ_x below we notice that the stress vanishes on the left and right faces, as required by the boundary condition. The other stress components also vanish on the appropriate boundaries. The values of σ_x are mostly negative, which means compressive stress, but there is a narrow region close to the base where the stress is tensile. It is interesting to reflect on the cause of this effect.

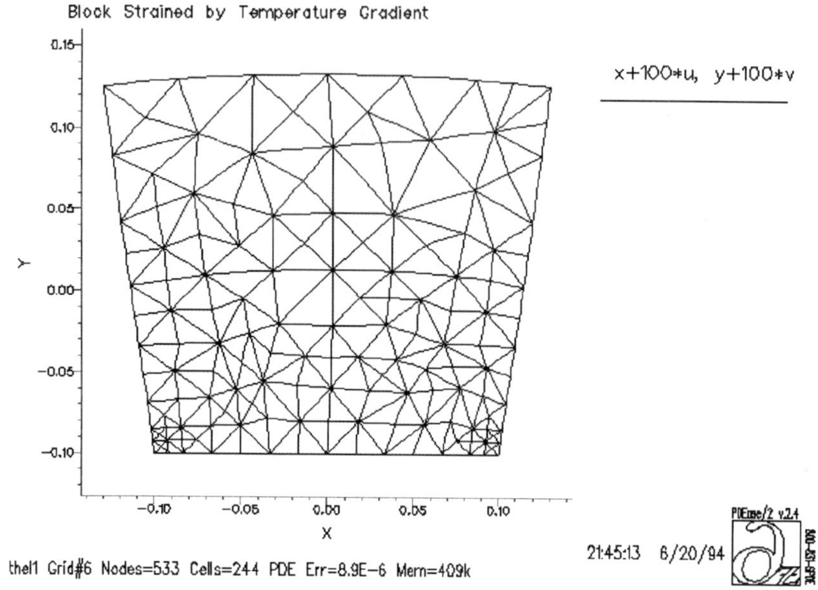
Block Strained by Temperature Gradient
x+100*u, y+100*v
0.15
0.10
0.05
0.00
−0.05
−0.10
Y
−0.10
−0.05
0.00
0.05
0.10
X
21:45:13 6/20/94
PDEase/2 v.2.4
thel1 Grid#6 Nodes=533 Cells=244 PDE Err=8.9E−6 Mem=409k

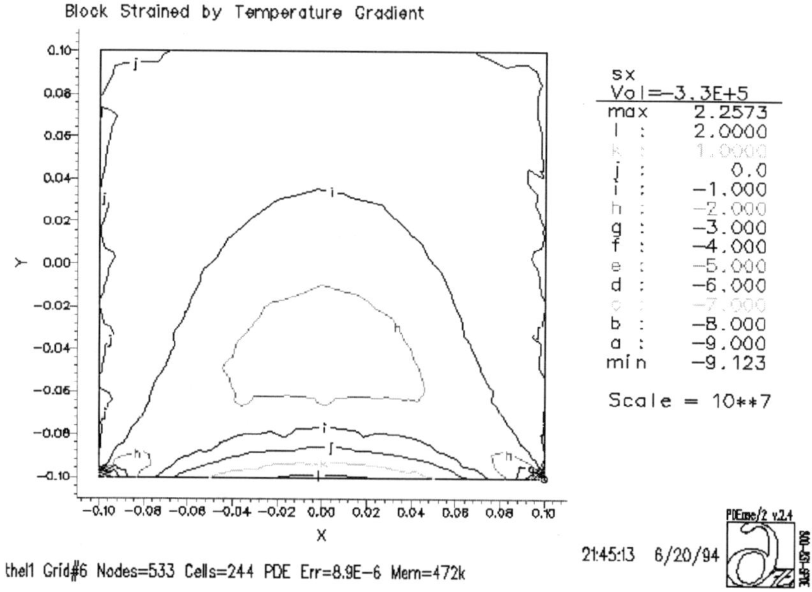
Block Strained by Temperature Gradient
sx
Vol=−3.3E+5
max 2.2573
l : 2.0000
k : 1.0000
j : 0.0
i : −1.000
h : −2.000
g : −3.000
f : −4.000
e : −5.000
d : −6.000
c : −7.000
b : −8.000
a : −9.000
min −9.123
Scale = 10**7
0.10
0.08
0.06
0.04
0.02
0.00
−0.02
−0.04
−0.06
−0.08
−0.10
Y
−0.10 −0.08 −0.06 −0.04 −0.02 0.00 0.02 0.04 0.06 0.08 0.10
X
21:45:13 6/20/94
PDEase/2 v.2.4
thel1 Grid#6 Nodes=533 Cells=244 PDE Err=8.9E−6 Mem=472k

Bi-Metallic Block

For simplicity we continue the line of hypothetical situations, assuming the upper half of our block to have twice as large thermal expansivity, the thermal conductivity and the elastic properties being the same. We also release the block from the rigid base. You only need to incorporate the following lines in the preceding descriptor.

```
                    region 1
start(-Lx,-Ly)
load(u)= 0          load(v)= 0
line to (Lx,-Ly) to (Lx,Ly) to (-Lx,Ly) to finish
                    region 2          alpha= 2e-5
start(-Lx,0)
line to (Lx,0) to (Lx,Ly) to (-Lx,Ly) to finish
```

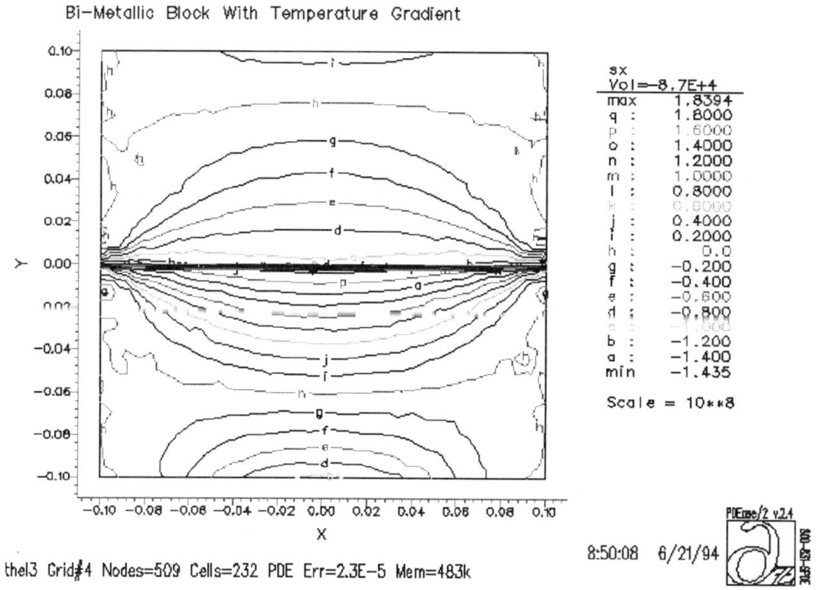

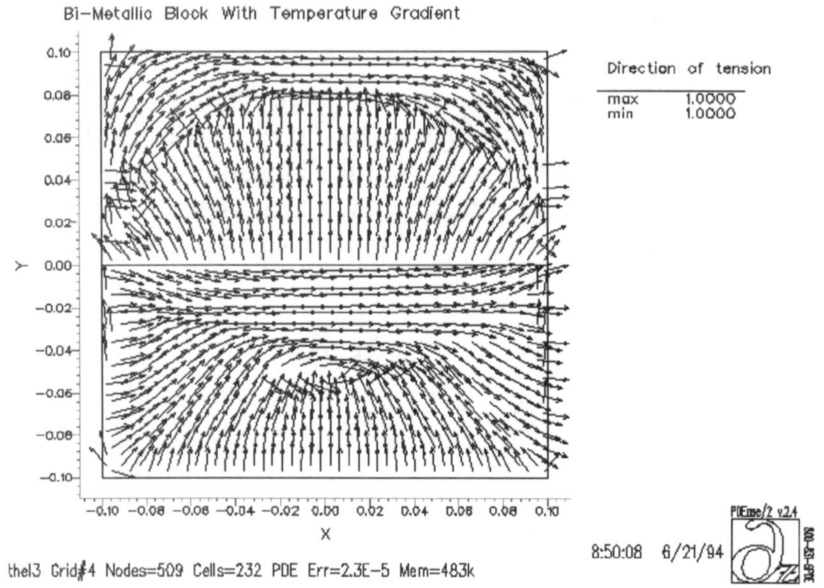

In this problem two plates of similar geometry deform by mutual forces at the interface. The *contour* plot on the preceding page displays the distribution of horizontal stress over the solution region. We may understand the main features of this plot by considering the way the upper plate expands, while being restrained by the lower plate. This interaction gives rise to forces and moments distorting the two parts of the block in different senses.

The last figure above illustrates how the interplay of forces is reflected in the principal directions of stress. A tensile axis on the lower side of the interface corresponds to a compressive axis on the upper side.

Three Simultaneous PDEs for dtemp, u and v

In the two preceding problems we could supply the exact expression for the temperature distribution. Now we shall deal with a situation where the solution for the temperature is unknown and has to be calculated along with the displacements.

Let us consider a modification of the preceding example, where only half of the upper face is at *dtemp*= 300, the rest of it being insulated.

The following list shows how the descriptor is to be modified, after the expression for *dtemp* has been eliminated. The *variables* must be declared in the order they have been entered in the *equations* segment. The first PDE expresses the conservation of energy (heat), the second one expresses mechanical equilibrium in the *x* (or *u*) direction, and so on.

```
  ...                                                         { thel11.pde }
  errlim= 5e-4
  ...
variables
  dtemp  u  v
definitions
  ...
  k= 45                                           { Thermal conductivity }
  ...
equations
  dx(-k*dx( dtemp))+ dy(-k*dy( dtemp))= 0
  dx( sx) + dy( sxy)= 0
  dx( sxy)+  dy( sy)= 0
boundaries
          region 1
  start(-Lx,-Ly)
  value( dtemp)= 0              load(u)= 0                 load(v)= 0
  line to (Lx,-Ly)
  natural( dtemp)= 0            line to (Lx,Ly)
  value( dtemp)= 300            line to (0,Ly)
  natural( dtemp)= 0            line to (-Lx,Ly) to finish
  ...
```

The figure below shows the temperature distribution, which here is independent of the state of stress. In a case of extremely high stress the thermal conductivity would depend on the stress components, but PDEase should be able to solve that problem as well.

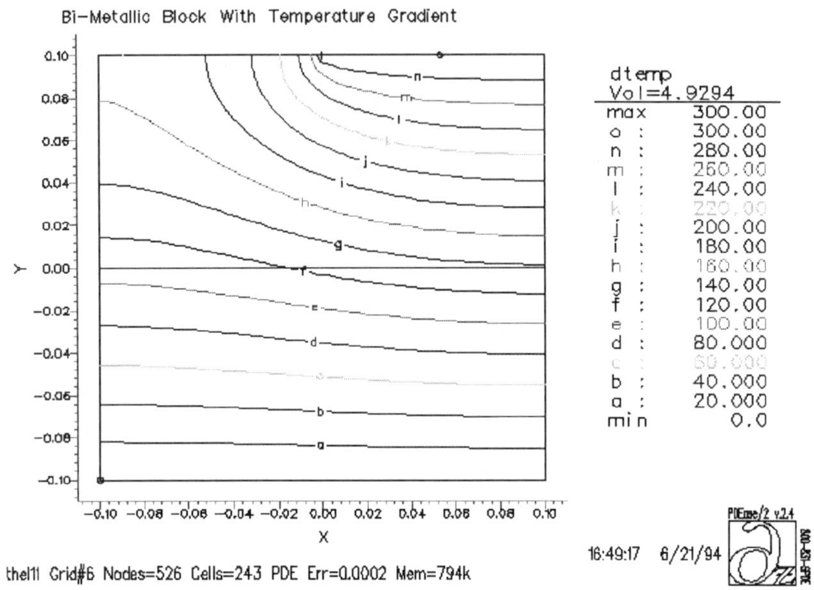
Bi-Metallic Block With Temperature Gradient
dtemp
Vol=4.9294
max 300.00
o : 300.00
n : 280.00
m : 260.00
l : 240.00
k : 220.00
j : 200.00
i : 180.00
h : 160.00
g : 140.00
f : 120.00
e : 100.00
d : 80.000
c : 60.000
b : 40.000
a : 20.000
min 0.0
X
Y
16:49:17 6/21/94
PDEase/2 v2.4
thel11 Grid#6 Nodes=526 Cells=243 PDE Err=0.0002 Mem=794k

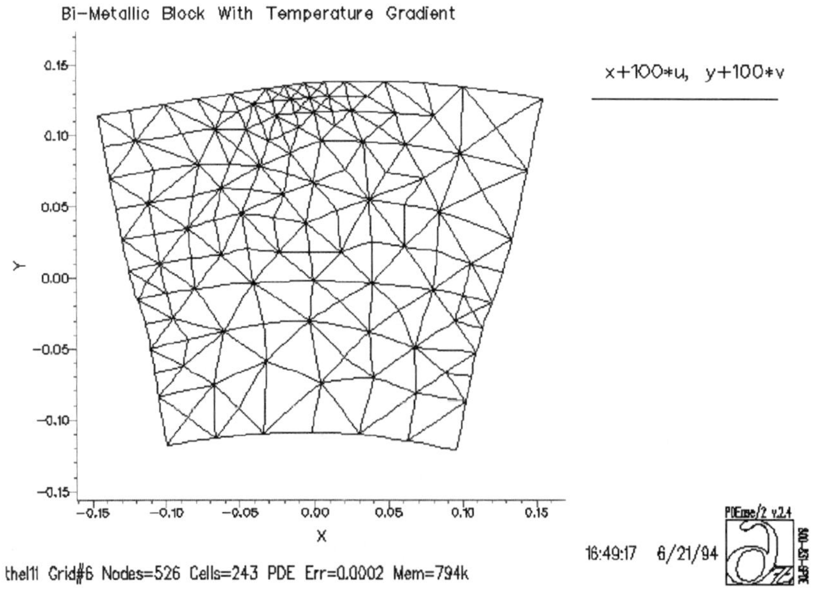
Bi-Metallic Block With Temperature Gradient
x+100*u, y+100*v
X
Y
16:49:17 6/21/94
PDEase/2 v2.4
thel11 Grid#6 Nodes=526 Cells=243 PDE Err=0.0002 Mem=794k

The preceding figure is a *grid* plot of the deformation, demonstrating the essential asymmetry of this problem. The plots of the stress distributions, however, are not drastically different from what we have just seen.

Axially Symmetric Heating of a Cylinder

We finally consider the thermo-elastic deformation of a cylindrical glass object, heated over the middle part of one face by a laser beam, while the opposite face is kept at constant temperature. In short, there is a given heating power per unit area over the central spot, the rest of the front face and the cylindrical surface is insulated, and the object is free to expand in all directions.

The essential parts of the descriptor are the following. The *definitions* of the stress components must be extended to the case of non-uniform temperature and the *equations* modified accordingly (p.207-209).

```
  ...
  alias(x) = 'Radius'           alias(y) = 'Z'
variables
  dtemp    u     w
definitions
  mm= 1e-3       r0= 0.3*mm       r1= 1*mm          L= 1*mm
  E= 70e9    mu= 0.25      k= 1.3       alpha= 1e-5                { Glass }
  Q= E/[(1-2*mu)*(1+mu)]      G= E/[2*(1+mu)]
    sr= Q*[ (1-mu)*dx(u)+ mu*u/x+ mu*dy(w)- (1+mu)*alpha*dtemp]
    st= Q*[ mu*dx(u)+ (1-mu)*u/x+ mu*dy(w)- (1+mu)*alpha*dtemp]
    sz= Q*[ mu*dx(u)+ mu*u/x+ (1-mu)*dy(w)- (1+mu)*alpha*dtemp]
    srz= G*[ dy(u)+ dx(w)]
  p_ang= 0.5* arctan( 2*srz/(sr-sz) )                            { Radians }
  sxp0= 0.5*(sr+sz)+ 0.5*(sr-sz)*cos(2*p_ang)+ srz*sin(2*p_ang)
  p_angl=  p_ang+ ustep(-sxp0)*pi/2
  sxp= 0.5*(sr+sz)+ 0.5*(sr-sz)*cos(2*p_angl)+ srz*sin(2*p_angl)
  syp= 0.5*(sr+sz)- 0.5*(sr-sz)*cos(2*p_angl)- srz*sin(2*p_angl)
  p_angle= p_angl * 180/pi                                       { Degrees }
  mises= sqrt( 0.5*[ (sr-st)**2+ (sr-sz)**2+ (st-sz)**2]+ 3*srz**2)
equations
```

```
  dx[-k*x*dx( dtemp)]+ x*dy[-k*dy( dtemp)]= 0
  x*dx[ Q*( (1-mu)*dx(u)+ mu*u/x+ mu*dy(w)- (1+mu)*alpha*dtemp )]+
             x*dy( G*(dy(u)+ dx(w) ))+ Q*(1-2*mu)* ( dx(u)- u/x)= 0
  x*dx( G*( dy(u)+ dx(w) ))+
     x*dy[ Q*( mu*dx(u)+ mu*u/x+ (1-mu)* dy(w)- (1+mu)*alpha*dtemp)]+
                                               G* ( dy(u)+ dx(w) )= 0
boundaries
         region 1
  start(0,0)
  value( dtemp)= 0   load(u)= 0      load(w)= 0    line to (r1,0)
  natural( dtemp)= 0      line to (r1,L) to (r0,L)
  natural(dtemp)= -x*1e5      line to (0,L)
  natural( dtemp)= 0  value(u)= 0      load(w)= 0    line to finish
monitors
plots
  grid( x+3e3*u, y+3e3*w)
  contour( dtemp)       contour( u)               contour( w)
  contour( sr)          contour( st)              contour( sz)
  contour( srz)         contour( p_angle)
  vector( cos(p_angl), sin(p_angl)) as 'Direction of tension'
  contour( sxp)         contour( syp)             contour( mises)
end
```

Although we are now dealing with three PDEs the calculation time remains reasonable. The *contour* plot on the next page illustrates the temperature distribution, demonstrating that the magnitude of the gradient is largest at the top and close to the axis.

The next figure indicates that the hoop stress is compressive at small radii and tensile further out. The *contour* plot of *sz* (not shown here) displays the pattern that we are accustomed to interpret as vanishing values. The principal axis also seems to be parallel throughout the object. This feature may be surprising in view of the distortion of the object, and it might be due to accidental cancellation. Other axially symmetric distributions of temperature on the upper circular face yield similar results in this respect. If, on the other hand, we apply axially symmetric heating on the cylindrical surface, this simple symmetric pattern is broken.

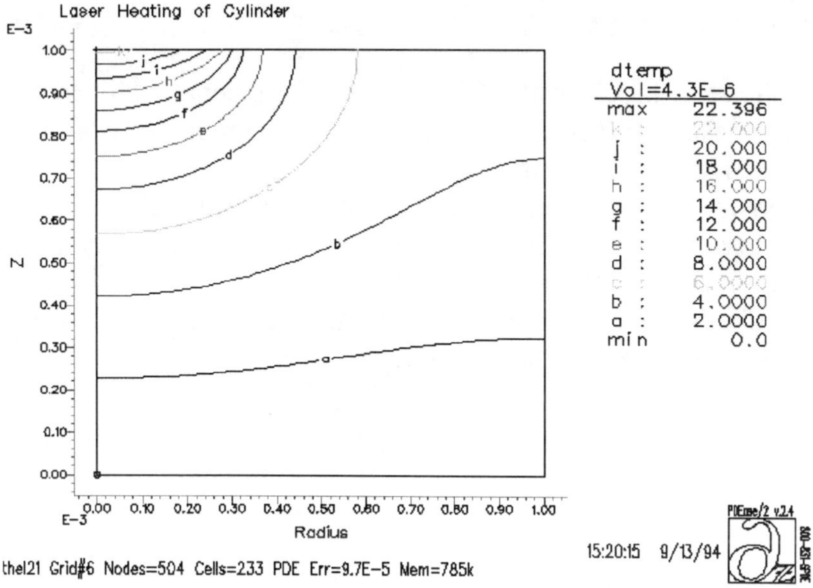
Laser Heating of Cylinder
E-3
1.00
0.90
0.80
0.70
0.60
Z 0.50
0.40
0.30
0.20
0.10
0.00
E-3 0.00 0.10 0.20 0.30 0.40 0.50 0.60 0.70 0.80 0.90 1.00
Radius
dtemp
Vol=4.3E-6
max 22.396
k : 22.000
j : 20.000
i : 18.000
h : 16.000
g : 14.000
f : 12.000
e : 10.000
d : 8.0000
c : 6.0000
b : 4.0000
a : 2.0000
min 0.0
15:20:15 9/13/94
PDEase/2 v2.4
thel21 Grid#6 Nodes=504 Cells=233 PDE Err=9.7E-5 Mem=785k

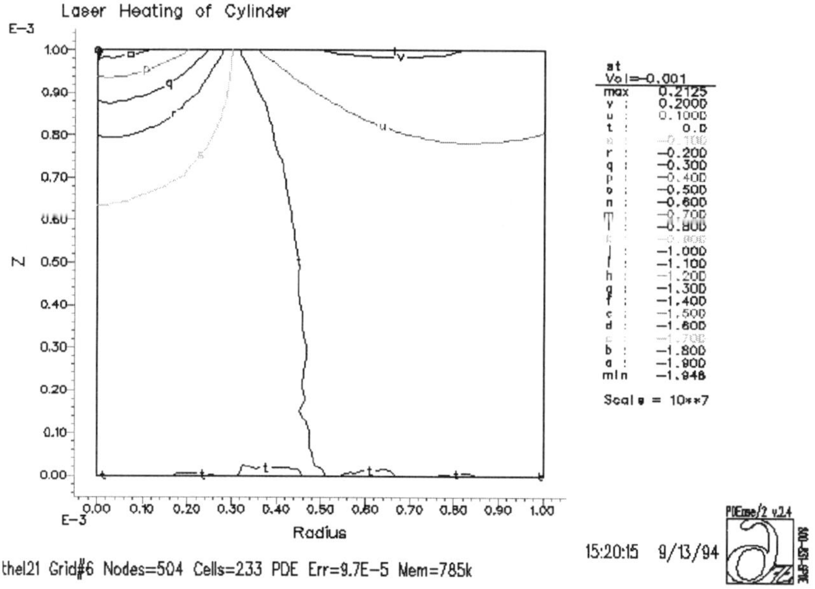
Laser Heating of Cylinder
E-3
1.00
0.90
0.80
0.70
0.60
Z 0.50
0.40
0.30
0.20
0.10
0.00
E-3 0.00 0.10 0.20 0.30 0.40 0.50 0.60 0.70 0.80 0.90 1.00
Radius
st
Vol=-0.001
max 0.2125
v : 0.2000
u : 0.1000
t : 0.0
r : -0.200
q : -0.300
p : -0.400
o : -0.500
n : -0.600
m : -0.700
l : -0.800
j : -1.000
i : -1.100
h : -1.200
g : -1.300
f : -1.400
e : -1.500
d : -1.600
b : -1.800
a : -1.900
min -1.948
Scale = 10**7
15:20:15 9/13/94
PDEase/2 v2.4
thel21 Grid#6 Nodes=504 Cells=233 PDE Err=9.7E-5 Mem=785k

Exercises

❑ Let the steel block of the first example be mechanically free from the rigid base, while still being at the same temperature as before. Study the distribution of stresses.
❑ Reverse the heat current through the bi-metallic block.
❑ Modify the first example by solving for the temperature distribution as well as for the two displacements.
❑ Modify the last example by applying heat to an annular region outside the radius *r0* on the top face.

Irrotational Flow of Liquids

Since the density of a liquid normally changes little within the range of pressures occurring in practical applications, we assume the density to be constant. In this chapter we also make more daring assumptions, i.e. that the liquid slips freely over solid surfaces and that viscous forces are vanishingly small compared to inertial forces. These assumptions are known to be useful, however, in some situations.

In the case of non-rotational flow ($\nabla \times \mathbf{u} = 0$) we may express the velocity in terms of a potential[7p115]

$$\mathbf{u} = \nabla \phi. \qquad \bullet$$

The conservation of mass, in combination with constant density, leads to the conservation of volume

$$\nabla \cdot \mathbf{u} = 0,$$

and hence the potential ϕ obeys the PDE

$$\nabla \cdot \mathbf{u} = \nabla^2 \phi = 0. \qquad \bullet$$

This in turn leads us to the Bernoulli equation of motion, which states

$$\frac{1}{2}\rho q^2 + p = constant \qquad \bullet$$

where ρ is the mass density, $q = \sqrt{u^2 + v^2}$ the magnitude of the velocity (with components $u,\ v$) and p the pressure.

Flow Through a Constricted Channel

Our first application of the above equations will be to the flow through a channel, limited by plane surfaces. The following descriptor defines the problem, which is a simple one since we only have one PDE for the velocity potential. After having solved for this potential we simply differentiate to obtain the components of velocity. Having obtained these components we then form the magnitude of the velocity q and finally the pressure p, as shown in the last line of the *definitions* segment.

In the *boundaries* segment we specify the input velocity $u0$ by a *natural* statement. For the output end we just fix a value for the potential *phi*. Its absolute value is of course arbitrary since only the derivatives will be used, but by specifying a constant value over this boundary we also stipulate that dy(*phi*) is to vanish, i.e. we force the liquid to exit horizontally.

```
 ...
 errlim= 1e-5            nodelimit= 500           vectorgrid= 20
 contours= 20            surfacegrid= 30          paint= off
variables
 phi                                                { Velocity potential }
definitions
 Lx= 1          Ly= 1                   coef= 2
 u0= 3                          { x component of velocity at left boundary }
 p0= 1e5                        { Atmospheric pressure at left boundary }
 dens= 1e3                                              { Mass density }
 u= dx(phi)      v= dy(phi)                      { Velocity components }
 speed2= u**2+v**2          speed= sqrt( speed2)
 p= p0+ 0.5*dens*(u0**2-speed2)                             { Pressure }
equations
 dxx( phi)+ dyy( phi)= 0
boundaries
        region 1
 start(0,Ly)       natural( phi)= -u0        line to (0,-Ly)
 natural( phi)= 0  line to (Lx,-Ly)  to (2*Lx,-Ly/coef)  to (3*Lx,-Ly/coef)
 value( phi)= 0   line  to (3*Lx,Ly/coef)
 natural( phi)= 0  line to (2*Lx,Ly/coef) to (Lx,Ly) to finish
monitors
plots
 contour(phi)            contour( speed)          contour( p)
 contour( p) zoom(1.5*Lx,0  ,Lx,Ly)
 vector(u,v) as 'Flow direction'
 surface( p) zoom(1.5*Lx,0  ,Lx,Ly)  interactive
end
```

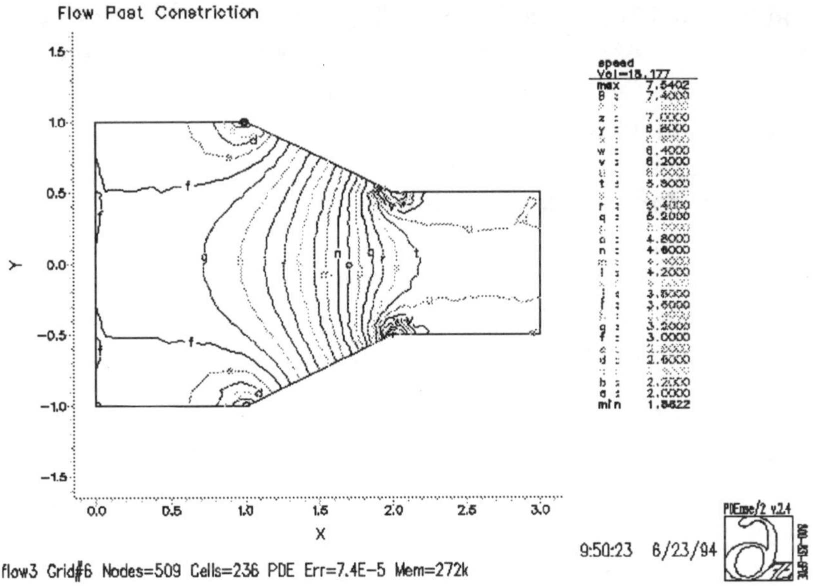
Flow Past Constriction
speed
Vol=15.177
max 7.5402
min 1.8622
Y
X
9:50:23 6/23/94
PDEase/2 v2.4
flow3 Grid#6 Nodes=509 Cells=236 PDE Err=7.4E-5 Mem=272k

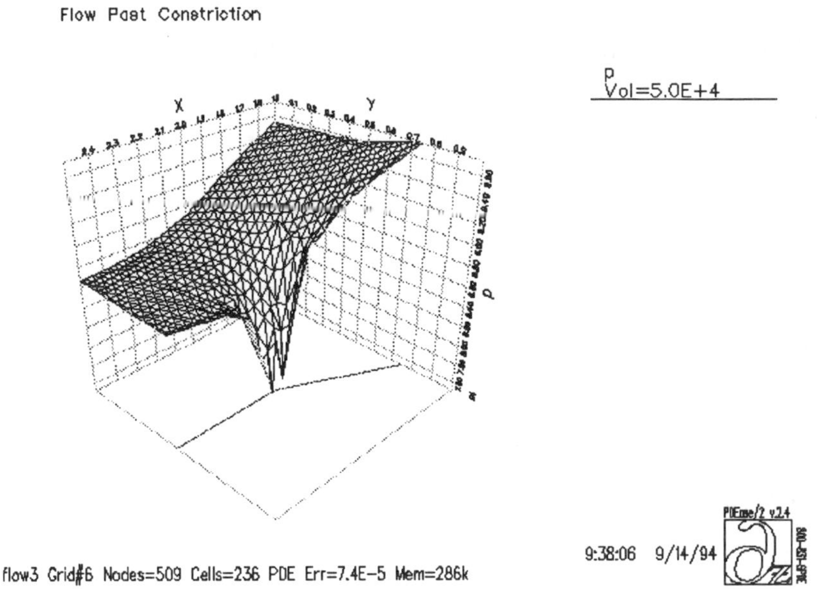
Flow Past Constriction
P
Vol=5.0E+4
X
Y
P
9:38:06 9/14/94
PDEase/2 v2.4
flow3 Grid#6 Nodes=509 Cells=236 PDE Err=7.4E-5 Mem=286k

From the first figure on the preceding page we see that the speed of the liquid increases by a factor of two from input to output, i.e. precisely by the constriction ratio. This is of course in accord with the conservation of mass, a principle which was incorporated into the PDE. Notice, however, that we did not explicitly introduce this constancy in the boundary conditions, although we could well have done so.

The above *surface* plot shows the pressure drop where the channel wall bends. From the corresponding *contour* plot we find that the pressure at the corner is only 74 % of the value at the input.

Square Obstacle in a Straight Channel

We shall next consider the flow around an obstacle, and in particular the forces exerted on it by the stream. In the descriptor, which is based on the first example, we introduce a bar of square cross-section across the channel. In order to be able to inspect the results in more detail we include the expression for the angle of flow (p.36).

```
...
errlim= 1e-7        nodelimit= 1000        vectorgrid= 20
contours= 20         surfacegrid= 30        paint= off
...
Lx= 1              Ly= 1       ax= 0.2          ay= 0.2
f_angle= sign(v)* arccos( u/speed)/pi*180
small= 1e-6*ax
...
            region 1
start(-Lx,Ly)
natural( phi)= -u0   line to (-Lx,-Ly)
natural( phi)= 0     line to (Lx,-Ly)
value( phi)= 0       line  to (Lx,Ly)
natural( phi)= 0 line to finish
start(-ax,-ay)                                                 { Cut-out }
natural( phi)= 0  line to (-ax,ay) to (ax,ay) to (ax,-ay) to finish
...
contour( p)       contour( p) zoom(0,0,  2*ax,2*ay)
vector( u, v)       vector( u, v) zoom(0,0,  2*ax,2*ay)
```

```
  contour( f_angle)              contour( f_angle) zoom(0,0,  2*ax,2*ay)
  contour( f_angle)  zoom(0.9*ax,-0.1*ay,  0.2*ax,0.2*ay)
  elevation( p) from (-ax-small,-ay) to (-ax-small,ay)
  elevation( p) from (ax+small,-ay) to (ax+small,ay)
end
```

The *contour* plot below demonstrates the symmetry of the pressure distribution with respect to both axes. Just to the left, and also to the right of the obstacle, we find pressure values slightly in excess of the input value, 1e5. The maximum value is close to the result (1.045) which we calculate directly from Bernoulli's equation for the points of stagnation.

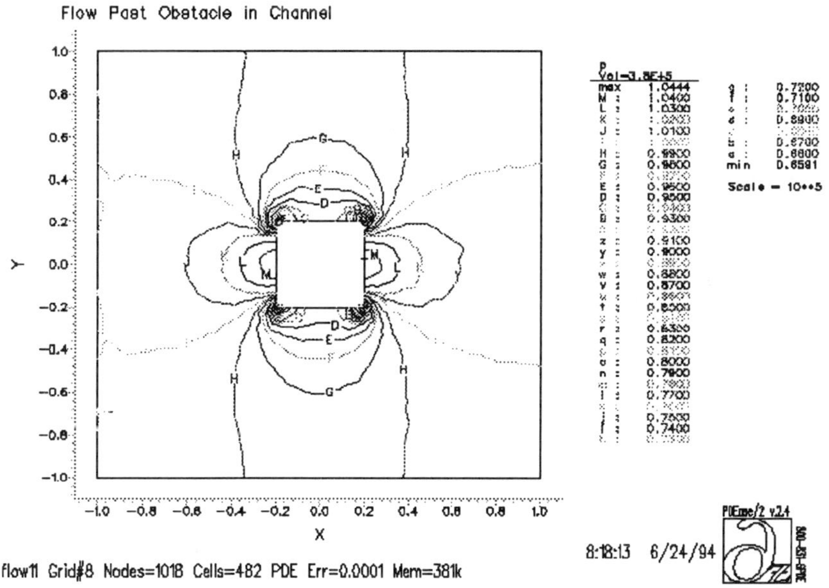

The next figure presents the direction of flow over the solution domain. We should expect angles up to 90^0 at the vertical faces, but the plot contains angles which are a few degrees larger. Exploring this curious feature by zooming we conclude that it must be due to round-off errors in the vicinity of the stagnation points.

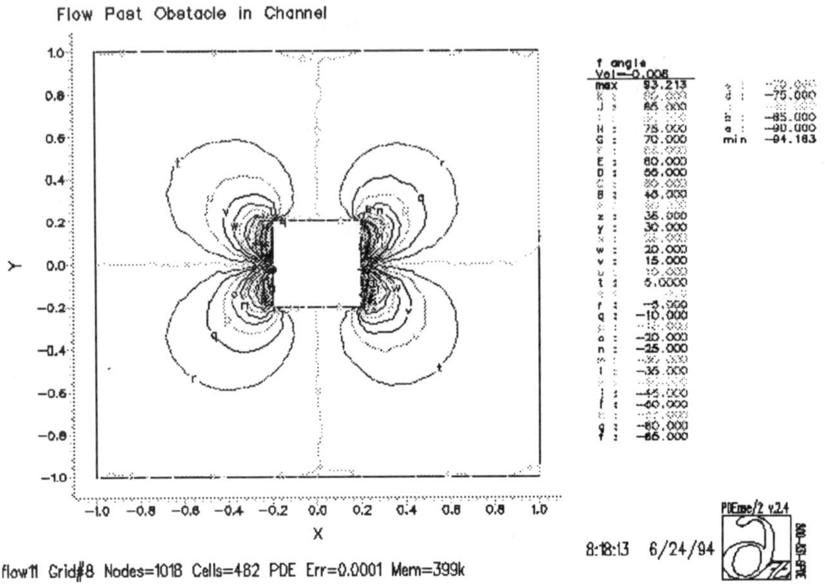

Before we leave this example we must find out the amount of drag on the obstacle by the stream. Since there is no viscosity, the relevant forces on the bottom and top faces vanish. From the symmetry of the pressure plot we already suspect that the upstream and downstream values are equal. We confirm this by two *elevation* plots, taken just outside the left and right boundaries (solution values do not always exist exactly on the boundaries). Thus, there is no drag on the obstacle in this hydrodynamic model. On the other hand, we find excess pressure on the left and right faces and a deficit on the bottom and top faces, so if the obstacle were elastic it would deform.

Lift on an Inclined Plate

We have just seen that there is no drag on a square bar, and for obvious reasons of symmetry there can be no lift either. Let us now turn to a situation where we all know from experience that a lifting force may occur, both in air and in water.

The geometry should be clear from the figure below. As before, we have a stream of liquid from left to right with constant velocity at the vertical boundaries. The obstacle is a rectangular plate at an angle of attack (*alpha*) with respect to the main stream. The new features of the descriptor follow.

The general expressions for the corner coordinates of the plate permit us to change the angle of attack at will. Here we modify the boundary conditions slightly by specifying the exit velocity to be equal to the input velocity.

```
  ...
  contours= 10
  ...
  Lx= 1              Ly= 1
  L= 0.5*Ly                    d= 0.2* L
  u0= 4
  alpha= 45* pi/180                                         { Angle of attack }
    si= sin( alpha)                  co= cos( alpha)
    x1= -d/2*si- L/2*co                    y1= -d/2*co+ L/2*si
    x2= d/2*si- L/2*co                     y2= d/2*co+ L/2*si
    x3= -x1    y3= -y1    x4= -x2    y4= -y2
  small= 1e-6*Lx
  ...
  region 1
  start(-Lx,Ly)    point value( phi)= 0
  natural( phi)= -u0   line to (-Lx,-Ly)
  natural( phi)= 0     line to (Lx,-Ly)
  natural( phi)= u0      line  to (Lx,Ly)
  natural( phi)= 0 line to finish
  start(x1,y1)                                                    { Cut-out }
  natural( phi)= 0 line to (x2,y2) to (x3,y3) to (x4,y4) to finish
plots
  contour( phi)             contour( speed)           contour( p)
  vector( u, v)             contour( f_angle)
    elevation( p) from (x1,y1) to (x2,y2)
    elevation( p) from (x2,y2) to (x3,y3)
    elevation( p) from (x4,y4-small) to (x3,y3-small)
    elevation( p) from (x1,y1-small) to (x4,y4-small)
  elevation( p) from (x1,y2+small) to (x3,y2+small)
  elevation( p) from (x1,y4-small) to (x3,y4-small)
end
```

The following *contour* plot of the pressure indicates that the forces on the major faces are symmetric in a reversed sense, but such that the total forces on opposite faces could well be equal. The four indented *elevation* plot commands are for computing the integrals of p over all four faces. The result is that the forces on opposite faces are in fact equal to within less than 0.1 %. Hence there does not seem to be any net force on the plate, neither drag nor lift.

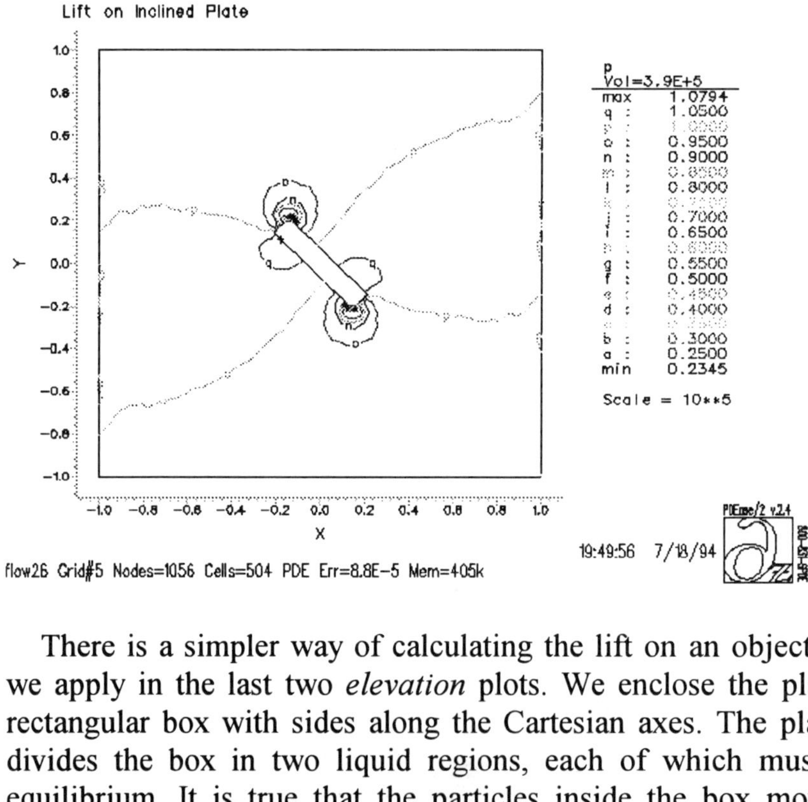

There is a simpler way of calculating the lift on an object, which we apply in the last two *elevation* plots. We enclose the plate in a rectangular box with sides along the Cartesian axes. The plate thus divides the box in two liquid regions, each of which must be in equilibrium. It is true that the particles inside the box move with different velocities, but the total momentum of this set of particles must be constant in steady flow. For this to pertain the total force on each volume of liquid must be zero. The force due to pressure on the top part, say, must be canceled by a force from the plate. The reaction force on the plate thus must be equal to the force on the box top. Using this method we need only two integrals to compute the lift. The conclusion remains the same: the top and bottom forces cancel, and there is no net vertical force on the plate.

An Airfoil of Triangular Cross-Section

If we are skeptical about the conclusion that no lift appears on an inclined slab in a stream, we would perhaps like to try a configuration more similar to an airplane wing. Let us investigate the simplest possible airfoil, formed by three plane surfaces normal to our screen. This model involves only three points, and by changing one or two of these we may easily obtain various thicknesses and angles of attack. In order to generate direction cosines for the normal vectors we also write expressions for the side lengths. These cosines are used in three *elevation* plots (indented), which directly yield the lift on each of the faces.

The last two *elevation* plots refer to the top and bottom surfaces of a box enclosing the airfoil. The difference of these two integrals should yield an auxiliary value of the lift.

```
 ...
 x1= -0.5                x2= -0.2            x3= 0.5
 y1= -0.0           y2= 0.1        y3= -0.2
   d12= sqrt[(x2-x1)**2+ (y2-y1)**2]                     { Side lengths... }
   d13= sqrt[(x3-x1)**2+ (y3-y1)**2]
   d23= sqrt[(x3-x2)**2+ (y3-y2)**2]
 small= 1e-3*Lx
 ...
 start(x1,y1)
 line to (x2,y2) to (x3,y3) to finish
 ...
   elevation( p* (x2-x1)/d12) from (x1,y1+small) to (x2,y2+small)
   elevation( p* (x3-x2)/d23) from (x2,y2+small) to (x3,y3+small)
   elevation( p* (x3-x1)/d13) from (x1,y1-small) to (x3,y3-small)
  elevation( p) from (x1,y2+small) to (x3,y2+small)
  elevation( p) from (x1,y3-small) to (x3,y3-small)
end
```

The figure below presents the pressure distribution, which shows that the minimum pressure is about one half of the maximum value. The integral (*Area*) values resulting from the *elevation* plots balance to within one percent or better, which indicates that the lift vanishes even in this case.

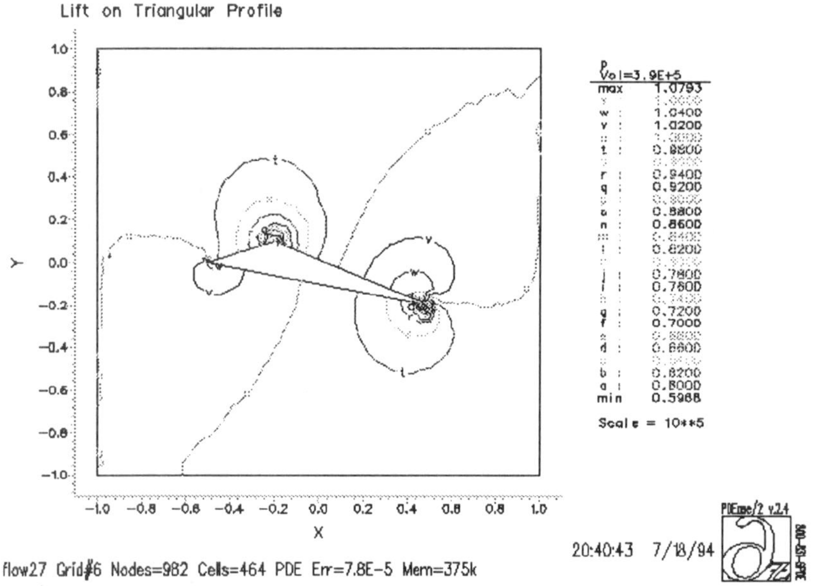

This and the preceding example show that there is neither drag nor lift on an object if the flow is irrotational. This may seem to contradict essential statements in textbooks, which disseminate the view that lift should be a consequence of the Bernoulli equation: the increased speed around the upper part of the triangular profile, say, should provide lower pressure, hence lift. This is true, but there are compensating forces close to the front and rear edges.

There is another version of the Bernoulli equation, however, valid for finite vorticity but vanishing viscosity, which may be derived by integrating the Navier-Stokes equation along a streamline[7p117]. This equation is better suited for studying the lift effect under more realistic conditions.

Flow Constriction in (ρ,z) Space

We finish this chapter by an example of axially symmetric flow, requiring cylindrical coordinates. The problem is analogous to the first one (p.230). The input section is a tube with circular cross-section, followed by a conical part connecting to a narrower tube. The flow is still assumed to be irrotational. The partial descriptor below indicates the changes necessary.

The PDE is similar to those used in other axially symmetric examples, e.g. with dielectrics (p.77). The boundary conditions are the same as in the channel constriction problem, except that the flow is now in a vertical direction on the screen.

```
select
  errlim= 1e-7            nodelimit= 1000         vectorgrid= 20
  contours= 20            surfacegrid= 30         paint= off
  alias(x)= 'Radius'      alias(y)= 'Z'
variables
  phi
definitions
  r0= 0.5         r1= 1.0          L= 1
  u0= 3           p0= 1e5          dens= 1e3
  u= dx(phi)              v= dy(phi)
  speed2= u**2+v**2   speed= sqrt( speed2)
  p= p0+ 0.5*dens*(u0**2-speed2)
  f_angle= sign(v)* arccos( u/speed)/pi*180
  small= 1e-6
equations
  (1/x)* dx[ x* dx(phi)]+ dyy( phi)= 0
boundaries
          region 1
  start(0,0)
  natural( phi)= -u0      line to (r1,0)
  natural( phi)= 0        line to (r1,L) to (r0,2*L) to (r0,3*L)
  value( phi)= 0          line to (0,3*L)
  natural( phi)= 0        line to finish
monitors
plots
  contour( phi)           contour( speed)
  contour( p)
```

```
  vector( u, v)          contour( f_angle)
end
```

The plot below shows that the final speed is as expected from mass conservation, but we could not have predicted the detailed variation against ρ and z on the basis of this principle only.

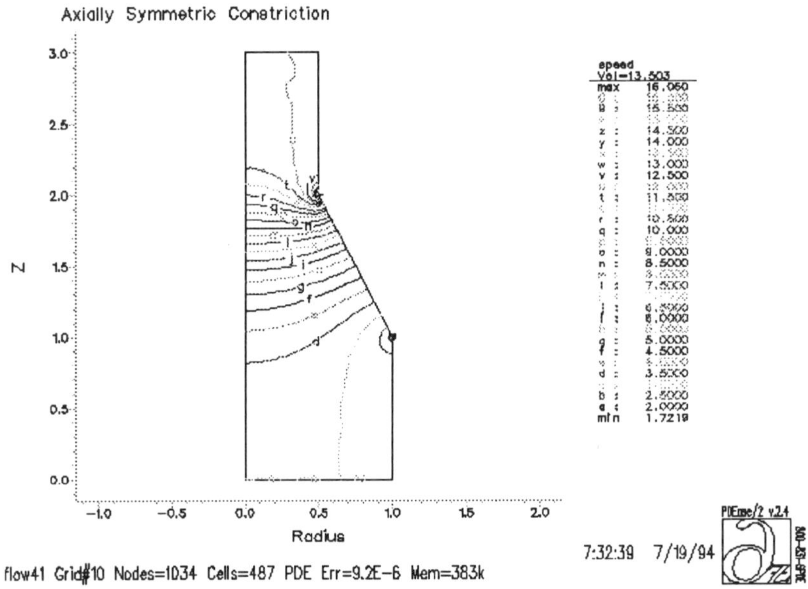

Exercises

❑ Change the boundary condition at the output end of the constricted channel (p.230) such that you specify the appropriate horizontal velocity. Compare the results to those of the original example. Then change the output speed in the boundary condition by 10 % and watch the consequences.

❑ Change the first example (p.230) such that the horizontal input velocity is made to vary across the channel according to the function *u*0= 3*[1-(*y*/*Ly*)**2], still keeping the value of the potential equal to zero at the output end.

❑ Use an input speed of 7 m/s in the first example and notice the minimum value of pressure resulting from the solution. Suggest a physical interpretation of the astonishing outcome.
❑ Change the angle of attack and the thickness of the inclined plate according to your own taste.
❑ Increase the angle of attack of the triangular profile to about twice the previous value.
❑ Expand the last descriptor to fit a symmetrical Venturi tube[7p120].

Viscous Flow of Liquids in (*x*,*y*) Space

In this chapter we deal with more realistic situations, where the fluid locally is at rest with respect to the solid objects in contact with it. Under such conditions curl(**u**) will in general be non-zero.

The classical laws of motion of a volume element of fluid yield the well-known Navier-Stokes equation[7p59]

$$\rho\frac{\partial \mathbf{u}}{\partial t}+\rho(\mathbf{u}\cdot\nabla)\mathbf{u}-\mathbf{F}+\nabla p-\eta\nabla^2\mathbf{u}=0 \qquad \bullet$$

where ρ is the constant mass density of the fluid, **u** the velocity vector, **F** the external force (e.g. gravity) per unit volume, p the pressure, and η the viscosity. In Cartesian coordinates (2D) this becomes

$$\rho\begin{Bmatrix}\dfrac{\partial u}{\partial t}\\ \dfrac{\partial v}{\partial t}\end{Bmatrix}+\rho\begin{Bmatrix}u\dfrac{\partial u}{\partial x}+v\dfrac{\partial u}{\partial y}\\ u\dfrac{\partial v}{\partial x}+v\dfrac{\partial v}{\partial y}\end{Bmatrix}-\begin{Bmatrix}F_x\\ F_y\end{Bmatrix}+\begin{Bmatrix}\dfrac{\partial p}{\partial x}\\ \dfrac{\partial p}{\partial y}\end{Bmatrix}-\eta\begin{Bmatrix}\dfrac{\partial^2 u}{\partial x^2}+\dfrac{\partial^2 u}{\partial y^2}\\ \dfrac{\partial^2 v}{\partial x^2}+\dfrac{\partial^2 v}{\partial y^2}\end{Bmatrix}=0 \qquad \bullet$$

where u and v are the components of **u**.

Thus we have only two equations for the three dependent variables u, v and p. Conservation of mass gives us the "continuity" equation

$$\nabla\cdot\mathbf{u}=\frac{\partial u}{\partial x}+\frac{\partial v}{\partial y}=0 \qquad \bullet$$

but unfortunately this does not contain the missing unknown p. Using this equation of continuity together with the equations of motion we may, however, generate a relation containing a second-order derivative in p. Applying the divergence operator to the Navier-Stoke equation we obtain[8]

$$\rho\frac{\partial}{\partial t}\nabla\cdot\mathbf{u}+\rho\nabla\cdot[(\mathbf{u}\cdot\nabla)\mathbf{u}]-\nabla\cdot\mathbf{F}+\nabla^2 p-\eta\nabla\cdot(\nabla^2\mathbf{u})=0$$

where the first term vanishes because of mass conservation. Further, we eliminate the last term using the identities
$\eta\nabla\cdot(\nabla^2\mathbf{u}) = \eta\nabla^2(\nabla\cdot\mathbf{u}) = \eta\nabla^2(0) = 0$.
The remaining terms of the modified Navier-Stokes equations are

$$\nabla^2 p + \rho\nabla\cdot[(\mathbf{u}\cdot\nabla)\mathbf{u}] - \nabla\cdot\mathbf{F} = 0$$

If the volume force **F** is constant in space, the divergence of it vanishes, leaving us with

$$\nabla^2 p + \rho\nabla\cdot[(\mathbf{u}\cdot\nabla)\mathbf{u}] = 0 \qquad \bullet$$

or in Cartesian coordinates

$$\frac{\partial^2 p}{\partial x^2} + \frac{\partial^2 p}{\partial y^2} + \rho\frac{\partial}{\partial x}\left(u\frac{\partial u}{\partial x} + v\frac{\partial u}{\partial y}\right) + \rho\frac{\partial}{\partial y}\left(u\frac{\partial v}{\partial x} + v\frac{\partial v}{\partial y}\right) = 0$$

Expanding the derivatives we obtain

$$\frac{\partial^2 p}{\partial x^2} + \frac{\partial^2 p}{\partial y^2} + \rho\left[\left(\frac{\partial u}{\partial x}\right)^2 + u\frac{\partial^2 u}{\partial x^2} + \frac{\partial v}{\partial x}\frac{\partial u}{\partial y} + v\frac{\partial^2 u}{\partial x \partial y}\right] +$$
$$\rho\left[\frac{\partial u}{\partial y}\frac{\partial v}{\partial x} + u\frac{\partial^2 v}{\partial x \partial y} + \left(\frac{\partial v}{\partial y}\right)^2 + v\frac{\partial^2 v}{\partial y^2}\right] = 0$$

In this expression the second-order derivatives cancel in pairs, which we can show using the equation of continuity, $\partial u/\partial x + \partial v/\partial y = 0$. Similarly we can prove that the squared terms are equal. The PDE for pressure finally takes the simple form

$$\frac{\partial^2 p}{\partial x^2} + \frac{\partial^2 p}{\partial y^2} + 2\rho\left[\left(\frac{\partial u}{\partial x}\right)^2 + \frac{\partial u}{\partial y}\frac{\partial v}{\partial x}\right] = 0$$

This equation, together with the Navier-Stokes vector equation, makes a total of three PDEs for calculating *u*, *v* and *p*. Although we derived the equation for *p* using mass conservation, it would be wrong to assume that any solution would satisfy $\nabla\cdot\mathbf{u}=0$. In fact, one can show[8] that this is true only in special cases.

It could never be wrong, however, to add a constant multiple of $\nabla \cdot \mathbf{u}$ to the equation for p, since the divergence eventually should vanish. Doing this we help the process of iteration, since we create extra pressure to eliminate any divergence remaining at a given stage. Hence we settle for the following form

$$\frac{\partial^2 p}{\partial x^2}+\frac{\partial^2 p}{\partial y^2}+2\rho\left[\left(\frac{\partial u}{\partial x}\right)^2+\frac{\partial u}{\partial y}\frac{\partial v}{\partial x}\right]+f_\nabla\left(\frac{\partial u}{\partial x}+\frac{\partial v}{\partial y}\right)=0\,,$$ •

where we may choose the factor f_∇ freely according to the problem at hand, such as to boost convergence.

Now that we have a PDE for pressure, we must find out what boundary conditions to use with it. This is easy enough where the pressure takes known values, but what about boundaries that just limit the fluid flow?

The alternative to *value(p)* is a *natural(p)* statement. In the latter case we need an expression for $\partial p/\partial n \equiv \mathbf{n}\cdot\nabla p$, where $\mathbf{n}$ is the outward normal ($|\mathbf{n}|=1$) at the boundary. The Navier-Stokes equation provides the answer rather directly:

$$\nabla p=\mathbf{F}+\eta\nabla^2\mathbf{u}-\rho\frac{\partial \mathbf{u}}{\partial t}-\rho(\mathbf{u}\cdot\nabla)\mathbf{u}$$

$$\mathbf{n}\cdot\nabla p=\mathbf{n}\cdot\mathbf{F}+\eta\mathbf{n}\cdot\nabla^2\mathbf{u}-\rho\mathbf{n}\cdot\frac{\partial \mathbf{u}}{\partial t}-\rho\mathbf{n}\cdot[(\mathbf{u}\cdot\nabla)\mathbf{u}]$$

The first term to the right we may neglect, since we shall not be concerned with any volume forces. The third term involves the time and may hence be neglected in the following applications. We are thus left with the boundary condition

$$natural(p)=\eta\mathbf{n}\cdot\nabla^2\mathbf{u}-\rho\mathbf{n}\cdot[(\mathbf{u}\cdot\nabla)\mathbf{u}]=$$

$$\eta\left[n_x\left(\frac{\partial^2 u}{\partial x^2}+\frac{\partial^2 u}{\partial y^2}\right)+n_y\left(\frac{\partial^2 v}{\partial x^2}+\frac{\partial^2 v}{\partial y^2}\right)\right]-$$ •

$$\rho\left[n_x\left(u\frac{\partial u}{\partial x}+v\frac{\partial u}{\partial y}\right)+n_y\left(u\frac{\partial v}{\partial x}+v\frac{\partial v}{\partial y}\right)\right]$$

In cases of straight-line boundaries **n** remains constant and the last term may be written

$$\rho\mathbf{n}\cdot\left[(\mathbf{u}\cdot\nabla)\,\mathbf{u}\right]=\rho(\mathbf{u}\cdot\nabla)(\mathbf{n}\cdot\mathbf{u})=\rho(\mathbf{u}\cdot\nabla)u_n$$

which thus vanishes on any straight and fixed (u_n=0) segment.

Steady Flow Through a Channel

As a first, simple example we study stationary flow between two parallel plates, driven by a prescribed pressure difference. The details should be evident from the following descriptor.

At this early stage we simplify by assuming a small Reynolds number, i.e. negligible inertial forces in comparison with viscous forces. We implement this assumption simply by putting the density equal to zero, thereby making all three PDEs linear in the dependent variables. As the results will show, we do not need the divergence term in the third equation; thus we also put *f_div* (=f_∇) equal to zero.

In the *variables* segment we list the variables in the order suggested by the derivation of the equations. For brevity we employ the symbol *del2* for the Laplace operator.

The *natural* boundary conditions are particularly simple in this case, since only n_y is different from zero on the horizontal boundaries. At the left and right boundaries we specify *natural*(u)= 0, which means dx(u)= 0 at the vertical faces.

This problem has an elementary solution. We enter the exact horizontal velocity component under the name *u_ex*, the vertical component being zero.

```
select
  errlim= 1e-10            nodelimit= 500
  contours= 10             vectorgrid= 30
  surfacegrid= 30          paint= off
variables
  u    v    p
definitions
  dens= 0                  f_div= 0                  visc= 1e-2
```

```
  Lx= 2e-2          Ly= 2e-2      p0= 1e5          delp= 1e-5
  Re= dens* u* Ly/ visc
  speed2= u**2+v**2                    speed= sqrt( speed2)
  f_angle= sign(v)* arccos( u/speed)/pi*180
  u_ex=  delp/(4*visc*Lx)*(Ly**2- y**2)
  div_u= dx(u)+dy(v)            curl_z= dx(v)-dy(u)       { Vorticity }
equations
  dens*[u*dx(u)+ v*dy(u)]+ dx(p)- visc* del2(u)= 0
  dens*[u*dx(v)+ v*dy(v)]+ dy(p)- visc* del2(v)= 0
  del2(p)+ 2*dens*[ dx(u)**2+ dy(u)*dx(v)]+
                                f_div*[ dx(u)+dy(v)]= 0
boundaries
        region 1
  start(-Lx,-Ly)            value(v)= 0
  value(u)= 0   natural(p)= -visc*del2(v)     line to (Lx,-Ly)
  natural(u)= 0    value(p)= p0    line to (Lx,Ly)
  value(u)= 0   natural(p)= visc*del2(v)     line to (-Lx,Ly)
  natural(u)= 0    value(p)= p0+delp   line to finish
monitors
  elevation( u) from (0,-Ly) to (0,Ly)
plots
  elevation( Re) from (0,-Ly) to (0,Ly)
  contour( u)               contour( v)                contour( p-p0)
  contour( speed)           contour(  u_ex)
  contour( speed- u_ex )  as 'Speed error'
  contour( f_angle)
  vector( u, v)             vector( u/speed, v/speed)
  contour( div_u)           contour( curl_z)
end
```

The first plot on the next page shows the solution for the horizontal component of velocity. The value is zero at the horizontal boundaries and takes a maximum at mid-distance. Several of the other plots demonstrate that the vertical component is very small.

The second plot below indicates an accuracy in u of better than 0.001 %, and this error may be further reduced if we choose a smaller value of *errlim*.

The last plot (not shown) of the descriptor illustrates that the z component of curl(**u**) (often called vorticity) is finite everywhere, except on the symmetry plane, where the function changes sign.

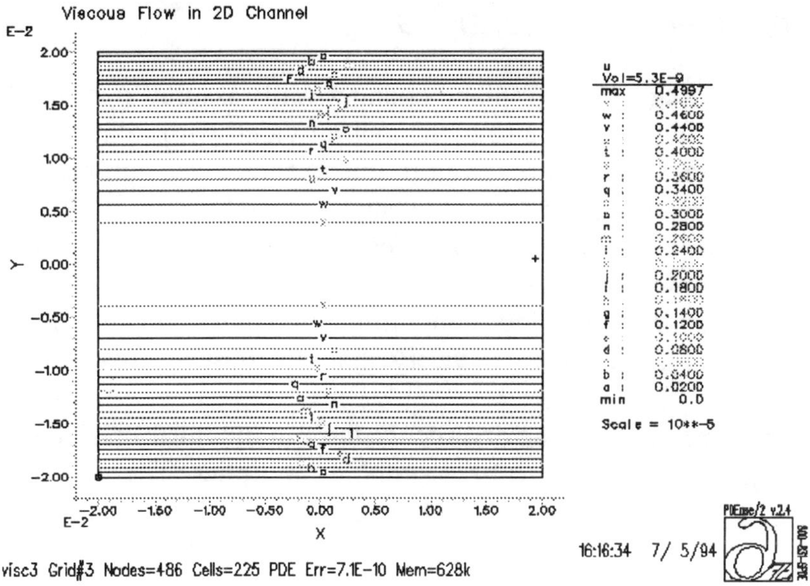

Viscous Flow in 2D Channel
u
Vol=5.3E-9
max 0.4997
w : 0.4600
v : 0.4400
t : 0.4000
r : 0.3600
q : 0.3400
o : 0.3000
n : 0.2800
l : 0.2400
j : 0.2000
i : 0.1800
g : 0.1400
f : 0.1200
d : 0.0800
b : 0.0400
a : 0.0200
min 0.0
Scale = 10**-5
X
Y
16:16:34 7/ 5/94
visc3 Grid#3 Nodes=486 Cells=225 PDE Err=7.1E-10 Mem=628k

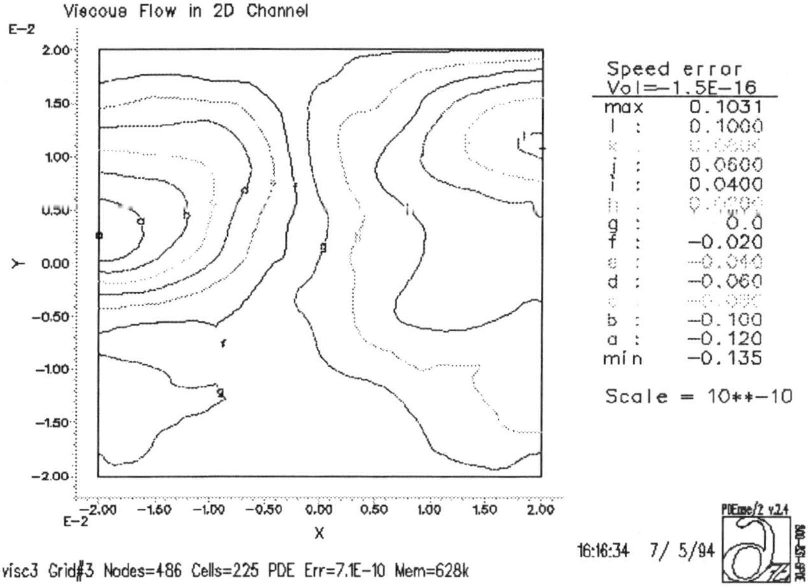

Viscous Flow in 2D Channel
Speed error
Vol=-1.5E-16
max 0.1031
l : 0.1000
j : 0.0600
i : 0.0400
g : 0.0
f : -0.020
d : -0.060
b : -0.100
a : -0.120
min -0.135
Scale = 10**-10
X
Y
16:16:34 7/ 5/94
visc3 Grid#3 Nodes=486 Cells=225 PDE Err=7.1E-10 Mem=628k

Channel With a Lateral Cavity

In the preceding example, plane channel walls assured parallel flow. Let us now study the case of a channel provided with a lateral cavity as shown in the next figure. Even in this example we shall restrict the calculations to vanishing *Re* by putting the density equal to zero.

We keep most of the preceding descriptor and modify it according to the list below. The exact solution *u_ex* is no longer applicable, and we also eliminate the plots containing this function. On the other hand we add three plots to verify that no mass is lost along the channel.

Here we need to use the divergence term in the third PDE, and trial runs suggest *f_div*=1e8 to be a suitable multiplier. The calculations may last up to an hour, depending on the speed of your PC system.

```
  ...
  errlim= 1e-7              nodelimit= 500
  ...
  dens= 0                   f_div= 1e8
  ...
                       region 1
  start(-1.5*Lx,0)       value(v)= 0
  value(u)= 0   natural(p)= -visc*del2(v) line to  (-0.5*Lx,0)
  natural(p)= -visc*del2(u)  line to (-0.5*Lx,-Ly)
  natural(p)= -visc*del2(v)  line to (0.5*Lx,-Ly)
  natural(p)= visc*del2(u)   line to (0.5*Lx,0)
  natural(p)= -visc*del2(v)   line to (1.5*Lx,0)
  natural(u)= 0   value(p)= p0   line to (1.5*Lx,Ly)
  value(u)= 0  natural(p)= visc*del2(v)     line to (-1.5*Lx,Ly)
  natural(u)= 0   value(p)= p0+delp   line to finish
  ...
  vector( u/speed, v/speed) zoom(-0.5*Lx,-Ly,  Lx,Ly)
  elevation( u) from (-1.4*Lx,0) to (-1.4*Lx,Ly)
  elevation( u) from (0,-Ly) to (0,Ly)
  elevation( u) from (1.4*Lx,0) to (1.4*Lx,Ly)
end
```

The figure below shows that the flow is mainly confined to the through part of the channel, the velocity being much lower in the adjacent cavity. We also notice that the minimum value, indicated in the plot by a small circle, is negative. The zero contour lies roughly in the middle of the cavity.

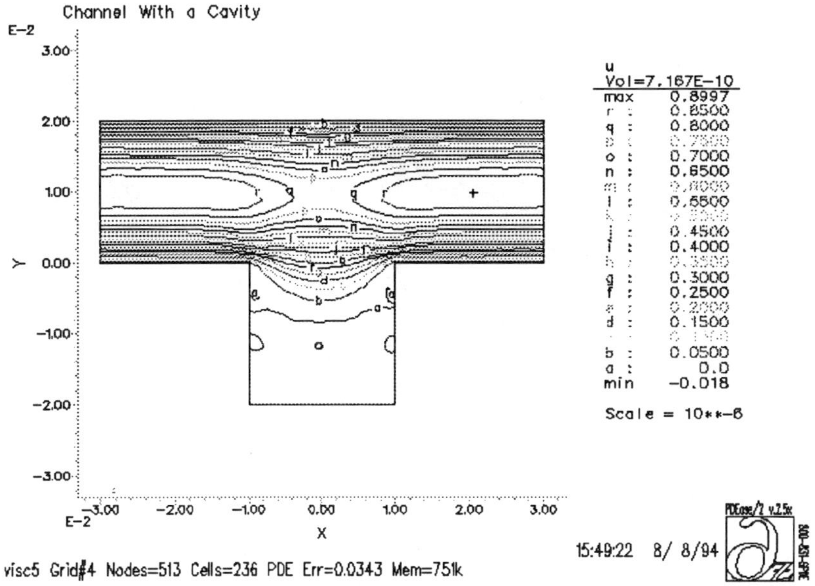

Several of the plots demonstrate that circulation occurs in the cavity, which causes the negative values of u. The next *vector* plot clearly shows the center of circulation. One may easily get the impression that the main stream "jumps" the cavity by inertia, but the present calculations do not take inertial forces into account, the density still being equal to zero.

Our assumption of a vanishing Reynolds number also leads to symmetry in absolute values with respect to the line x=0.

The three *elevation* plots across the channel confirm that no mass is lost along the stream.

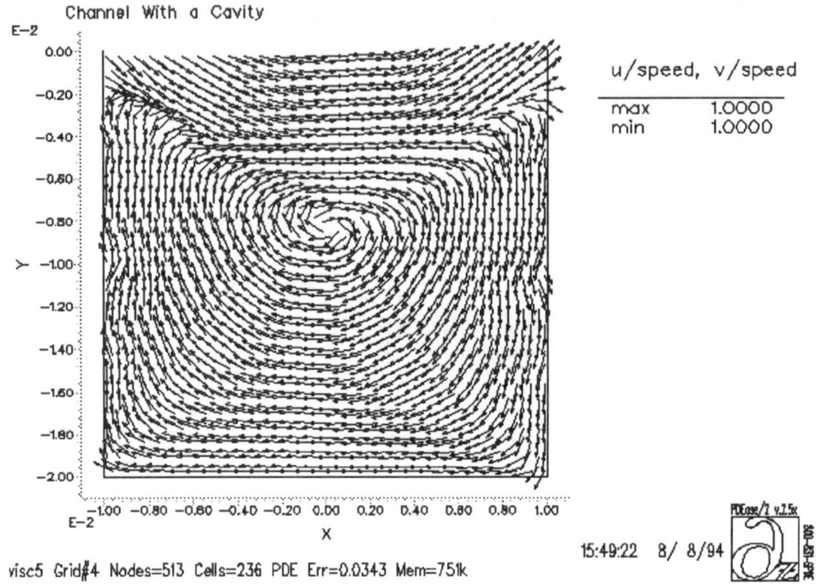

Uniform Input Velocity at Various Reynolds Numbers

The first example involved uniform pressure at the left and right boundaries of a straight channel and resulted in identical velocity profiles over all cross-sections. Let us now consider the case of a uniform horizontal velocity, *u*, imposed at the left boundary.

We use the *stages* device to make a preliminary calculation at vanishing Reynolds number (*dens*=0). In later stages we increase the density in steps up to a realistic value. This permits us to study the flow pattern at several values of *Re*.

The Navier-Stokes equations are particularly difficult to solve in this case. To obtain the accuracy required we need *converge* and *subconverge* to control the iterative calculations.

Although there is no exact solution to this problem, we may use a reference function *u_ref* for the velocity profile, analogous to that in the first example. The leading coefficient makes the flux equal to that injected at the front. In an *elevation* plot we compare the final velocity distribution to *u_ref*.

On the entrance boundary we simply impose the horizontal velocity $u0$ and ambient pressure. The fact that we have u=0 on the solid boundaries creates a discontinuity at the left corners. We alleviate this condition by densifying the grid, simply by putting extra points near the corners. The resulting denser grid permits u to increase more steeply from zero to the constant value $u0$.

Furthermore, we assume that the velocity becomes horizontal and varies little with x close to the exit boundary, which would be true for a long channel.

```
  stages= 6
  errlim= 1e-3                    nodelimit= 300
  converge= 1e-6                  subconverge= 1e-6
  ...
  u0= 1e-2        visc= 1e-2          f_div= 1e8
  dens= 0.2*(stage-1)* 1e3
  Lx= 5e-2        Ly= 2e-2            d= 5e-2*Ly          p0= 1e5
  u_ref=  3*u0/2*[1- (y/Ly)**2]       { Equal mass flux }
  ...
initial values
  u= u0           p= p0
  ...
       region 1
  start(-Lx,-Ly)      value(v)= 0
  value(u)= 0   natural(p)= -visc*del2(v)    line to (Lx,-Ly)
  natural(u)= 0   natural(p)= visc*del2(u)    line to (Lx,Ly)
  value(u)= 0   natural(p)= visc*del2(v)    line to (-Lx,Ly)
  value(u)= u0   value(p)= p0   line to (-Lx,Ly-d)
                                      to (-Lx,-Ly+d) to finish
  ...
  elevation( u) from (-0.9*Lx,-Ly) to (-0.9*Lx,Ly)
  elevation( u) from (0,-Ly) to (0,Ly)
  elevation( u, u_ref) from (0.9*Lx,-Ly) to (0.9*Lx,Ly)
end
```

This problem may take several hours to solve, and it had better be planned as an overnight run.

The first *contour* plot of *u* below (for *Re*=0) reveals a more complex pattern than in the first example. The speed profile is uniform on the left boundary and gradually changes to become parabolic with a central maximum near the exit. The *vector* plots (not shown) suggest how the flow redistributes to reproduce this final pattern, known from the first example.

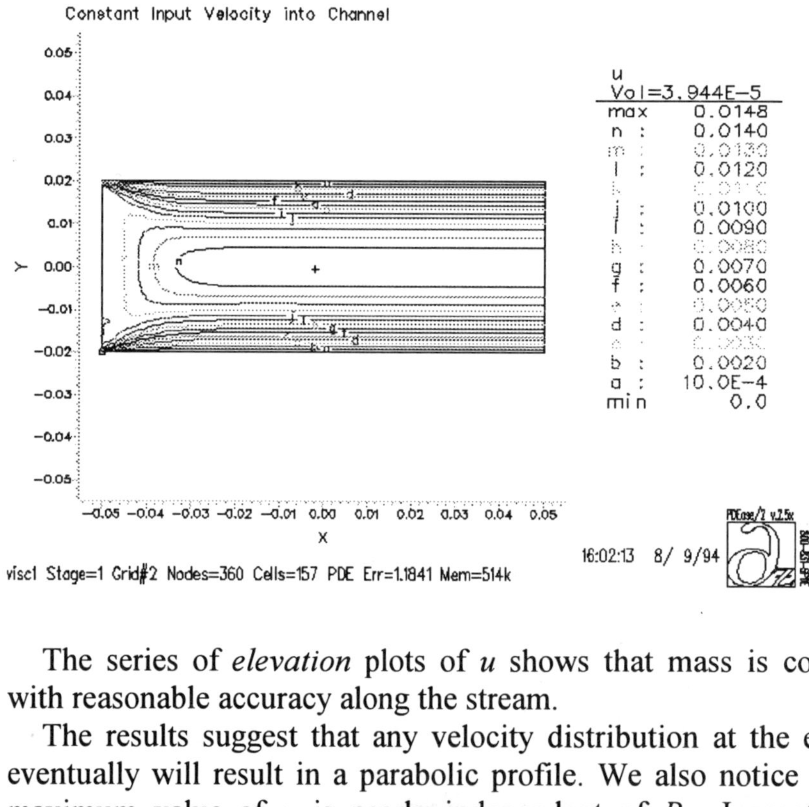

The series of *elevation* plots of *u* shows that mass is conserved with reasonable accuracy along the stream.

The results suggest that any velocity distribution at the entrance eventually will result in a parabolic profile. We also notice that the maximum value of *u* is nearly independent of *Re*. Inspecting the series of *contour* plots for various values of *Re*, we are now in a position to find out how far the deviation from parabolic behavior persists. To make our task simpler we concentrate on the innermost contour (*n*), where *u* is 140 against a maximum of about 150 (in appropriate units). We first notice that for *Re*=0 this contour begins at a distance of about 0.8**Ly* downstream and that it shifts by almost constant intervals to the right as *Re* increases. If we summarize the results we find the following approximate relation

$$\frac{X}{D} = 0.4 + \frac{Re}{30},$$

where X is the length from the entrance to the contour of interest and D is the total width of the channel. Apart from the constant corresponding to zero Reynolds number, this is similar to a relation found for flow through a tube[7p15].

The next figure of the direction of flow illustrates how the initially uniform velocity transforms into another parallel stream, but with parabolic speed profile.

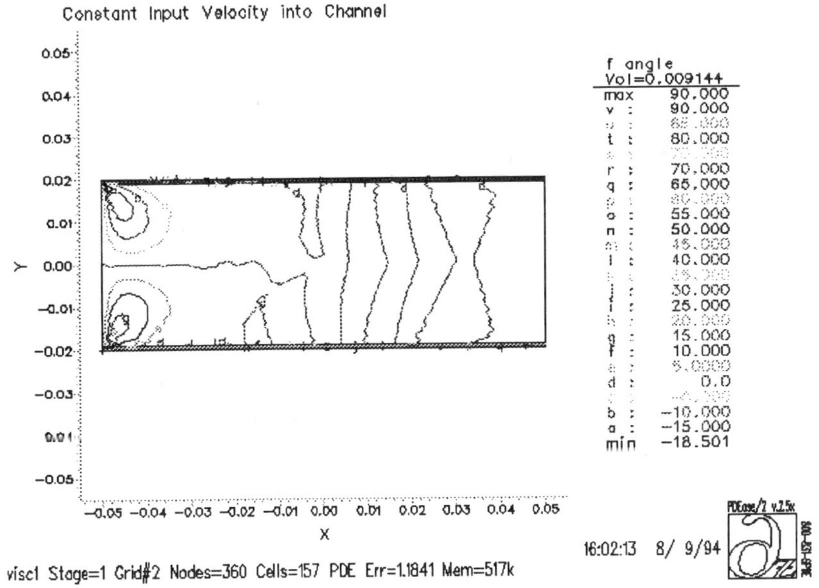

The figure below is a *contour* plot of $\mathrm{curl}_z(\mathbf{u})$, which exhibits four distinct regions of different signs. Those extending to the horizontal boundaries are of the same character as before, and the others are due to the particular velocity pattern at the input end.

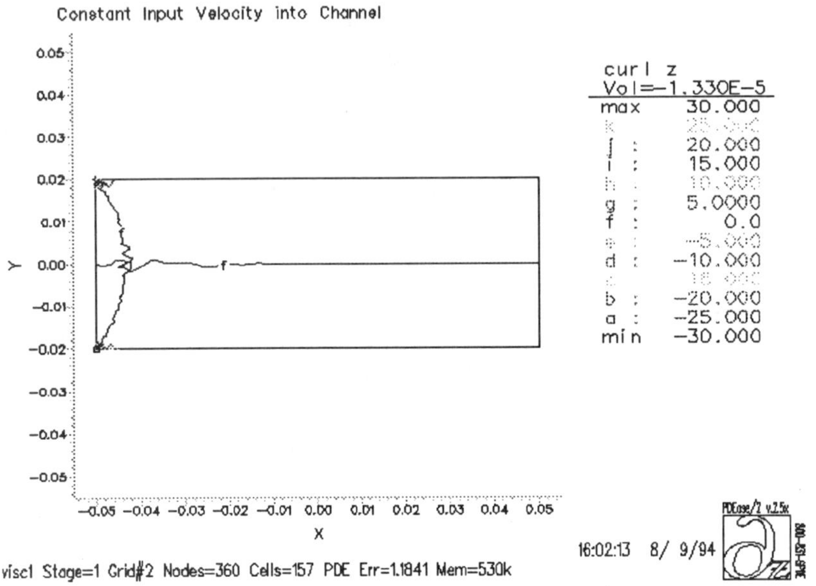

Tangential Input Velocity

We now specify a constant vertical (y) velocity at the left face, while keeping the other three sides closed by fixed walls. In practice, we could impose this lateral velocity by an endless tape, driven over rollers at constant velocity past the left face of the channel. This only requires a few changes of the preceding descriptor, as suggested by the list below. We eliminate the expression for *u_ref* and the *elevation* plots.

```
  ...
  stages= 2                                nodelimit= 500
  ...
  v0= 0.1          f_div= 1e8              dens= (stage-1)*1e3
  Lx= 2e-2         Ly= 2e-2                p0= 1e5
  visc= 1e-2                               Re= dens* v0* Ly/ visc
initial values
  p= p0
  ...
```

```
                 region 1
  start(-Lx,-Ly)    value(v)= 0
  value(u)= 0       natural(p)= -visc*del2(v)    line to (Lx,-Ly)
  value(u)= 0       natural(p)= visc*del2(u)     line to (Lx,Ly)
  value(u)= 0       natural(p)= visc*del2(v)     line to (-Lx,Ly)
  value(u)= 0       value(v)= v0       value(p)= p0   line to finish
  ...
  vector( u/speed, v/speed) zoom(0,0,  Lx,Ly)
end
```

As before the calculation proceeds in two stages, the first one corresponding to a vanishing value of *Re*. The second stage is for *Re* ≅ 200, and results of the latter run are shown below. The *contour* plot of the angle consists of curves which branch out from a common point, which means that the liquid circulates around a singularity. This phenomenon appears near the center and in a rudimentary way in the upper-right corner. The fuzzy curves indicate horizontal flow. They suggest that the flow pattern at three corners have particular properties.

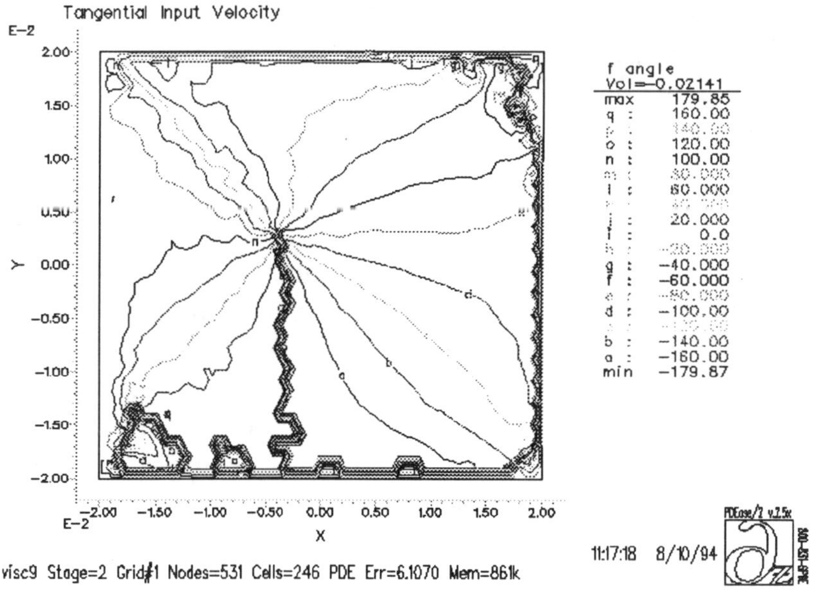

The *vector* plot below indicates in less detail, but more directly, in what way the corner regions are peculiar. The zoomed *vector* plot (not shown) confirms that there is reverse circulation in the upper right corner.

The plot of $\text{curl}_z(\mathbf{u})$ suggests that there are at least three distinct regions, separated by curves corresponding to the value zero. A zoomed plot, excluding a narrow strip at the left face, would probably shed more light on this feature.

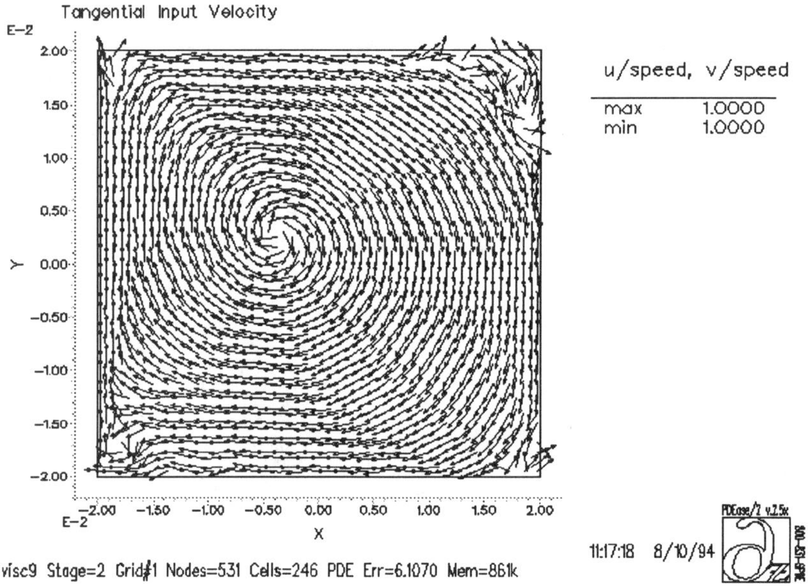

The contour plot of $\text{div}(\mathbf{u})$ yields the irregular contours of value zero which we habitually associate with a vanishing function.

The two stages effectively yield results for vanishingly small *Re* as well as for $Re \cong 200$. Comparing the salient features we find the flow pattern at low *Re* to be symmetrical with respect to the line y=0. At high *Re*, this symmetry becomes broken, and the main center of circulation shifts towards larger x and larger y.

Exercises

❑ Show that the function *u_ex* in the first example satisfies the PDEs and the boundary conditions.

❑ Verify the calculation of *curl_z* in the first example using the contour plot of *u* and the integral definition of curl(**u**).

❑ Try a finite density value, *dens*= 1e3, in the first example. Then increase the driving pressure in suitable steps. For how high Reynolds numbers can you obtain solutions?

❑ Calculate the total viscous force on the liquid in the first example, using an elevation plot close to one of the horizontal boundaries. Compare the integral to the driving force exerted by the pressure.

❑ Modify the example of the long (5e-2) channel with uniform input velocity (p.251). Impose a parabolic pressure distribution at the input end, such that the pressure vanishes at the center and takes its largest values at the channel walls. Does the flow profile eventually become parabolic? Use *stages* with different densities to investigate how the entry length varies with *Re*. Is the maximum speed independent of *Re*?

Viscous Flow in (*x*,*y*) Space Past an Obstacle

Throughout the preceding chapter the solution domain was a rectangular box. Now we shall introduce solid objects into the stream, which will locally accelerate the fluid. Obstacles create interesting effects of several kinds. In particular, questions of forces on the obstacle arise, both in the main direction of the stream (drag) and transverse to it (lift).

Flow Around a Square Bar

We shall now study the flow in a channel partially blocked by a square bar across the stream. Hoping to reduce calculation times, which become considerable in this chapter, we exploit the symmetry of the problem, limiting the solution to one half of the real domain. In the following descriptor we thus use the auxiliary boundary conditions *natural*(u)=0 and *natural*(p)=0 on the symmetry plane, i.e. specify that the normal derivatives of u and p are to vanish, which is a way of expressing the symmetry required.

On the solid boundaries, the *natural*(p) statements become particularly simple. As discussed earlier (p.245) the second term vanishes whenever the boundary is a straight line segment and is immobile (u_n=0).

Here we no longer bother to specify the atmospheric pressure $p0$, since any constant pressure value may always be added to the solution afterwards.

As in the preceding chapter, we study the flow at several Reynolds numbers. We provide a first approximation by specifying zero density (Re=0), then increasing the density in steps.

```
select
  stages= 3
```

```
  errlim= 1e-2              nodelimit= 500
  converge= 1e-6            subconverge= 1e-6
  ...
  delp= 1.0                 f_div= 1e8          visc= 1e-2
  dens= 0.5*(stage-1)*1e3
  Lx= 2e-2                  Ly= 1e-2            Re= dens* u* Ly/ visc
  ax= 0.2e-2                ay= 0.2e-2
  ...
            region 1
  start(-Lx,0)      value(v)= 0
  natural(u)= 0  natural(p)= 0    line to (-ax,0)
  value(u)= 0  natural(p)= visc*del2(u)  line to (-ax,ay)
  natural(p)= -visc*del2(v)    line to (ax,ay)
  natural(p)= -visc*del2(u)     line to (ax,0)
  natural(u)= 0  natural(p)= 0   line to (Lx,0)
  natural(u)= 0  value(p)= 0 line to (Lx,Ly)
  value(u)= 0   natural(p)= visc*del2(v) line to (-Lx,Ly)
  natural(u)= 0  value(p)= delp line to   finish
monitors
  contour(u)
plots
  elevation( Re) from (-Lx/2,0) to (-Lx/2,Ly)
  contour( u)               contour( v)                contour( p)
  elevation( p) from (-Lx,Ly/2) to (Lx,Ly/2)
  contour( f_angle)       vector( u, v)
  vector( u, v) zoom(-2*ax,0,  4*ax,4*ay)
  vector( u/speed, v/speed) zoom(-2*ax,0,  ax,ay)
  vector( u/speed, v/speed) zoom(ax,0,  ax,ay)
  vector( u/speed, v/speed) zoom(ax,0,  ax/5,ay/5)
  contour( div_u)          contour( curl_z)
  elevation( u) from (-0.9*Lx,0) to (-0.9*Lx,Ly)
  elevation( u) from (0,0) to (0,Ly)
  elevation( u) from (0.9*Lx,0) to (0.9*Lx,Ly)
end
```

The first *contour* plot below displays the velocity component u for Re=0. The maximum value occurs in the constriction, just above the obstacle. At vanishing Re the flow pattern evidently is symmetric with respect to x= 0.

The *vector* plot below, zoomed on a minute region of the wake, shows signs of incipient circulation.

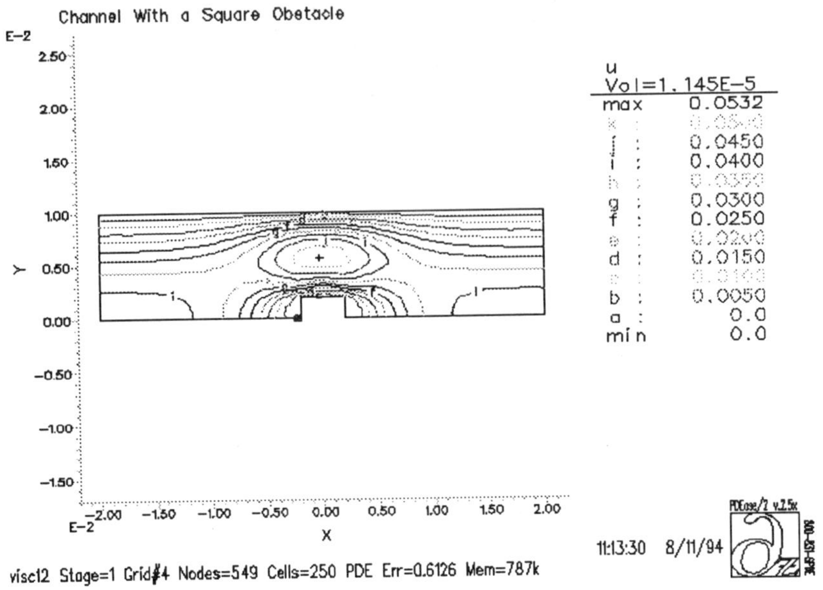
Channel With a Square Obstacle
E-2
2.50
2.00
1.50
1.00
Y 0.50
0.00
-0.50
-1.00
-1.50
-2.00 -1.50 -1.00 -0.50 0.00 0.50 1.00 1.50 2.00
E-2
X
u
Vol=1.145E-5
max 0.0532
j : 0.0450
i : 0.0400
g : 0.0300
f : 0.0250
d : 0.0150
b : 0.0050
a : 0.0
min 0.0
11:13:30 8/11/94
visc12 Stage=1 Grid#4 Nodes=549 Cells=250 PDE Err=0.6126 Mem=787k

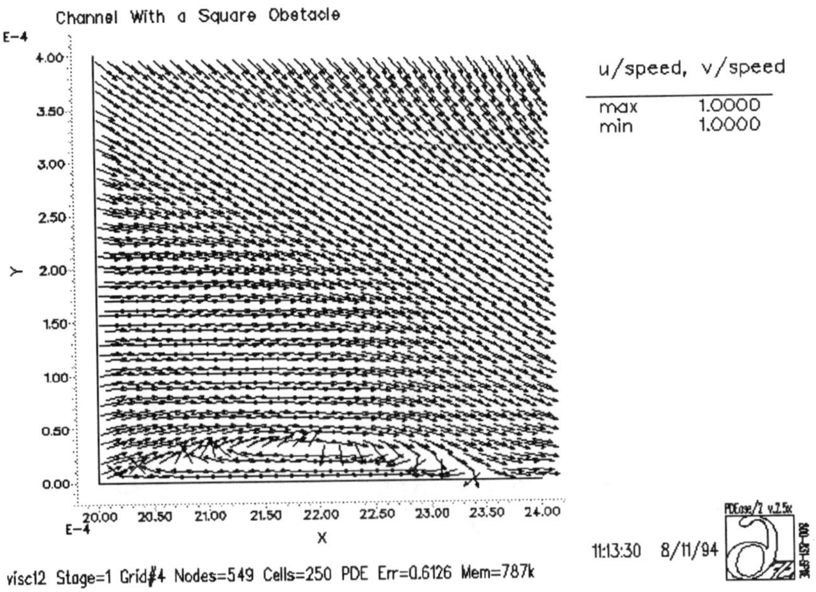
Channel With a Square Obstacle
E-4
4.00
3.50
3.00
2.50
Y 2.00
1.50
1.00
0.50
0.00
20.00 20.50 21.00 21.50 22.00 22.50 23.00 23.50 24.00
E-4
X
u/speed, v/speed
max 1.0000
min 1.0000
11:13:30 8/11/94
visc12 Stage=1 Grid#4 Nodes=549 Cells=250 PDE Err=0.6126 Mem=787k

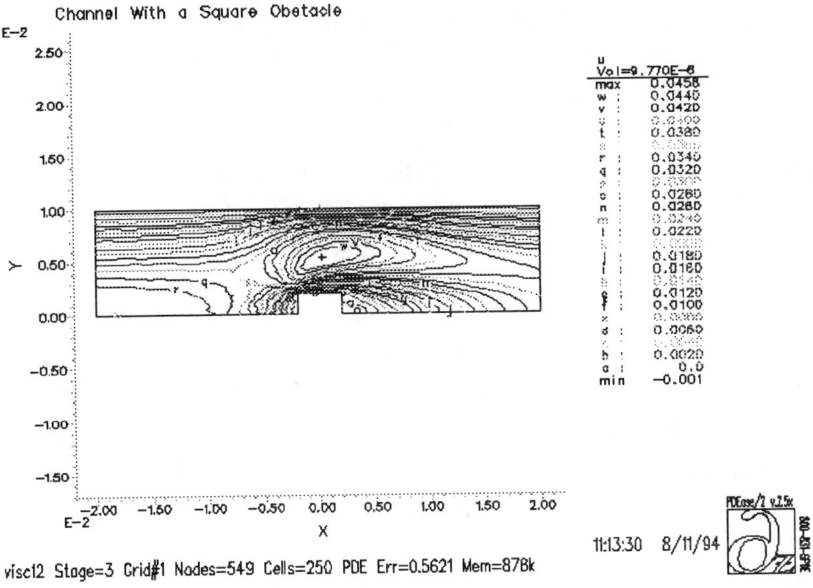
Channel With a Square Obstacle
E-2
Y
X
u
Vol=9.770E-6
max 0.0458
w : 0.0440
v : 0.0420
t : 0.0380
r : 0.0340
q : 0.0320
o : 0.0280
n : 0.0260
l : 0.0220
j : 0.0180
i : 0.0160
g : 0.0120
f : 0.0100
d : 0.0060
b : 0.0020
a : 0.0
min -0.001
11:13:30 8/11/94
PDEase/2 v.2.5x
visc12 Stage=3 Grid#1 Nodes=549 Cells=250 PDE Err=0.5621 Mem=878k

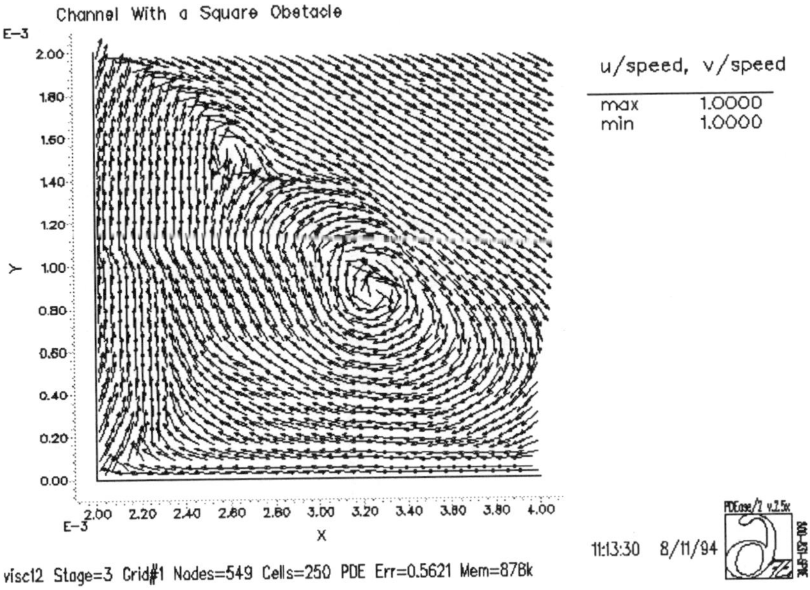
Channel With a Square Obstacle
E-3
Y
X
u/speed, v/speed
max 1.0000
min 1.0000
11:13:30 8/11/94
PDEase/2 v.2.5x
visc12 Stage=3 Grid#1 Nodes=549 Cells=250 PDE Err=0.5621 Mem=878k

The last two figures, corresponding to Re≅24, exhibit a pronounced asymmetry in the distribution of u and also circulation over an appreciable region of the wake.

In the above series of calculations it is interesting to note how the maximum velocity varies with density at constant pressure difference.

We could have stepped the input pressure instead of the density to obtain results for a range of Reynolds numbers. Increasing the driving pressure, however, changes the dependent variables more profoundly than the corresponding change in density, thus making the calculations longer. Doubling the density at small *Re* hardly changes u and v at all, whereas these values would be doubled if we changed the pressure. Hydrodynamic theory[7p90] shows that it is sufficient to know solutions at various values of *Re* for a given domain geometry in order to predict the flow behavior at other domain sizes, different velocities and driving pressures. Hence, we may solve for a fixed size and fixed pressure, if we cover the required range of *Re* by suitable values of density. Starting at zero density PDEase only needs to solve a system of linear PDEs to generate a first approximation, for later use with finite density.

Drag on a Square Bar

The liquid flowing through the channel in the preceding example acts on the obstacle in its way by a force in the main direction of the stream. There are two distinct contributions to this force: the pressure on the left and right faces of the bar and the viscous drag on the bottom and top faces.

In the descriptor defining half of the domain we might calculate half of the drag on the bar by three *elevation* plots, one for each of the faces. The integrated pressure difference for the left and right faces contributes to the horizontal force, and on the top face we have the viscous drag, which we obtain by integrating $\eta \partial u / \partial y$ along a line on the interface between solid and liquid.

We need not even take the force integrals over lines immediately around the obstacle. As already discussed (p.236) we may equally well integrate over an extended box, as long as the channel itself, to obtain the drag on the bar. Since we know the forces due to the pressure difference over this box, only one line integral (for the viscous force) would be needed; the contribution from the integral along the symmetry line vanishes.

There is a feature of the present version of PDEase which should be noted: Interpolated values of a dependent variable and its derivatives do not always exist on the boundary itself, which is the reason why we occasionally must integrate along a line displaced a short distance into the liquid. We choose a displacement (*dl*) so small that it cannot be expected to change the integral value appreciably. The list below indicates what to add to the previous descriptor to calculate the drag.

Our experience from the preceding example shows that it would be wise to leave more space for the downstream flow pattern; hence we extend the right part of the domain by 50 %.

Most of the previous plots are no longer needed and may be removed to save disk space and to avoid confusion.

```
   ...
   dl= 1e-6
   ...
   value(u)= 0   natural(p)= -visc*del2(u)   line to (ax,0)
   natural(u)= 0  natural(p)= -visc*del2(v) line to (1.5*Lx,0)
   natural(u)= 0  value(p)= 0 line to (1.5*Lx,Ly)
   ...
   contour(u)
   elevation[ visc* dy(u)] from (-Lx,ay+dl) to (1.5*Lx,ay+dl)
end
```

The *elevation* plot shown below gives us the viscous drag, expressed by the "*Area*" variable. The force due to pressure on the box is 0.2e-2, which is of the same order of magnitude. The increase of the total drag up to the maximum density is about 15 %. We also note that the maximum velocity, obtained by the *contour* plot decreases over the same range by roughly 12 %. These two facts are connected, since any drag on an obstacle also means that a force of equal magnitude acts on the liquid, impeding its flow.

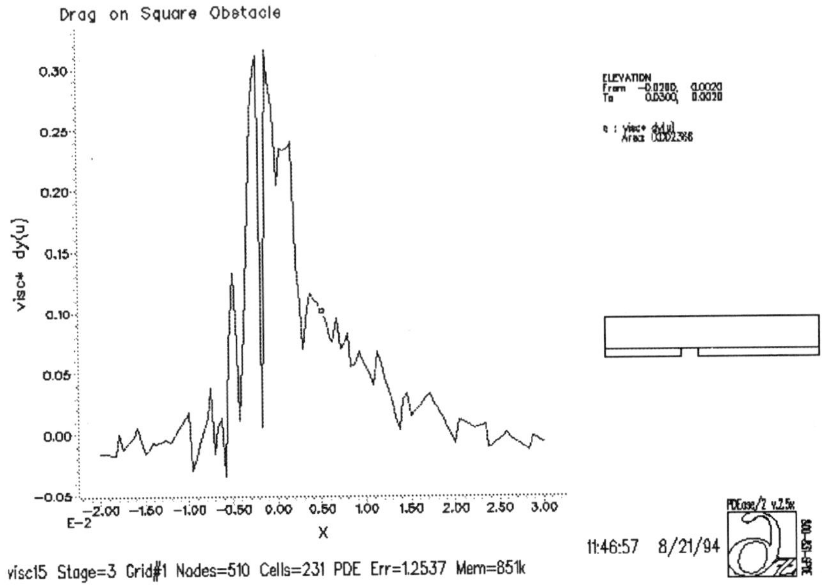

Drag and Lift on an Inclined Slab

In the previous two examples the obstacle was symmetrical with respect to the middle plane of the channel, and we only needed to consider the force along the main stream. Now we introduce a sloping obstacle, on which the stream may exert a vertical force. We could compute the drag as well as the lift by a set of *elevation* plots, taken just outside the surface of the slab. In view of the more complicated geometry, however, we would need tangential as well as normal unit vectors, and the expressions to integrate would be rather long. It is more convenient to calculate the horizontal and vertical forces on a box enclosing our obstacle, as we also did in a previous example.

The coordinates $x1$ etc. refer to the corners of the plate, and $n1x$ etc. are the components of the unit vectors normal to the sides, taken in the order they appear in the descriptor, i.e. clockwise.

The first *stage* of the solution pertains to zero density (*Re*=0), and the second one corresponds to Re≅26.

By the eight last *elevation* plots we thus calculate the various terms of drag and lift, the signs being chosen such that the "*Area*" results just may be added to yield the force components. All forces are for a unit length of the slab in the *z* direction.

```
  ...
  stages= 2
  ...
  delp= 0.2                  f_div= 1e8
  dens= (stage-1)*1e3                    visc= 1e-2
  Lx= 2e-2                   Ly= 2e-2
  L= 0.5*Ly          d= 0.2* L          Re= dens* u* Ly/ visc
  alpha= 45* pi/180      si= sin( alpha)     co= cos( alpha)
  x1= -d/2*si- L/2*co              y1= -d/2*co+ L/2*si
  x2= d/2*si- L/2*co               y2= d/2*co+ L/2*si
  x3= -x1            y3= -y1              x4= -x2              y4= -y2
    n1x= cos(-alpha)                     n1y= sin(-alpha)                   { #1-#2 }
    n2x= cos(-alpha-pi/2)                n2y= sin(-alpha-pi/2)
    n3x= cos(-alpha-pi)                  n3y= sin(-alpha-pi)
    n4x= cos(-alpha-3*pi/2)              n4y= sin(-alpha-3*pi/2)
  ...
             region 1
  start(-Lx,-Ly)      value(v)= 0
  value(u)= 0   natural(p)= -visc*del2(v)    line to (1.5*Lx,-Ly)
  natural(u)= 0    value(p)= 0    line to (1.5*Lx,Ly)
  value(u)= 0   natural(p)= visc*del2(v)    line to (-Lx,Ly)
  natural(u)= 0    value(p)= delp   line to finish
  start(x1,y1)                                                            { Cut-out }
  value(u)= 0    value(v)= 0
  natural(p)= visc*[ n1x*del2(u)+ n1y*del2(v)]  line to (x2,y2)
  natural(p)= visc*[ n2x*del2(u)+ n2y*del2(v)]  line to (x3,y3)
  natural(p)= visc*[ n3x*del2(u)+ n3y*del2(v)]  line to (x4,y4)
  natural(p)= visc*[ n4x*del2(u)+ n4y*del2(v)]  line to finish
monitors
  contour(u)
plots
  elevation( Re) from (-0.5*Lx,-Ly) to (-0.5*Lx,Ly)
  contour( u)
    elevation( u) from (-0.9*Lx,-Ly) to (-0.9*Lx,Ly)                    { div_u ... }
    elevation( u) from (0,-Ly) to (0,Ly)
    elevation( u) from (0.9*Lx,-Ly) to (0.9*Lx,Ly)
```

```
  elevation( p) from (x1,y4) to (x1,y2)                              { drag, p }
  elevation( -p) from (x3,y4) to (x3,y2)
  elevation[ visc*dy(u)] from (x1,y2) to (x3,y2)                     { drag, visc }
  elevation[ -visc*dy(u)] from (x1,y4) to (x3,y4)
    elevation( p) from (x1,y4) to (x3,y4)                            { lift, p }
    elevation( -p) from (x1,y2) to (x3,y2)
    elevation( -visc*dx(v)) from (x1,y4) to (x1,y2)                  { lift, visc }
    elevation( visc*dx(v)) from (x3,y4) to (x3,y2)
end
```

The following figure presents a *contour* plot of the *x*-component of velocity for *Re*=0. We notice that the maximum value of *u* occurs about half-way between an edge of the slab and the channel wall and that the two extreme points are centered on the middle of the slab. The plot also confirms that *u* vanishes on the solid surfaces, as required.

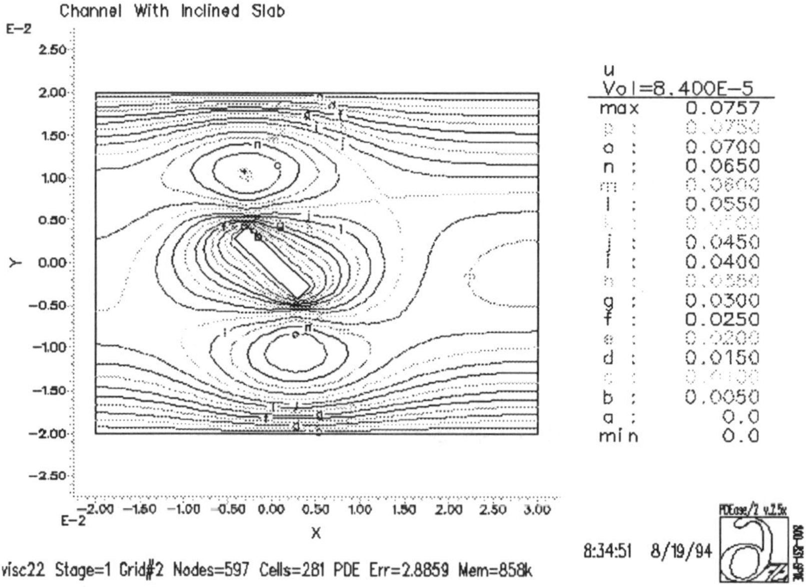

The series of three *elevation* plots demonstrates that the flux through cross-sections of the channel is constant to high accuracy, i.e. that mass is conserved.

The force integrals demonstrate that the contributions from pressure and from viscous drag are of comparable magnitude under the present conditions.

Comparing results for different densities we find that the maximum value of u is about 15 % lower at Re=26. The drag, on the other hand, increases by about 12 % as we go to the higher Reynolds number.

The calculations show that the lift is smaller than the drag in this geometry, although the lift increases more as Re becomes higher. The slab is thus far from the ideal airfoil.

Flow Past a Cylinder

The case of a circular cylinder across a stream is rather similar to the example with the square bar (p.256). The difference resides essentially in the *natural*(p) boundary condition on the curved surface, where we must use both components (n_x,n_y) of the normal vector, as shown below. Here we must also include the second term, the one proportional to density (p.244).

This descriptor and others involving curved contours require PDEase version 2.5 or later, and the run takes hours on a DX66 processor. For a slower system it should be arranged to run overnight. It is possible to shorten the calculation time in this and similar cases to follow, by choosing *errlim* to be, say, 50 % higher than the *PDE Err* found on the plots in this book. This will of course impair the accuracy and the definition, but the solution proceeds much faster in the initial phase than close to the end.

```
stages= 2
...
delp= 2.0                        f_div= 1e8
dens= 1e3*(stage-1)              visc= 1e-2
Lx= 2e-2     Ly= 1e-2            r0= 0.2e-2            Re= dens* u* Ly/ visc
...
                  region 1
start(-Lx,0)        value(v)= 0
natural(u)= 0  natural(p)= 0    line to (-r0,0)
```

```
  value(u)= 0  natural(p)= visc*[ -(x/r0)*del2(u) -(y/r0)*del2(v)]-
        dens*[ -(x/r0)*(u*dx(u)+v*dy(u)) -(y/r0)*(u*dx(v)+v*dy(v))]
                          arc to (0,r0) to (r0,0)
  natural(u)= 0   natural(p)= 0    line to (1.5*Lx,0)
  natural(u)= 0  value(p)= 0     line to (1.5*Lx,Ly)
  value(u)= 0  natural(p)= visc*del2(v)- dens*[ u*dx(v)+ v*dy(v)]
                          line to (-Lx,Ly)
  natural(u)= 0  value(p)= delp    line to   finish
  ...
  vector( u/speed, v/speed) zoom(-2*r0,0,  2*r0,2*r0)
  vector( u/speed, v/speed) zoom(0,0,  3*r0,2*r0)
end
```

The following plots are results from the second *stage*, corresponding to $Re \cong 55$. The circulating region in the wake is a well-known phenomenon, which also has been observed experimentally.

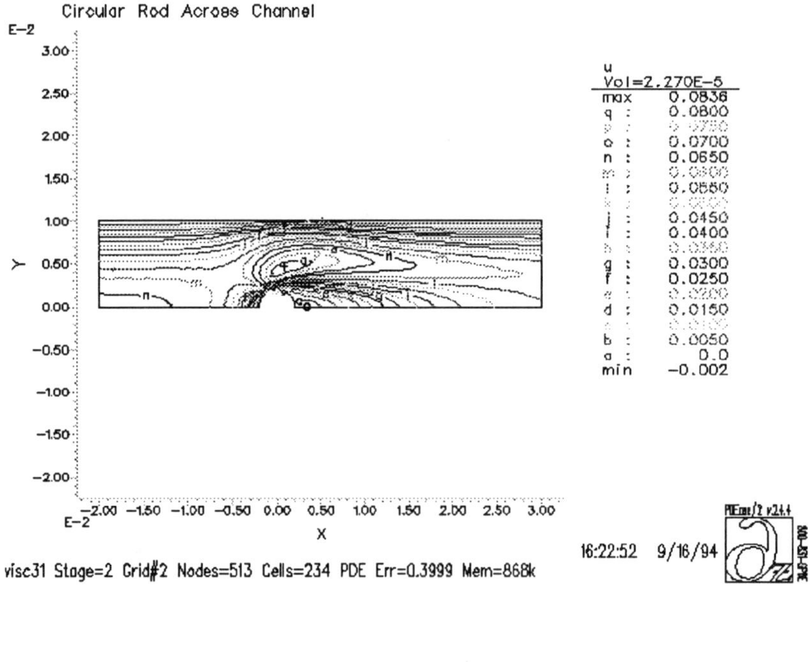

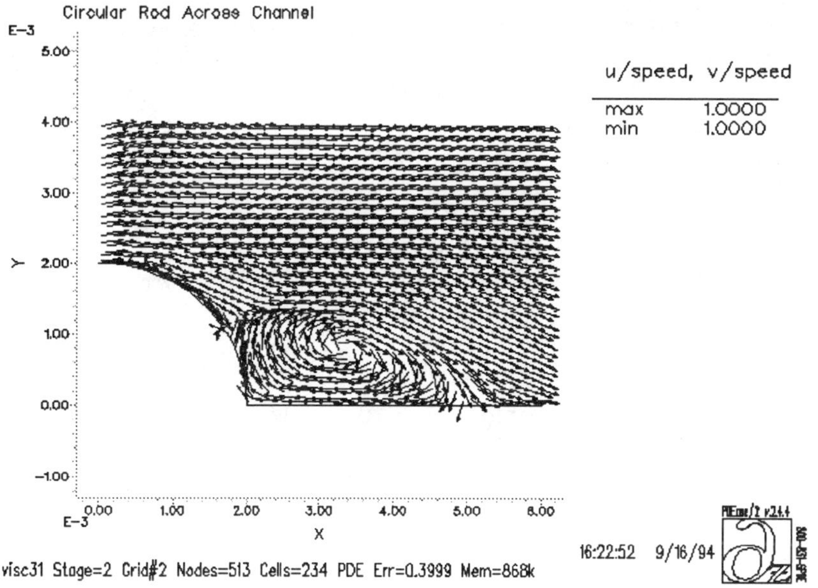

Rotating Shaft Across a Channel

We finish this chapter by a variation of the preceding problem. The rod is now rotating around its axis of symmetry while moving to the left along the channel. Thus the channel walls are at rest, and we assume the liquid at the left boundary of the channel to be at rest as well. Seen from the origin of the *xy* system, which we fix at the center of the shaft, the left boundary and the walls move to the right at a constant speed *u*0. The tangential speed due to the rotation we denote as a multiple of *u*0. Since the symmetry with respect to the line *y*=0 now is broken, we must solve over the entire problem domain. The amendments to the preceding descriptor are as follows.

```
...
f_div= 1e8        u0= 2e-2            ur0= 2*u0
...
                  region 1
start(-Lx,-Ly)      value(v)= 0
value(u)= u0   natural(p)= -visc*del2(v)- dens*[  -(u*dx(v)+ v*dy(v))]
```

```
                    line to (1.5*Lx,-Ly)
  natural(u)= 0  natural(p)= 0    line to (1.5*Lx,Ly)
  value(u)= u0   natural(p)= visc*del2(v)- dens*[ u*dx(v)+ v*dy(v)]
                    line to (-Lx,Ly)
  value(u)= u0   value(p)= 0   line to   finish
  start(r0,0)                                                  { Cut-out }
  value(u)= -(y/rad)*ur0     value(v)= (x/rad)*ur0
  natural(p)= visc*[ -(x/r0)*del2(u) -(y/r0)*del2(v)]-
              dens*[ -(x/r0)*(u*dx(u)+v*dy(u)) -(y/r0)*(u*dx(v)+v*dy(v))]
                    arc to (0,-r0) to (-r0,0) to (0,r0) to finish
  ...
  vector( u, v) zoom(-3*r0,-3*r0,  6*r0,6*r0)
  vector( u/speed, v/speed) zoom(-3*r0,-3*r0,  6*r0,6*r0)
  surface( speed) interactive
end
```

The figures below show a zoomed *vector* plot of the direction of flow close to the shaft and a *surface* plot of the speed, both at $Re \cong 20$.

It is well known that a force will act on the shaft, perpendicular to the axis of rotation and to the main direction of the stream[7p159] (the Magnus effect).

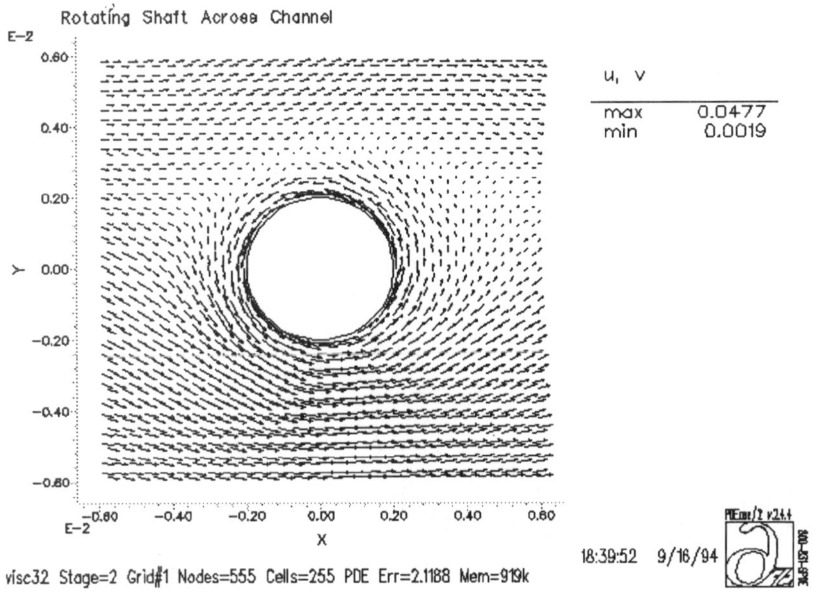

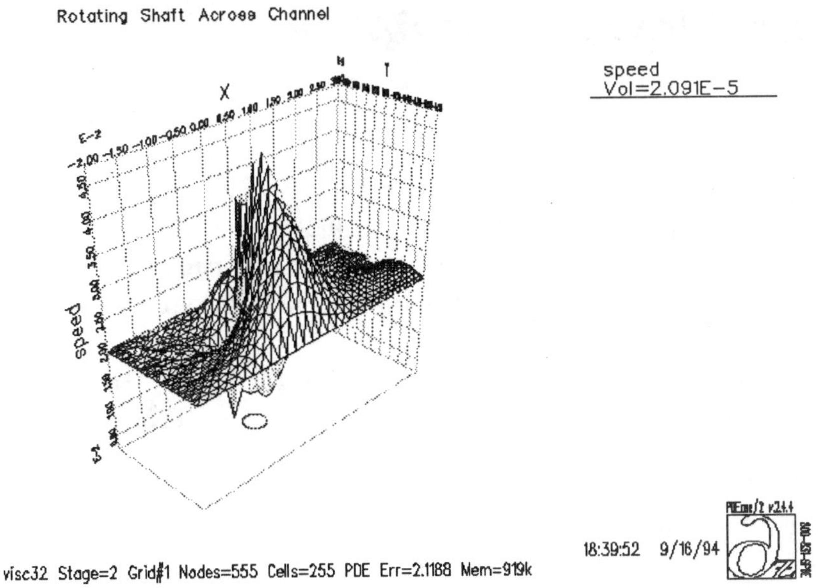

Exercises

❑ Investigate how the drag on the square bar varies with flow speed.

❑ Modify the example with the square bar by adding two line integrals for viscous drag along the entire channel, one just below the upper boundary, the other one just above the obstacle. Study the force balance on the liquid.

❑ Investigate how the drag and lift forces vary with the angle (*alpha*) of attack.

❑ Repeat the calculations of drag and lift, using a ten times thinner slab.

❑ The solution for pressure is quite noisy in case of the circular cylinder, mostly due to the high coefficient we use for the divergence term. Nevertheless, try to find the drag on that obstacle.

❑ Determine the lift (Magnus effect) on the shaft under clockwise and anti-clockwise rotation and at different *Re*. Does the effect exist at small *Re*?

Viscous Flow in (ρ,z) Space

We finish hydrodynamics by a few examples of steady, axially symmetric flow. The theory underlying the PDEs (p.242) is the same as in previous chapters. The only modification necessary is to transform the three equations and the *natural* boundary condition to cylindrical coordinates.

We start with the Navier-Stokes equation for steady flow and vanishing volume forces

$$\rho(\mathbf{u}\cdot\nabla)\cdot\mathbf{u}+\nabla p-\eta\nabla^2\mathbf{u}=0\ ,$$

Recasting these equations is straightforward, except for the last term, which is awkward to transform[9p602].

$$\rho_0\left(u\frac{\partial}{\partial\rho}+w\frac{\partial}{\partial z}\right)\begin{Bmatrix}u\\ w\end{Bmatrix}+\begin{Bmatrix}\dfrac{\partial p}{\partial\rho}\\ \dfrac{\partial p}{\partial z}\end{Bmatrix}-\eta\begin{Bmatrix}\dfrac{1}{\rho}\dfrac{\partial}{\partial\rho}\left(\rho\dfrac{\partial u}{\partial\rho}\right)-\dfrac{u}{\rho^2}+\dfrac{\partial^2 u}{\partial z^2}\\ \dfrac{1}{\rho}\dfrac{\partial}{\partial\rho}\left(\rho\dfrac{\partial w}{\partial\rho}\right)+\dfrac{\partial^2 w}{\partial z^2}\end{Bmatrix}=0$$

Here we denote the density by ρ_0 to distinguish it from the radial coordinate. The axial component of velocity is w. We recognize most of the last column vector from previous PDEs for axially symmetric dielectric (p.77) and heat conduction problems (p.108). After operating on the first column vector we obtain the final equations

$$\rho_0\begin{Bmatrix}u\dfrac{\partial u}{\partial\rho}+w\dfrac{\partial u}{\partial z}\\ u\dfrac{\partial w}{\partial\rho}+w\dfrac{\partial w}{\partial z}\end{Bmatrix}+\begin{Bmatrix}\dfrac{\partial p}{\partial\rho}\\ \dfrac{\partial p}{\partial z}\end{Bmatrix}-\eta\begin{Bmatrix}\dfrac{1}{\rho}\dfrac{\partial}{\partial\rho}\left(\rho\dfrac{\partial u}{\partial\rho}\right)-\dfrac{u}{\rho^2}+\dfrac{\partial^2 u}{\partial z^2}\\ \dfrac{1}{\rho}\dfrac{\partial}{\partial\rho}\left(\rho\dfrac{\partial w}{\partial\rho}\right)+\dfrac{\partial^2 w}{\partial z^2}\end{Bmatrix}=0 \qquad \bullet$$

The third PDE,

$$\nabla^2 p+\rho_0\nabla\cdot[(\mathbf{u}\cdot\nabla)\cdot\mathbf{u}]=0\ ,$$

requires some work, caused by the convoluted second term. We first transform the expression in square brackets:

$$(\mathbf{u}\cdot\nabla)\cdot\mathbf{u}=\left(u\frac{\partial}{\partial\rho}+w\frac{\partial}{\partial z}\right)\begin{Bmatrix}u\\ w\end{Bmatrix}=\begin{Bmatrix}u\dfrac{\partial u}{\partial\rho}+w\dfrac{\partial u}{\partial z}\\ u\dfrac{\partial w}{\partial\rho}+w\dfrac{\partial w}{\partial z}\end{Bmatrix},$$

where u is the radial component of the velocity **u,** etc.. It now remains to take the divergence of this column vector. For any vector **v** we have

$$\nabla\cdot\mathbf{v}=\frac{1}{\rho}\frac{\partial}{\partial\rho}\left(\rho v_\rho\right)+\frac{\partial v_z}{\partial z},$$

and hence, for the components of the column vector,

$$\nabla\cdot\left[(\mathbf{u}\cdot\nabla)\cdot\mathbf{u}\right]=\nabla\cdot\begin{Bmatrix}u\dfrac{\partial u}{\partial\rho}+w\dfrac{\partial u}{\partial z}\\ u\dfrac{\partial w}{\partial\rho}+w\dfrac{\partial w}{\partial z}\end{Bmatrix}=$$

$$\frac{1}{\rho}\frac{\partial}{\partial\rho}\left[\rho u\frac{\partial u}{\partial\rho}+\rho w\frac{\partial u}{\partial z}\right]+\frac{\partial}{\partial z}\left[u\frac{\partial w}{\partial\rho}+w\frac{\partial w}{\partial z}\right]=$$

$$\frac{1}{\rho}\left[u\frac{\partial u}{\partial\rho}+\rho\left(\frac{\partial u}{\partial\rho}\right)^2+\rho u\frac{\partial^2 u}{\partial\rho^2}+\quad w\frac{\partial u}{\partial z}+\rho\frac{\partial w}{\partial\rho}\frac{\partial u}{\partial z}+\rho w\frac{\partial^2 u}{\partial\rho\partial z}\right]+$$

$$\frac{\partial u}{\partial z}\frac{\partial w}{\partial\rho}+u\frac{\partial^2 w}{\partial\rho\partial z}+\left(\frac{\partial w}{\partial z}\right)^2+w\frac{\partial^2 w}{\partial z^2}=$$

$$\frac{u}{\rho}\frac{\partial u}{\partial\rho}+\left(\frac{\partial u}{\partial\rho}\right)^2+u\frac{\partial^2 u}{\partial\rho^2}+\frac{w}{\rho}\frac{\partial u}{\partial z}+\frac{\partial w}{\partial\rho}\frac{\partial u}{\partial z}+w\frac{\partial^2 u}{\partial\rho\partial z}+$$

$$\frac{\partial u}{\partial z}\frac{\partial w}{\partial\rho}+u\frac{\partial^2 w}{\partial\rho\partial z}+\left(\frac{\partial w}{\partial z}\right)^2+w\frac{\partial^2 w}{\partial z^2}.$$

Just as in the case of Cartesian coordinates we find that the second-order derivatives cancel in pairs. We again make use of the equation of continuity, $\nabla\cdot\mathbf{u}=0$, which now takes the form

$$\nabla\cdot\mathbf{u}=\frac{1}{\rho}\frac{\partial}{\partial\rho}(\rho u)+\frac{\partial w}{\partial z}= \frac{u}{\rho}+\frac{\partial u}{\partial\rho}+\frac{\partial w}{\partial z}=0 \qquad \bullet$$

Differentiating with respect to ρ, then multiplying by u we obtain the relation

$$u\frac{\partial^2 u}{\partial\rho^2}+u\frac{\partial^2 w}{\partial\rho\partial z}=\frac{u^2}{\rho^2}-\frac{u}{\rho}\frac{\partial u}{\partial\rho}$$

and differentiating with respect to z, then multiplying by w

$$w\frac{\partial^2 u}{\partial\rho\partial z}+w\frac{\partial^2 w}{\partial z^2}=-\frac{w}{\rho}\frac{\partial u}{\partial z}$$

The left parts of these equations both occur in the expression on the preceding page, and we replace them by the right parts to obtain

$$\nabla\cdot[(\mathbf{u}\cdot\nabla)\cdot\mathbf{u}]=\left(\frac{\partial u}{\partial\rho}\right)^2+2\frac{\partial u}{\partial z}\frac{\partial w}{\partial\rho}+\left(\frac{\partial w}{\partial z}\right)^2+\frac{u^2}{\rho^2}= \left(\frac{\partial u}{\partial\rho}+\frac{\partial w}{\partial z}\right)^2+\frac{u^2}{\rho^2}$$

Using $\nabla\cdot\mathbf{u}=0$ again we find that the above expression collapses into $2u^2/\rho^2$ and hence the third PDE becomes

$$\nabla^2 p+2\rho_0\frac{u^2}{\rho^2}=0$$

Finally we express the Laplacian in cylindrical coordinates and add the divergence term as before

$$\frac{1}{\rho}\frac{\partial}{\partial\rho}\left(\rho\frac{\partial p}{\partial\rho}\right)+\frac{\partial^2 p}{\partial z^2}+2\rho_0\frac{u^2}{\rho^2}+f_\nabla\left(\frac{u}{\rho}+\frac{\partial u}{\partial\rho}+\frac{\partial w}{\partial z}\right)=0 \qquad \bullet$$

This is thus the third PDE we need for the descriptor.

It now only remains to adapt the *natural* boundary condition to the system for axial symmetry. This is almost immediate, since we already have employed an expression for $\nabla^2\mathbf{u}$. The condition must now be written

$$natural(p) = \eta \mathbf{n} \cdot \nabla^2 \mathbf{u} - \rho \mathbf{n} \cdot [(\mathbf{u} \cdot \nabla)\mathbf{u}] =$$

$$\eta\left[n_\rho\left(\frac{1}{\rho}\frac{\partial}{\partial\rho}\left(\rho\frac{\partial u}{\partial\rho}\right) - \frac{u}{\rho^2} + \frac{\partial^2 u}{\partial z^2}\right) + n_z\left(\frac{1}{\rho}\frac{\partial}{\partial\rho}\left(\rho\frac{\partial w}{\partial\rho}\right) + \frac{\partial^2 w}{\partial z^2}\right)\right] -$$

$$\rho_0\left[n_\rho\left(u\frac{\partial u}{\partial\rho} + w\frac{\partial u}{\partial z}\right) + n_z\left(u\frac{\partial w}{\partial\rho} + w\frac{\partial w}{\partial z}\right)\right]$$

The last term, proportional to the density, vanishes on straight-line, fixed boundaries and may thus be omitted in the following examples. When writing a descriptor we should multiply all the PDEs by the factor ρ. As we already discussed (p.209) this implies that all *natural* boundary conditions also must be multiplied by ρ. No modifications of the *value* boundary conditions are necessary, however.

Tube With Uniform Driving Pressure

In accord with our usual policy we apply the equations to the simplest possible case, that of Poiseuille flow[7p11] through a tube. In the limit of low *Re* values we may put the density equal to zero. The following descriptor hardly contains any surprises.

```
  ...
  errlim= 1e-12           nodelimit= 200
  converge= 1e-6          subconverge= 1e-6
  contours= 10            vectorgrid= 20
  surfacegrid= 30         paint= off
  alias(x)= 'Radius'      alias(y)= 'Z'
variables
  u    w    p
definitions
  f_div= 0                dens= 0                    visc= 1e-2
  delp= 1e-4              Re= 1e3* w* r1/ visc
  r1= 1e-2                L1= 1e-2
  speed2= u**2+w**2            speed= sqrt( speed2)
  f_angle= sign(w)* arccos( u/speed)/pi*180
  div_u= u/x+dx(u)+dy(w)       curl_phi= dy(u)-dx(w)
  w_ex= delp/(4*2*L1*visc)* (r1**2-x**2)
```

```
initial values
equations
  x*dens* [u*dx(u)+ w*dy(u)]+ x*dx(p)-
                      visc* [dx(x*dx(u))- u/x +x*dyy(u)]= 0
  x*dens* [u*dx(w)+ w*dy(w)]+ x*dy(p)- visc* [dx(x*dx(w))+x*dyy(w)]= 0
  dx(x*dx(p))+ x*dyy(p)+ 2*dens* u**2/x+
                                          f_div*[ u+ x*dx(u)+ x*dy(w)]= 0
boundaries
            region 1
  start(0,-L1)      value(u)= 0
  natural(w)= 0   value(p)= delp     line to (r1,-L1)
  value(w)= 0
  natural(p)= visc*[ dx(x*dx(u))- u/x +x*dyy(u)]  line to (r1,L1)
  natural(w)= 0   value(p)= 0     line to (0,L1)
  natural(w)= 0   natural(p)= 0    line to finish
monitors
  contour(w)              delay= 2
  elevation(w) from (0,0) to (r1,0)
plots
  elevation( Re) from (0,0) to (r1,0)
  contour( u)             contour( w)              contour( p)
  elevation(w, w_ex) from (0,0) to (r1,0)
  elevation(w- w_ex) from (0,0) to (r1,0)
  contour( f_angle)       vector( u, w)
  contour( div_u)         contour( curl_phi)
      elevation( 2*pi*x*w) from (0,-0.9*L1) to (r1,-0.9*L1)
      elevation( 2*pi*x*w) from (0,0) to (r1,0)
      elevation( 2*pi*x*w) from (0,0.9*L1) to (r1,0.9*L1)
end
```

The expression *w_ex* is the well-known exact solution which we shall use for testing the numerical accuracy.

The *vector* plot below displays the parabolic velocity profile we have already seen in the case of a 2D channel, only in this case we are looking at a cross-section of axially symmetric flow. The *elevation* plot of the difference shows that our solution is good to within one part in 10^7.

The second figure below shows that the vorticity $\text{curl}_\varphi(\mathbf{u})$ is finite over all the domain, except on the axis of symmetry.

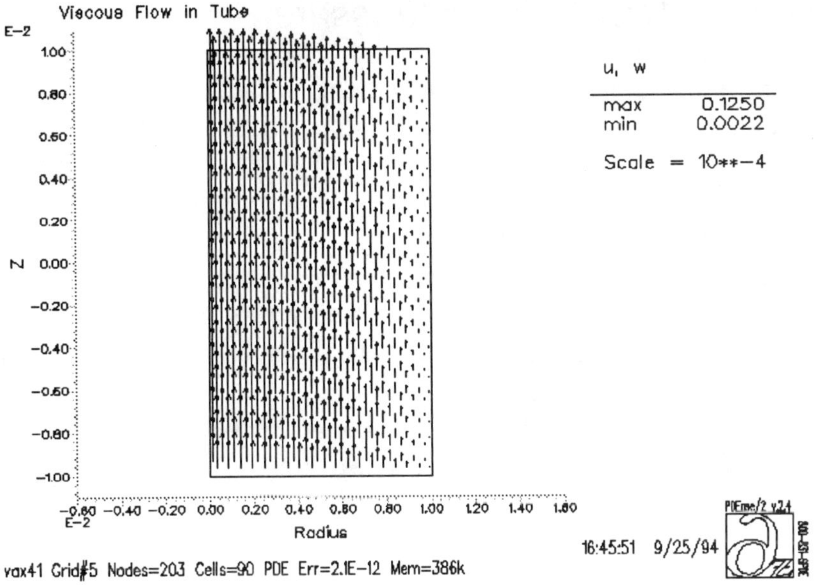
Viscous Flow in Tube
E−2
1.00
0.80
0.60
0.40
0.20
Z 0.00
−0.20
−0.40
−0.60
−0.80
−1.00
−0.60 −0.40 −0.20 0.00 0.20 0.40 0.60 0.80 1.00 1.20 1.40 1.60
E−2
Radius
u, w
max 0.1250
min 0.0022
Scale = 10**−4
16:45:51 9/25/94
PDEase/2 v2.4
vax41 Grid#5 Nodes=203 Cells=90 PDE Err=2.1E−12 Mem=386k

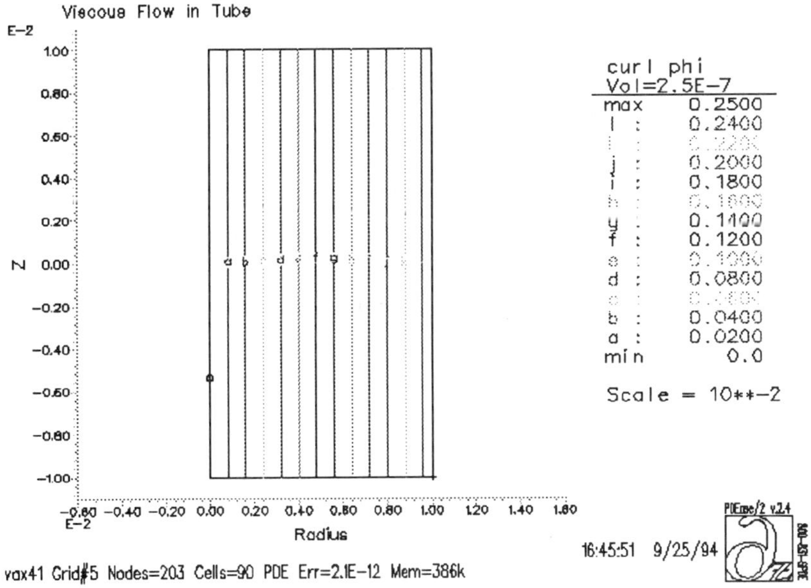
Viscous Flow in Tube
E−2
1.00
0.80
0.60
0.40
0.20
Z 0.00
−0.20
−0.40
−0.60
−0.80
−1.00
−0.60 −0.40 −0.20 0.00 0.20 0.40 0.60 0.80 1.00 1.20 1.40 1.60
E−2
Radius
curl phi
Vol=2.5E−7
max 0.2500
l : 0.2400
j : 0.2000
i : 0.1800
g : 0.1400
f : 0.1200
d : 0.0800
b : 0.0400
a : 0.0200
min 0.0
Scale = 10**−2
16:45:51 9/25/94
PDEase/2 v2.4
vax41 Grid#5 Nodes=203 Cells=90 PDE Err=2.1E−12 Mem=386k

Tube With Non-Uniform Pressure

The preceding example was convenient because it only involved parallel flow. We shall now engage more terms in the equations by imposing parabolic pressure (*delp*) at the front end, as we did in a problem for a channel (p.257). Only a few modifications of the previous descriptor are necessary.

In this example we limit the study to the case of *Re*=0. Because of the divergence term in the third PDE the errors appear to be rather high, and a reasonable compromise between accuracy and calculation time is to request *errlim*=2.0.

The *elevation* plot at the end compares the velocity near the exit with a suitably scaled parabolic profile *w_ref*.

```
  ...
  errlim= 2.0              nodelimit= 200
  converge= 1e-6           subconverge= 1e-6
  ...
  f_div= 1e10              dens= 0             visc= 1e-2
  dpmax= 1e-4              delp= dpmax*(x/r1)**2
  Re= dens* w* r1/ visc        w_ref= dpmax/(24*visc*L1)*(r1**2-x**2)
  ...
  elevation( w, w_ref) from (0,0.9*L1) to (r1,0.9*L1)
end
```

The figure on the next page shows contours of constant vertical velocity. It reminds us of the pattern obtained with the channel (p.257), but there are differences in details.

The *vector* plot below demonstrates how the complex input flow finally becomes nearly parabolic. The input speed has a maximum close to the cylindrical boundary, and there is even reversal of the velocity near the axis.

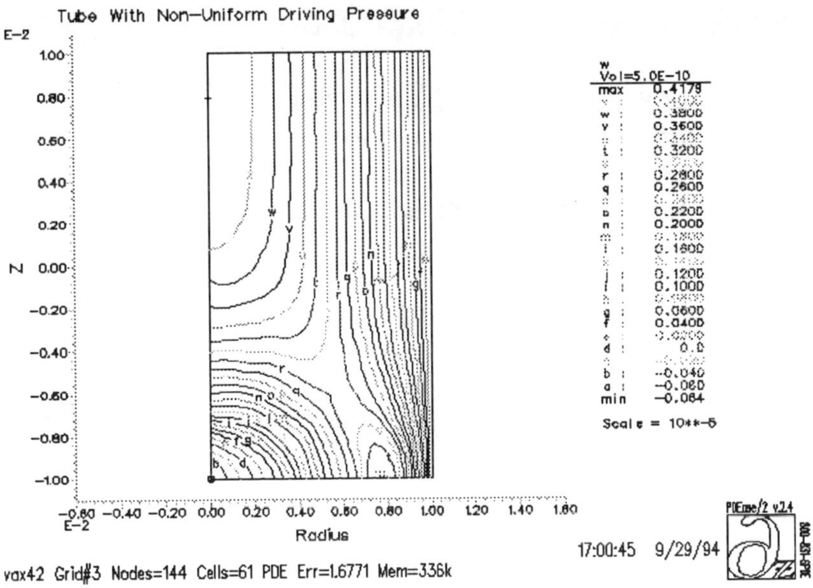

Tube With Non-Uniform Driving Pressure
E-2
Z
Radius
w
Vol=5.0E-10
max 0.4179
w : 0.3800
v : 0.3600
t : 0.3200
r : 0.2800
q : 0.2600
o : 0.2200
n : 0.2000
l : 0.1600
j : 0.1200
i : 0.1000
g : 0.0600
f : 0.0400
d : 0.0
b : -0.040
a : -0.060
min -0.064
Scale = 10**-5
PDEase/2 v2.4
17:00:45 9/29/94
vax42 Grid#3 Nodes=144 Cells=61 PDE Err=1.6771 Mem=336k

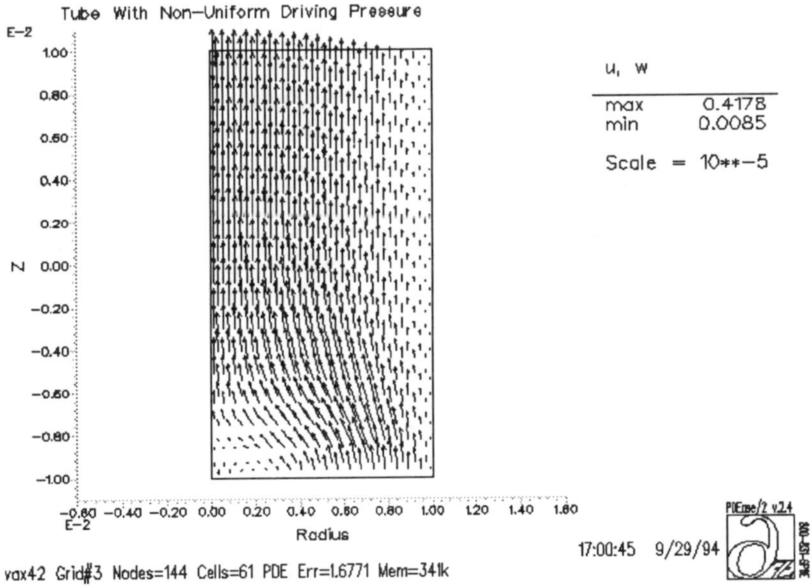

Tube With Non-Uniform Driving Pressure
E-2
Z
Radius
u, w
max 0.4178
min 0.0085
Scale = 10**-5
PDEase/2 v2.4
17:00:45 9/29/94
vax42 Grid#3 Nodes=144 Cells=61 PDE Err=1.6771 Mem=341k

Tube With a Venturi Constriction

Let us now study the flow through a Venturi tube[7p120], a device often used for determining speed from measurements of pressure. The geometry is evident from the figures. In this example we limit ourselves to *Re*=0, but we note that the second term in *natural*(*p*) vanishes on all fixed, linear boundary segments (even for *Re*≠0).

```
...
errlim= 20
...
f_div= 1e10            delp= 1e-4        dens= 0            visc= 1e-2
r1= 1e-2          L1= 1e-2          r2= 2e-2
nx= 1/sqrt(2)            ny= nx
w_ref= delp/(200*visc*L1)*(r2**2-x**2)
...
                    region 1
start(0,-4*L1)      value(u)= 0
natural(w)= 0   value(p)= delp    line to (r2,-4*L1)
value(w)= 0
natural(p)= visc*[ dx(x*dx(u))- u/x +x*dyy(u)]  line to (r2,-2*L1)
natural(p)= nx*visc*[ dx(x*dx(u))- u/x +x*dyy(u)]
                    +ny*visc*[ dx(x*dx(w))+x*dyy(w)]  line to (r1,-L1)
natural(p)= nx*visc*[ dx(x*dx(u))- u/x +x*dyy(u)]  line to (r1,L1)
natural(p)= -nx*visc*[ dx(x*dx(u))- u/x +x*dyy(u)]
                    -ny*visc*[dx(x*dx(w))+x*dyy(w)]  line to (r2,2*L1)
natural(p)= visc*[ dx(x*dx(u))- u/x +x*dyy(u)]  line to (r2,4*L1)
natural(w)= 0  value(p)= 0    line to (0,4*L1)
natural(w)= 0   natural(p)= 0    line to finish
...
elevation( 2*pi*x*w) from (0,-0.9*L1) to (r2,-0.9*L1)                { div... }
elevation( 2*pi*x*w) from (0,0) to (r1,0)
elevation( 2*pi*x*w) from (0,0.9*L1) to (r1,0.9*L1)
elevation( 2*pi*x*w) from (0,3.9*L1) to (r2,3.9*L1)
  elevation( w, w_ref) from (0,3.9*L1) to (r2,3.9*L1)               { Exit w }
  elevation( p) from (0.1*r1,-4*L1) to (0.1*r1,4*L1)              { Pressure }
end
```

The first figure below shows the symmetrical velocity pattern typical of the low-*Re* regime. The maximum occurs on the axis, as we could expect.

The *contour* plot of the angle of flow again brings out the symmetry with respect to z=0. The extreme values of the angle shows that the liquid follows the conical surfaces, although the speed goes to zero near the wall.

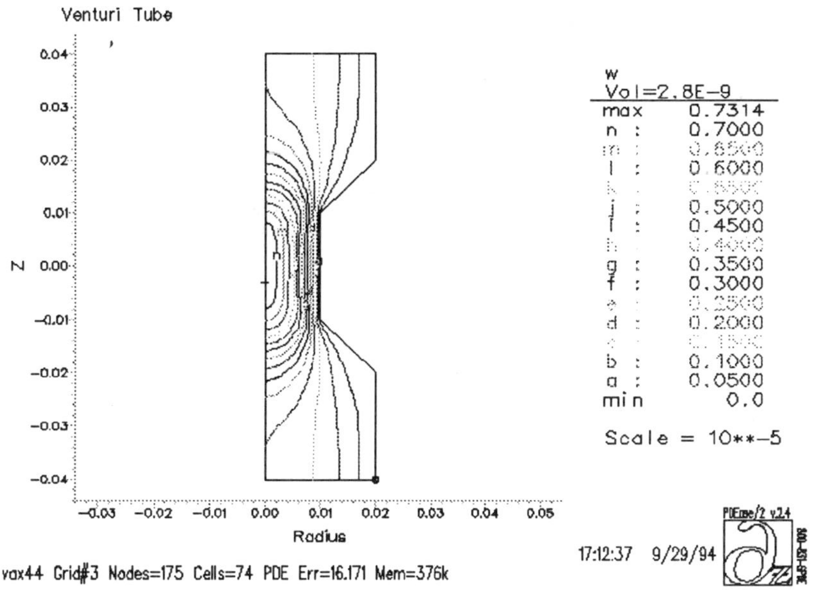

The solution for pressure is not as well defined as for the velocity components. The cause is no doubt the large divergence term that we have to add to the pressure equation. The last figure illustrates the strong pressure fall along the axis in the region of the constriction. This should be compared to the case of irrotational flow (p.232), where we obtain a symmetric pressure distribution, independent of the direction of flow.

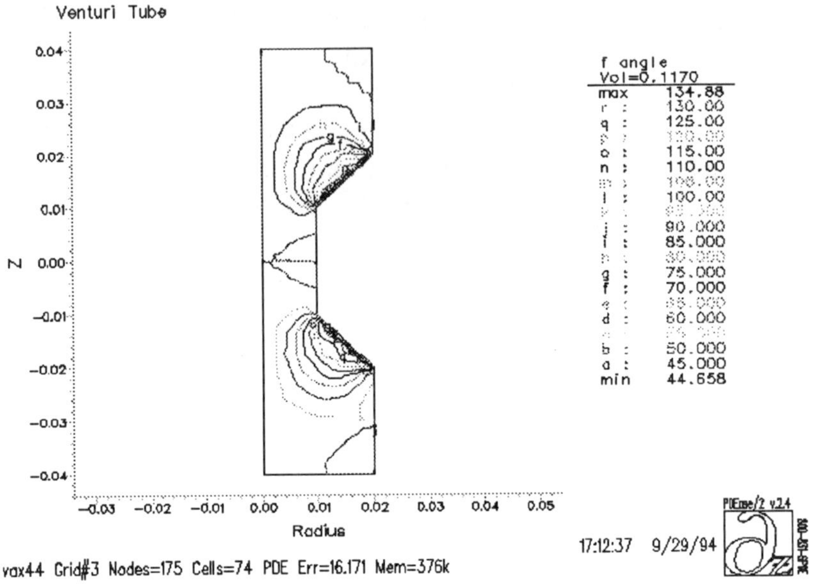
Venturi Tube
Z
Radius
f angle
Vol=0.1170
max 134.88
r : 130.00
q : 125.00
o : 115.00
n : 110.00
l : 100.00
j : 90.000
i : 85.000
g : 75.000
f : 70.000
d : 60.000
b : 50.000
a : 45.000
min 44.658
17:12:37 9/29/94
PDEase/2 v2.4
vax44 Grid#3 Nodes=175 Cells=74 PDE Err=16.171 Mem=376k

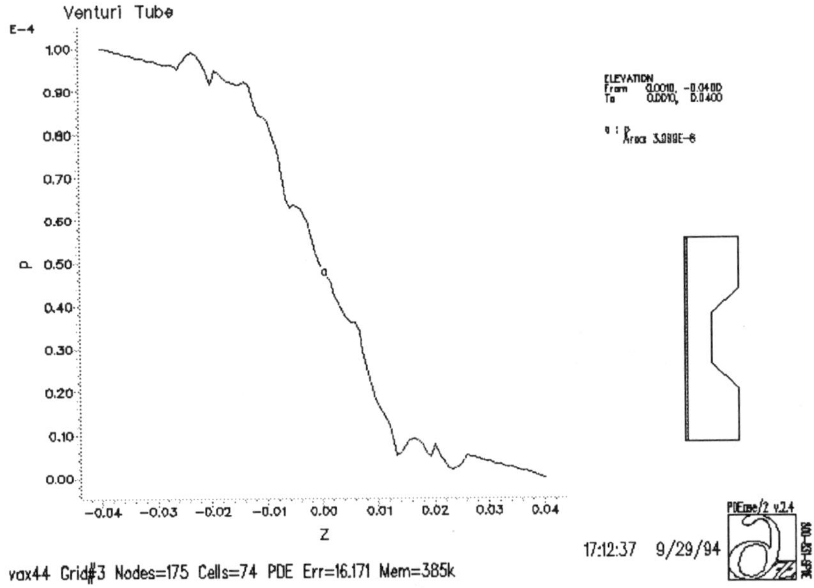
Venturi Tube
E-4
P
Z
ELEVATION
From 0.0010, -0.0400
To 0.0010, 0.0400
a : p
Area 3.088E-6
17:12:37 9/29/94
PDEase/2 v2.4
vax44 Grid#3 Nodes=175 Cells=74 PDE Err=16.171 Mem=385k

Exercises

❑ Show that the solution *w_ex* with *u*= 0 for the simple tube satisfies the PDE and the boundary conditions.

❑ Modify the first example by putting *delp*=1.0. Use two stages with *dens*=(*stage*-1)*1e3 to treat flow at finite Reynolds number.

❑ Modify the descriptor for the Venturi tube by putting *delp*=1.0. Use two stages with *dens*=(*stage*-1)*1e3 to treat flow at finite *Re*. Select *errlim*=1e3 and *ngrid*=10. Also increase the length of the outlet tube to twice the previous value, thus making room for the downstream pattern.

Waves in Fluids

The PDE for wave propagation in gases and liquids becomes simple if viscosity is neglected. Pressure (p) is then the only form of stress we need to take into account, and the velocity ($\mathbf{v}$) is the response to that stress.

In the limit of small compression, i.e. small pressure variation relative to the average value, all equations become linear. The following relation expresses mass conservation

$$\rho\nabla\cdot\mathbf{v} = -\frac{\partial\rho}{\partial t} = -\left(\frac{\partial\rho}{\partial p}\right)_S\frac{\partial p}{\partial t} = -\left(\frac{\partial V^{-1}}{\partial p}\right)_S\frac{\partial p}{\partial t} = -\frac{1}{VB_S}\frac{\partial p}{\partial t}$$

which may be written

$$\nabla\cdot\mathbf{v} = -\frac{1}{B_S}\frac{\partial p}{\partial t} \qquad \bullet$$

where B_S is the adiabatic bulk modulus.

Newton's law of motion gives us the additional equation

$$\rho\frac{d\mathbf{v}}{dt} = -\nabla p \qquad \bullet$$

where the left side expresses the acceleration of a volume element and the right side the force acting on it.

The above PDEs, with the appropriate boundary conditions, are sufficient to define a solution. The last one is a vector equation, which means that in total we have four PDEs, and if we are to find time-sinusoidal solutions by the complex method we have to face 8 equations. In addition, they are not 2nd order PDEs in the space variables, which PDEase assumes to be the case.

We may resort to an ingenious device, however, converting the PDEs into a single equation. The key to this simplification is the velocity potential ϕ, defined by

$$\mathbf{v} = -\nabla\phi \qquad \bullet$$

Substituting this expression for **v** into Newton's law of motion, we obtain

$$-\nabla p = \rho \frac{d\mathbf{v}}{dt} = \rho \frac{d}{dt}(-\nabla \phi) \cong \nabla\left(-\rho \frac{\partial \phi}{\partial t}\right)$$

from which we deduce that

$$p = \rho \frac{\partial \phi}{\partial t},$$ ●

apart from an arbitrary constant. With this expression for p we have

$$-\frac{1}{B_s}\frac{\partial p}{\partial t} = -\frac{1}{B_s}\frac{\partial}{\partial t}\left(\rho \frac{\partial \phi}{\partial t}\right) = -\frac{\rho}{B_s}\frac{\partial^2 \phi}{\partial t^2}$$

In terms of the velocity potential the equation of mass conservation may be written

$$-\frac{1}{B_S}\frac{\partial p}{\partial t} = \nabla \cdot \mathbf{v} = \nabla \cdot (-\nabla \phi) = -\nabla^2 \phi$$

and combining these two equations we obtain the 2nd order PDE

$$\nabla^2 \phi - \frac{\rho}{B_s}\frac{\partial^2 \phi}{\partial t^2} = 0$$ ●

Complex Solution

For the time-sinusoidal solutions we use the same complex form as for thermal waves (p.130)

$$\phi = \phi_0 \exp(i\omega t) \equiv (\phi_r + i\phi_i)\exp(i\omega t).$$

If we introduce this expression into the above PDE, the time derivative produces the factor $(i\omega)^2 = -\omega^2$ and the exponential factors cancel. We may separate the result into a real and an imaginary equation.

$$\begin{cases} \nabla^2 \phi_r + \dfrac{\rho\omega^2}{B_s} \phi_r = 0 \\ \nabla^2 \phi_i + \dfrac{\rho\omega^2}{B_s} \phi_i = 0 \end{cases}$$ •

This system of PDEs may be directly solved using PDEase. Once we have a solution, we obtain the velocity field **v** in the complex form

$$\mathbf{v} = -\nabla\phi_r - i\nabla\phi_i$$
$$\mathbf{v}_r = -\nabla\phi_r, \qquad \mathbf{v}_i = -\nabla\phi_i$$ •

and the pressure excess by

$$p = i\rho\omega(\phi_r + i\phi_i) = -\rho\omega\phi_i + i\rho\omega\phi_r$$
$$p_r = -\rho\omega\phi_i, \qquad p_i = \rho\omega\phi_r$$ •

Needless to say, all velocity components as well as the pressure take real values at all times. To convert any of these quantities into real form we must multiply the complex expression by $\exp(i\omega t)$ and take the real part of the result, as follows.

$$\begin{aligned} \mathbf{v} &= \mathrm{Re}\{(\mathbf{v}_r + i\,\mathbf{v}_i)(\cos(\omega t) + i\sin(\omega t))\} = \\ &\quad \mathbf{v}_r \cos(\omega t) - \mathbf{v}_i \sin(\omega t) = \\ &\quad -\nabla\phi_r \cos(\omega t) + \nabla\phi_i \sin(\omega t) \end{aligned}$$ •

$$\begin{aligned} p &= \mathrm{Re}\{(p_r + i p_i)(\cos(\omega t) + i\sin(\omega t))\} = \\ &\quad p_r \cos(\omega t) - p_i \sin(\omega t) = \\ &\quad -\rho\omega\phi_i \cos(\omega t) - \rho\omega\phi_r \sin(\omega t) \end{aligned}$$ •

Plane Acoustic Wave

We consider a domain of length L in the x direction and height h in the y direction. In the first examples there will be no variation with y, but we prepare a more general descriptor for later 2D applications.

In this first study we choose a frequency more or less at random. On three of the boundaries we require that the outward normal velocity be zero, which concerns both real and imaginary parts. Thus most of the boundary conditions are *natural*( )= 0, which is the default with PDEase. We specify it here only for clarity.

```
title 'Acoustic wave in 1D'                                  { acw1.pde }
select
  errlim= 1e-4         nodelimit= 1000
  contours= 10         surfacegrid= 30      paint= off
variables
  phir   phii                              { Real and imaginary potential }
definitions
  omega= 250                                        { SI units assumed }
  L= 10        h= 1        dens= 1.176        Bs= 1.4e5              { Air }
  vxr= -dx(phir)      vxi= -dx(phii)                { Velocity components }
  vyr= -dy(phir)      vyi= -dy(phii)
  preal= -dens* omega* phii                                  { Pressure }
  pimag= dens* omega* phir
equations
  del2(phir)+ dens* omega**2/ Bs*phir= 0
  del2(phii)+ dens* omega**2/ Bs*phii= 0
boundaries
          region 1
  start (0,-h/2)
  natural(phir)= 0   natural(phii)= 0    line to (L,-h/2)
  natural(phir)= 0   natural(phii)= 0    line to (L,h/2)
  natural(phir)= 0   natural(phii)= 0    line to (0,h/2)
  natural(phir)= 1   natural(phii)= 0    line to finish
monitors
  elevation(vxr) from (0,0) to (L,0)
plots
  elevation(phir, phii) from (0,0) to (L,0)
  elevation(vxr, vxi) from (0,0) to (L,0)
  elevation(preal, pimag) from (0,0) to (L,0)
  contour(phir)                    contour(phii)
end
```

The resulting curve (not shown here) for the real part of v_x starts at 1.0 as required by the boundary conditions, and it remains of comparable magnitude throughout the domain. The imaginary part, on the other hand, is zero everywhere. This is a consequence of the vanishing normal velocities specified by the boundary conditions.

We shall now interpret the curves for v_x and p. As is evident from the plots, only the real part of the potential, *phir*, is different from zero. Hence, the time dependence takes a particularly simple form, i.e.

$$v_x = v_{xr}\cos(\omega t)$$

$$p = -p_i \sin(\omega t)$$

These relations permit us to assess the time-variation of the quantities, the plots providing the spatial variation.

The following figure shows curves of *preal* (vanishing) and *pimag*. As already discussed, *pimag* multiplied by $-\sin(\omega t)$ gives us the time variation as well.

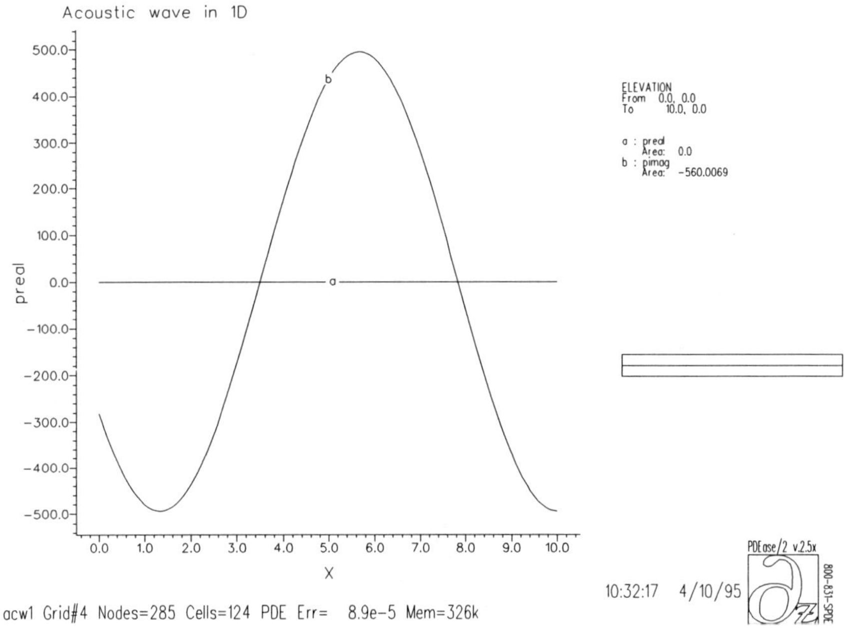

Resonance

In the preceding example the velocity inside the cavity turned out to have about the same magnitude as that imposed at the left end. We shall now scan the frequency over a range to investigate whether this remains true in general.

We modify the above descriptor as follows, scanning *omega* by *stages* and retaining only one plot per stage.

```
select                                              { acw2.pde }
  stages= 20
  errlim= 3e-4   nodelimit= 1000
...
definitions
  omega= 200.7+ 2*stage
...
plots
  elevation( vxr) from (0,0) to (L,0)
end
```

Running the new file we discover that the amplitudes initially increase, to reach a maximum at *stage*=8. This anomalous increase, known as a resonance, makes the imposed perturbation at the left end negligible compared to the oscillation inside the cavity.

The figure to the right shows a plot of the velocity at resonance. The quality of this plot (a derivative) may be improved by an extra run at the resonant frequency, using a smaller value of *errlim*.

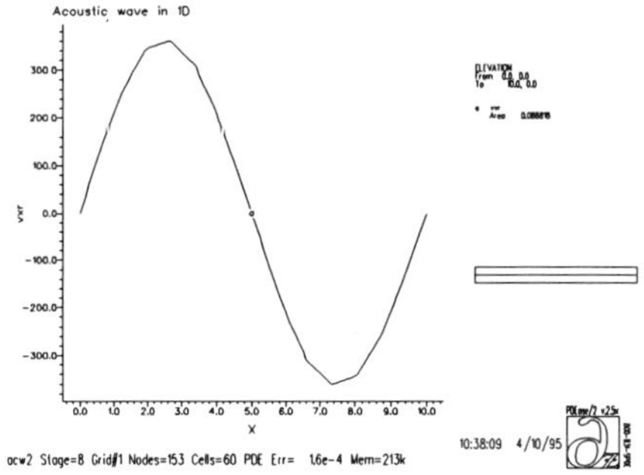

Constricted Cavity

We now proceed to waves propagating in two dimensions. As before, we drive the sound wave from the left but add a constriction half-way along the cavity, as is evident from the next figure. The boundary conditions are the same as in previous examples. Here we simplify by specifying the boundary conditions only once, knowing that PDEase assumes similar *natural* conditions to apply for the following sections of the boundary. We can now request plots liberally, since there is only one *stage*.

```
select                                                            { acw11.pde }
  errlim= 3e-4          nodelimit= 1000
  contours= 10          surfacegrid= 30         paint= off
variables
  phir  phii            { Real and imaginary potential }
definitions
  omega= 200
  L= 10      h= 5       dens= 1.176     Bs= 1.4e5                 { Air }
  vxr= -dx(phir)        vxi= -dx(phii)
  vyr= -dy(phir)        vyi= -dy(phii)
  preal= -dens* omega* phii        pimag= dens* omega* phir
equations
  del2(phir)+ dens* omega**2/ Bs*phir= 0
  del2(phii)+ dens* omega**2/ Bs*phii= 0
boundaries
          region 1
  start (0,-h/2)
  natural(phir)= 0   natural(phii)= 0
  line to  (0.45*L,-0.5*h)  to (0.45*L,-0.1*h)  to (0.55*L,-0.1*h)
    to (0.55*L,-0.5*h)  to  (L,-0.5*h)  to  (L,0.5*h)
    to (0.55*L,0.5*h)  to (0.55*L,0.1*h)  to (0.45*L,0.1*h)
    to (0.45*L,0.5*h)  to (0,0.5*h)
  natural(phir)= 1   line to finish
monitors
  elevation( vxr) from (0,0) to (L,0)
plots
  elevation( vxr) from (0,0) to (L,0)
  contour( vxr)         contour(vxi)
  contour( vyr)         contour(vyi)
```

```
  contour( preal)        contour( pimag)
  vector( vxr, vyr)      surface( pimag)  interactive
end
```

The following *vector* plot gives a snapshot of the air velocity at a time where the input velocity reaches its maximum. The range of values obtained from the table to the right indicates that the frequency is far from resonance.

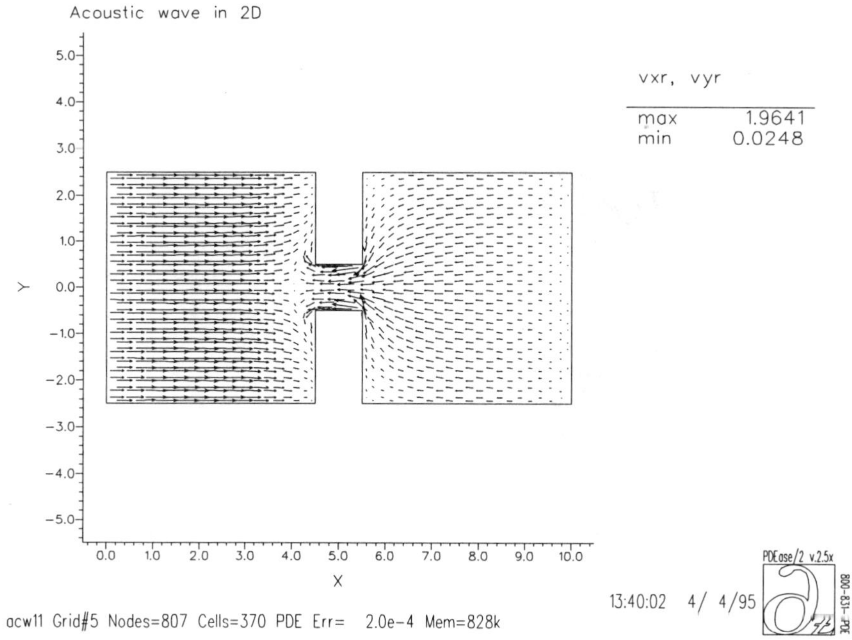

We now proceed to scan over a frequency range, searching for resonant states. The modified descriptor lines are as follows.

```
  ...                                        { acw12.pde }
  stages= 20
  ...
  omega= 258+ 0.5*stage
  ...
plots
```

```
  elevation ( vxr, vxi) from (0,0) to (L,0)
  surface( vxr)
end
```

When running this new file you will notice that the amplitude reaches very large values. The figure below shows some unusual characteristics, i.e. that the velocity is positive in the lateral cavities and negative only in the constricted part.

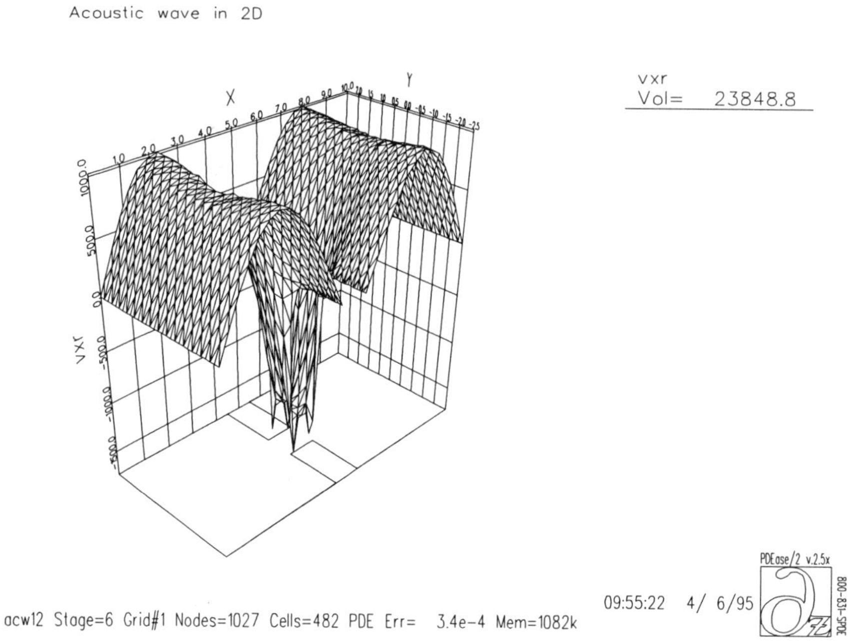

The *elevation* plot already indicates that the imaginary part of the velocity *vxi* is zero. A closer inspection by *contour* plots at the resonant frequency confirms this. Again this is a trivial consequence of the special boundary conditions.

Cylindrical Resonator

We shall next study a different kind of cavity, consisting of a channel leading to a cylindrical vessel. As before, we excite the oscillations from the left end. The descriptor is as follows.

```
select                                                        { acw14.pde }
  errlim= 3e-4          nodelimit= 1000
  contours= 10          surfacegrid= 30         paint= off
variables
  phir      phii                          { Real and imaginary potential }
definitions
  omega= 70
  L= 10      h= 4       dens= 1.176     Bs= 1.4e5                  { Air }
  vxr= -dx(phir)     vxi= -dx(phii)
  vyr= -dy(phir)     vyi= -dy(phii)
  preal= -dens* omega* phii         pimag= dens* omega* phir
equations
  del2(phir)+ dens* omega**2/ Bs*phir= 0
  del2(phii)+ dens* omega**2/ Bs*phii= 0
boundaries
                     region 1
  start (0,-0.5*h)
  natural(phir)= 0        natural(phii)= 0
  line to  (L,-0.5*h)
  arc to  (L+ 4*h, 0) to (L,0.5*h)
  line to (0,0.5*h)
  natural(phir)= 1    line to finish
monitors
  elevation( vxr, vxi) from (0,0) to (L+4*h,0)
plots
  elevation( vxr, vxi) from (0,0) to (L+4*h,0)
  contour( vxr)           contour(vxi)
  contour( vyr)           contour(vyi)
  contour( preal)         contour( pimag)
  vector( vxr, vyr)       surface( pimag)  interactive
end
```

The plot below shows contours of constant *vxr*, and as we see from the adjoining table the amplitudes in the interior are about as large as on the left boundary. We also notice that the wave propagates through the channel as a plane wave, to spread slightly at the entrance to the cylinder.

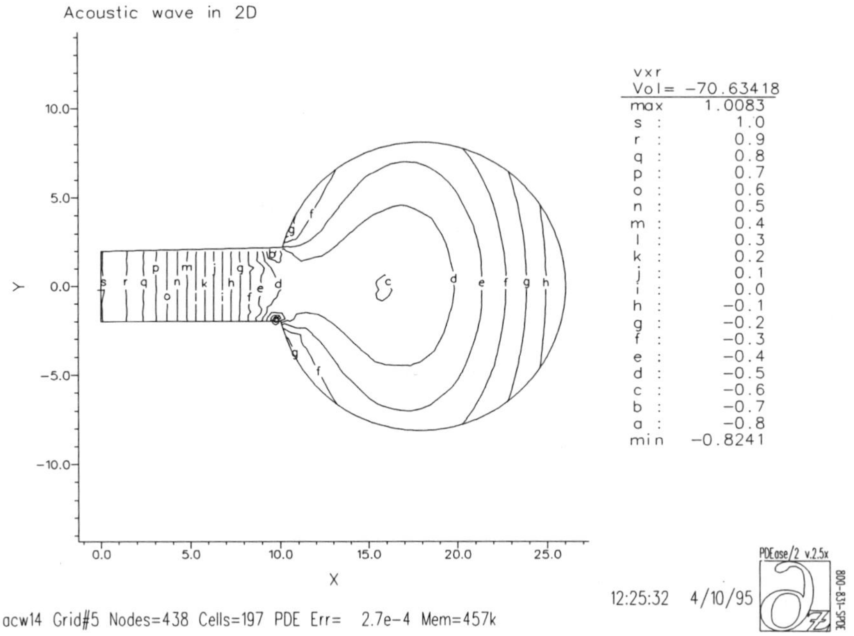

We shall now modify the preceding descriptor as indicated below, in order to search for resonances.

```
...                                               { acw15.pde }
select
  stages= 20
  ...
definitions
  omega= 120.4+ 1*stage
  ...
plots
```

```
  contour( vxr)
end
```

The next plot resulted from a similar descriptor, containing only one frequency close to resonance . The *max* and *min* values show that the maximum velocity amplitude is more than a thousand times that imposed at the input end. The imaginary part of the velocity potential is still zero over the entire domain.

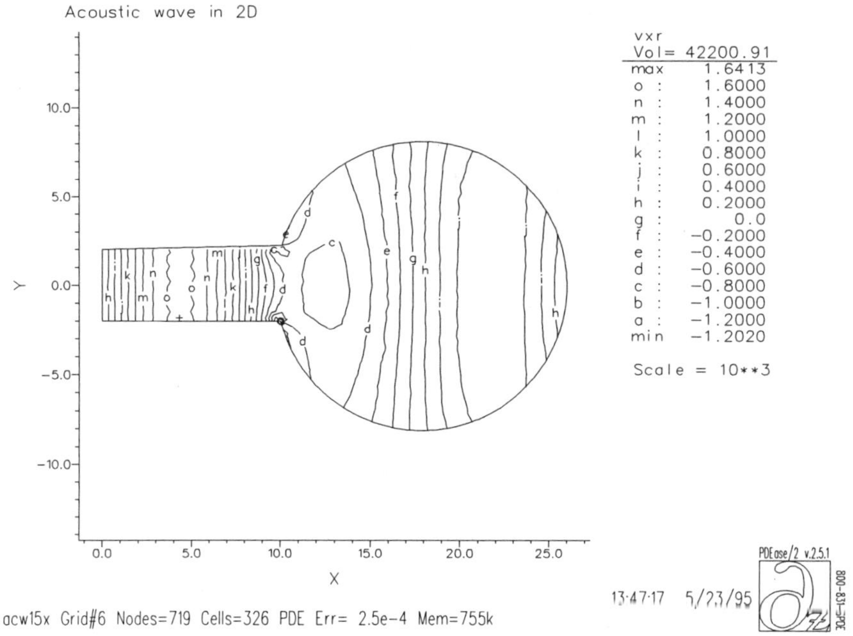

Since we now concentrate on the frequency yielding the largest amplitudes, we may bring in the whole arsenal of plots from previous descriptors without generating too many picture files. The next figure gives an example of what we may observe, in this instance the imaginary part of the pressure. The real part vanishes, as usual. It would be wrong, however, to assume that either the real or the imaginary part of the dependent variables and their derived functions always must zero. In the problems we have considered so far, the situation was exceptional, in the sense that the boundary conditions for *phii* involved vanishing outward derivatives. In the next section we shall see an example of a more general situation where both real and imaginary parts play a role.

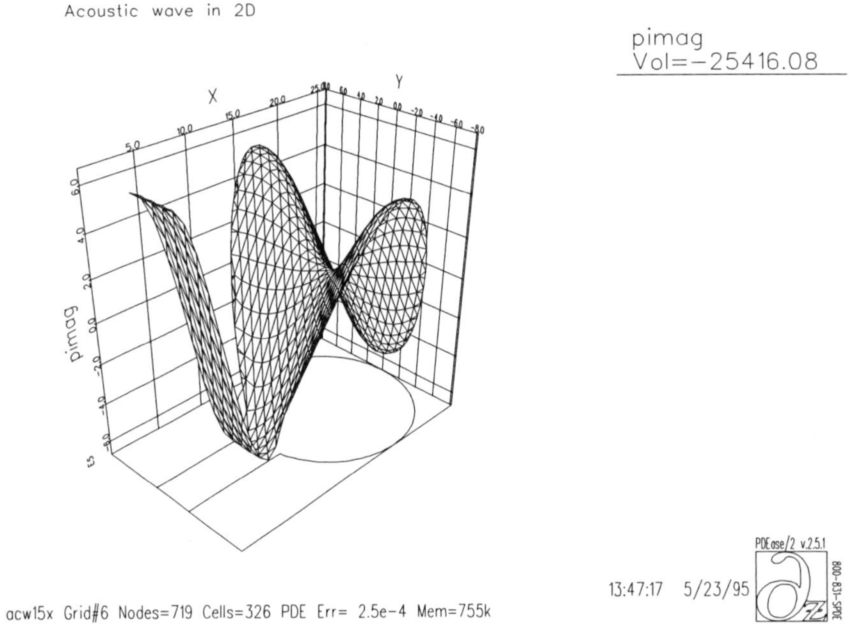

Full Complex Solution

As an example of a more general type of problem, we shall now revisit the simple descriptor ACW1.PDE, which we change by specifying a varying phase angle for the imposed oscillation. We also make the solution domain square by the following amendments.

```
...
definitions                                                  { acw21.pde }
  omega= 250
  L= 10     h= 10     dens= 1.176     Bs= 1.4e5              { Air }
...
```

```
        region 1
  start (0,-h/2)
  natural(phir)= 0   natural(phii)= 0    line to (L,-h/2)
  natural(phir)= 0   natural(phii)= 0    line to (L,h/2)
  natural(phir)= 0   natural(phii)= 0    line to (0,h/2)
  natural(phir)= 1+y/h   natural(phii)= 1-y/h    line to finish
monitors
  elevation( vxr) from (0,0) to (L,0)
plots
  contour(phir)          contour(phii)
  contour( vxr)          contour( vxi)
  vector( vxr, vyr)      vector( vxi, vyi)
  surface( vxi)
end
```

It is evident from the first plots that the potential now has real and imaginary parts of comparable magnitude, in contrast to what we have seen in previous examples. The *surface* plot of *vxi* below illustrates the consequences regarding the velocity.

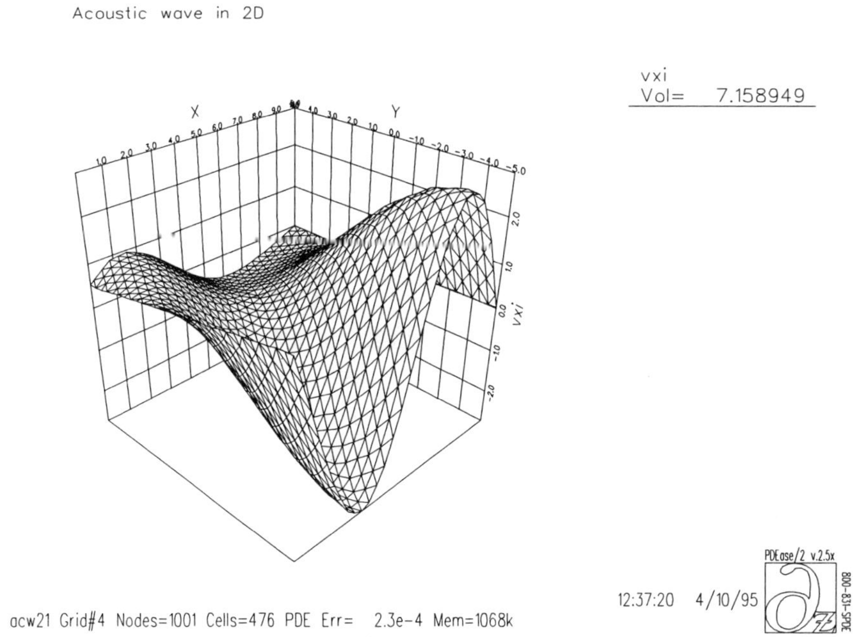

The additional plot below (*vxr*) shows a novel pattern with curved zero lines. The corresponding plot of *vxi* appears somewhat similar, but upside-down, which means that the zero lines of the two plot will not coincide. In other words, when we multiply *vxr* and *vxi* by their cosine and sine as discussed, we never find zero velocity along a curve. This is in contrast to previous problems, where only *vxr* was non-zero, which yielded vanishing absolute values along lines.

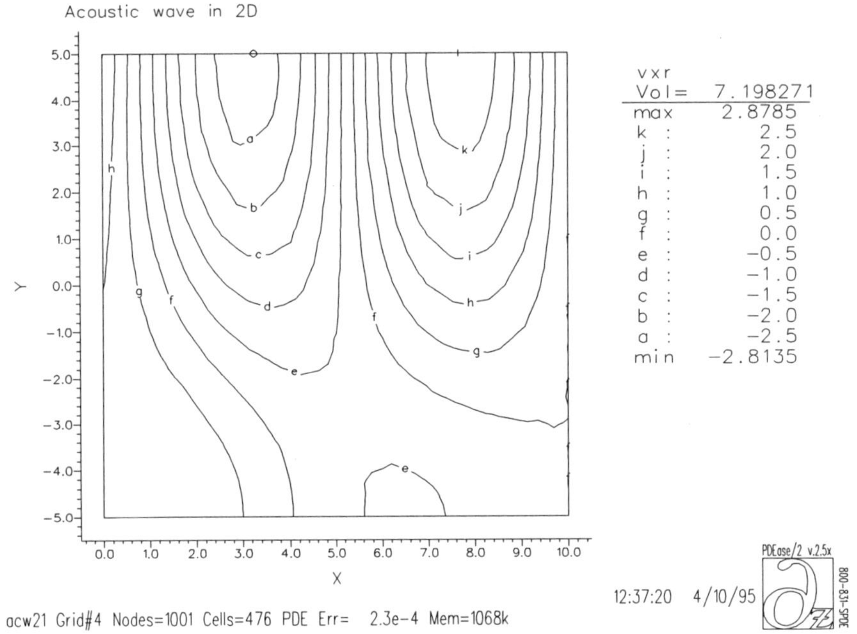

We have seen that forced oscillations may produce lines of zero velocity, if the boundary conditions make the real (or the imaginary) part of the potential vanish. When both parts are non-zero, lines of vanishing magnitude generally do not occur.

In a resonant state, however, the amplitudes on the boundaries are negligible compared to those in the interior of the domain. Hence, we are effectively back at zero boundary conditions, and lines of vanishing oscillation may again appear.

Exercises

❑ Add the necessary plots to ACW1.PDE to check that the boundary conditions are satisfied.
❑ Modify ACW2.PDE to search for one resonance at a lower, another at a higher frequency.
❑ Use ACW12.PDE to search for a resonance in the region of $230 < \omega < 240$.
❑ Set $\omega = 413.8$ in ACW12.PDE and explore the corresponding resonant state, including the pressure field.
❑ Modify ACW15.PDE to discover a new resonance at about $\omega = 85$.
❑ Add definitions of the absolute values of v_x and p to ACW21.PDE and plot these quantities to search for zero lines.
❑ Scan the frequency using ACW21.PDE in modified form to find a resonance at about $\omega = 217$.

Waves Along a Beam

Previously we have only been concerned with static deformation of solid objects. We shall now study dynamic deformation in the form of sinusoidal waves, generated by periodic deformation at the boundary.

In the chapter on elementary elasticity (p.157) we started with the equations of equilibrium

$$\begin{cases} \dfrac{\partial \sigma_x}{\partial x} + \dfrac{\partial \tau_{xy}}{\partial y} + F_x = 0 \\ \dfrac{\partial \tau_{xy}}{\partial x} + \dfrac{\partial \sigma_y}{\partial y} + F_y = 0 \end{cases}$$

where F_x, F_y are volume forces of external origin. These equations express the total x and y forces on a volume element of the material, but this force is not generally zero in the dynamic case. According to Newton's equations of motion we should now write

$$\begin{cases} \dfrac{\partial \sigma_x}{\partial x} + \dfrac{\partial \tau_{xy}}{\partial y} + F_x = \rho \dfrac{\partial^2 u}{\partial^2 t} \\ \dfrac{\partial \tau_{xy}}{\partial x} + \dfrac{\partial \sigma_y}{\partial y} + F_y = \rho \dfrac{\partial^2 v}{\partial^2 t} \end{cases}$$

where we have included the mass of the volume element and its acceleration along the axes. In the present chapter we shall assume the volume forces to be zero.

As we have already seen (p.159) the components of plane stress take the form (at constant T)

$$\begin{Bmatrix} \sigma_x \\ \sigma_y \end{Bmatrix} = \frac{E}{1-\mu^2} \begin{Bmatrix} \varepsilon_x + \mu \varepsilon_y \\ \mu \varepsilon_x + \varepsilon_y \end{Bmatrix} = E_2 \begin{Bmatrix} \dfrac{\partial u}{\partial x} + \mu \dfrac{\partial v}{\partial y} \\ \mu \dfrac{\partial u}{\partial x} + \dfrac{\partial v}{\partial y} \end{Bmatrix},$$

with the abbreviation $E_2 = E/(1-\mu^2)$, and

$$\tau_{xy} = G\gamma_{xy} = G\left(\frac{\partial v}{\partial x} + \frac{\partial u}{\partial y}\right).$$

If we substitute these expressions, the equations of motion become 2nd-order PDEs in the dependent variables u and v.

Complex Solution

For time-sinusoidal solutions we use the same complex form as for thermal waves (p.130)

$$\begin{Bmatrix} u \\ v \end{Bmatrix} = \begin{Bmatrix} u_0 \\ v_0 \end{Bmatrix} \exp(i\omega t) \equiv \begin{Bmatrix} u_r + iu_i \\ v_r + iv_i \end{Bmatrix} \exp(i\omega t).$$

If we introduce these expressions into the stresses, the exponential appears as a common factor. When these stresses are substituted in the equations of motion, the exponential factors on both sides cancel. The 2nd-order time derivative, however, produces the extra factor $(i\omega)^2 = -\omega^2$ on the right. In the case of thermal waves we separated the resulting complex equation into its real and imaginary parts. If we proceed in a similar way with the first equation of motion, we obtain

$$\frac{\partial}{\partial x}\left(E_2\left[\frac{\partial u_r}{\partial x} + \mu\frac{\partial v_r}{\partial y}\right]\right) + \frac{\partial}{\partial y}\left(G\left[\frac{\partial v_r}{\partial x} + \frac{\partial u_r}{\partial y}\right]\right) + \rho\omega^2 u_r = 0 \qquad \bullet$$

for the real part. The imaginary part becomes similar, with the subscript i replacing r.

The real part of the second equation of motion takes the form

$$\frac{\partial}{\partial x}\left(G\left[\frac{\partial v_r}{\partial x} + \frac{\partial u_r}{\partial y}\right]\right) + \frac{\partial}{\partial y}\left(E_2\left[\mu\frac{\partial u_r}{\partial x} + \frac{\partial v_r}{\partial y}\right]\right) + \rho\omega^2 v_r = 0 \qquad \bullet$$

and the imaginary part may again be written down just by changing the subscripts.

While the real and imaginary variables appeared mixed in the equation for a thermal wave, each of the above PDEs contains only real variables. This means that we may solve for the real parts u_r, v_r and the imaginary parts u_i, v_i separately. The two independent systems of PDEs are similar, but the boundary conditions may be different.

Forced Shear Vibration

Perhaps the simplest mode of vibration is that where one end of a steel beam is driven by a time-sinusoidal, transverse oscillation, the boundary conditions chosen being compatible with transverse displacement. If we choose to drive the oscillation at an arbitrary frequency, say ω=33000, a snapshot of the beam might look as shown below. The amplitude would generally not be larger than that applied at the left face.

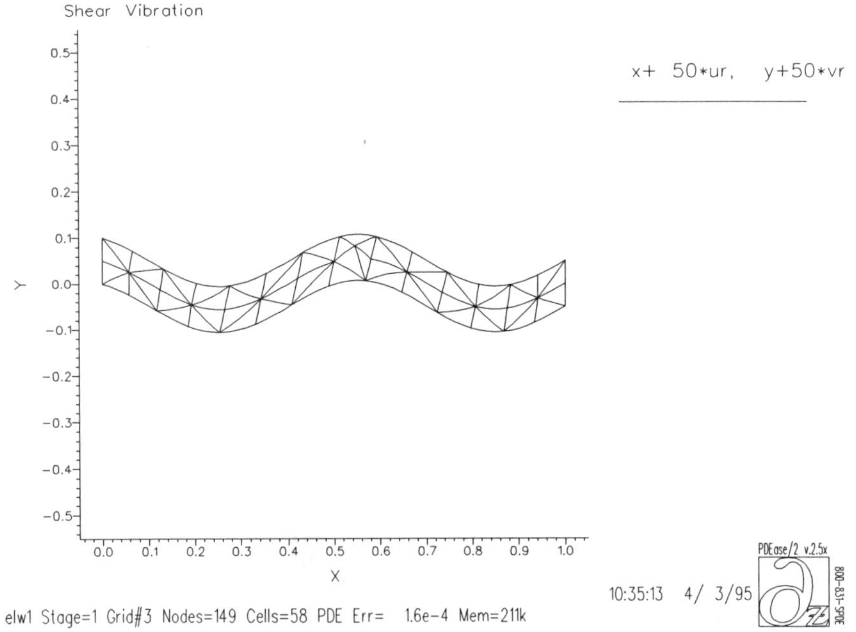

If we vary the angular frequency over a wide interval, however, we may find that the amplitude occasionally rises to very high values.

Initially we do not know much about the modes of motion; hence we use the *stages* device to scan over a range of ω to detect possible resonant states.

PDEase versions 2.5 and later permit us to write the PDEs of equilibrium in terms of stresses, and the program will look into the *definitions* segment to substitute expressions for strains and finally displacements.

We specify zero longitudinal displacement and zero transverse load on the horizontal boundaries. On the left end we impose transverse oscillation by *vr*= 1e-3, and the right end of the bar we fix by $ur = vr = 0$.

This problem could give rise to many figures, but we limit ourselves to one *elevation* plot per stage, showing the transverse displacement *vr* of the bar.

```
title                                                          { elw1.pde }
  'Shear Vibration'
select
  stages= 20
  errlim= 2e-4   nodelimit= 1000
  contours= 10   surfacegrid= 30   paint= off
variables
  ur  vr
definitions
  omega= 19000+ 100*stage                     { Scan angular frequency }
  L= 1    h= 0.1
  dens= 7.8e3      mu = 0.3       E = 200e9                      {Steel}
  c = E/(1-mu**2)      G = E/2/(1+mu)
  exr= dx(ur)          eyr= dy(vr)                { x-, y- and xy-strains }
  exyr= dx(vr)+ dy(ur)
  sxr= c*(exr+ mu*eyr)   syr= c*(mu*exr+ eyr)                { Stresses }
  sxyr= G*exyr
initial values
equations
  dx( sxr)+ dy( sxyr)+ dens*omega**2*ur= 0
  dx( sxyr)+ dy( syr)+ dens*omega**2*vr= 0
boundaries
          region 1
  start (0,-h/2)
  value(ur)= 0    load(vr)= 0    line to (L,-h/2)
```

```
  value(ur)= 0   value(vr)= 0    line to (L,h/2)                    { Fixed }
  value(ur)= 0    load(vr)= 0    line to (0,h/2)
  value(ur)= 0   value(vr)= 1e-3  line to finish                    { Driven }
monitors
  elevation(vr) from (0,0) to (L,0)
plots
  elevation( vr) from (0,0) to (L,0)
end
```

As already explained in the case of temperature waves, a plot of v_r provides a snapshot of the bar at a certain point in time, i.e. at the moment when the applied displacement reaches it maximum positive value (1e-3). The following figure presents the transverse displacement *vr* at a frequency close to resonance. Evidently, the maximum displacement is as large as a hundred times the input value.

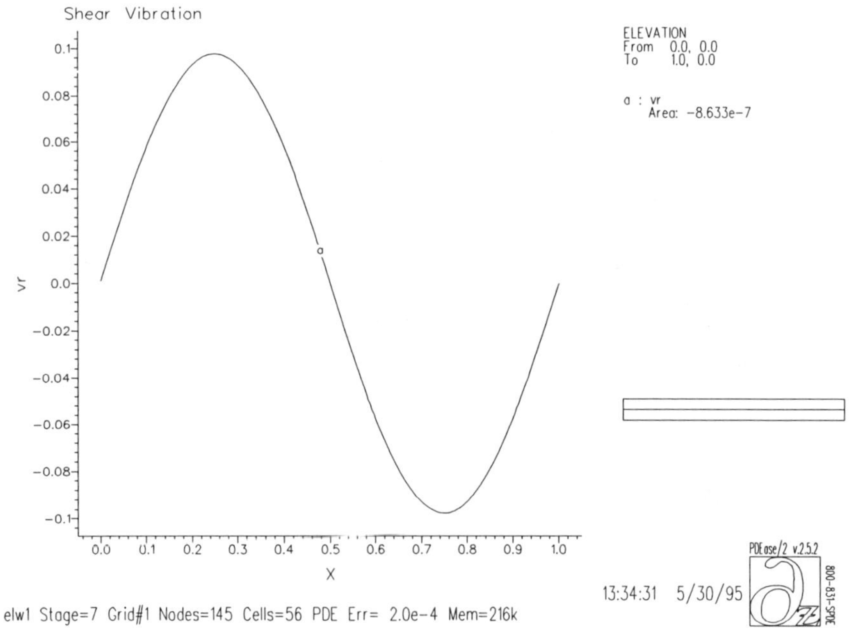

Longitudinal Vibration

Alternatively, we could excite a beam by a time-sinusoidal axial displacement at the left end. The oscillations would then be mainly longitudinal, but in view of Hooke's law the transverse dimensions would also oscillate. The motion along the z-axis strictly leads us into a three-dimensional problem, but we assume that the thickness perpendicular to the figure is small compared to a typical wavelength, so we may take the z-strain to be independent of z.

Let us now modify the preceding descriptor ELW1.PDE as follows. We investigate the range close to the resonance by *stages*. The horizontal boundaries remain unloaded, but the right end is fixed. We could have fixed the vertical position by a *point value* command, but this is not necessary here. We again drive the vibration from the left end, in this case by oscillating it longitudinally while applying no vertical load. Since the motion is now mainly axial, we add a *contour* plot to view the extreme points of *ur*.

```
                                                        { elw2.pde }
  omega= 47460+ 20*stage
  ...
        region 1
  start (0,-h/2)
  load(ur)= 0      load(vr)= 0     line to (L,-h/2)
  value(ur)= 0     load(vr)= 0     line to (L,h/2)          { Fixed }
  load(ur)= 0      load(vr)= 0     line to (0,h/2)
  value(ur)= 1e-3  load(vr)= 0     line to finish           { Driven }
monitors
  elevation(ur) from (0,0) to (L,0)
plots
  elevation(ur) from (0,0) to (L,0)
  contour(ur)
end
```

The figure below shows the situation near resonance. The plot of *ur* shows two maxima and one minimum, and searching for contour *j* we may identify the node lines, corresponding to zero displacement. Multiplying *ur* by $\cos(\omega t)$ we obtain the instantaneous value of the real part of the displacement. The amplification by resonance is about 470, which means that the oscillation driving the wave from the left is too weak to be visible on the plot. We could also solve for the imaginary part by similar PDEs, with the appropriate boundary conditions.

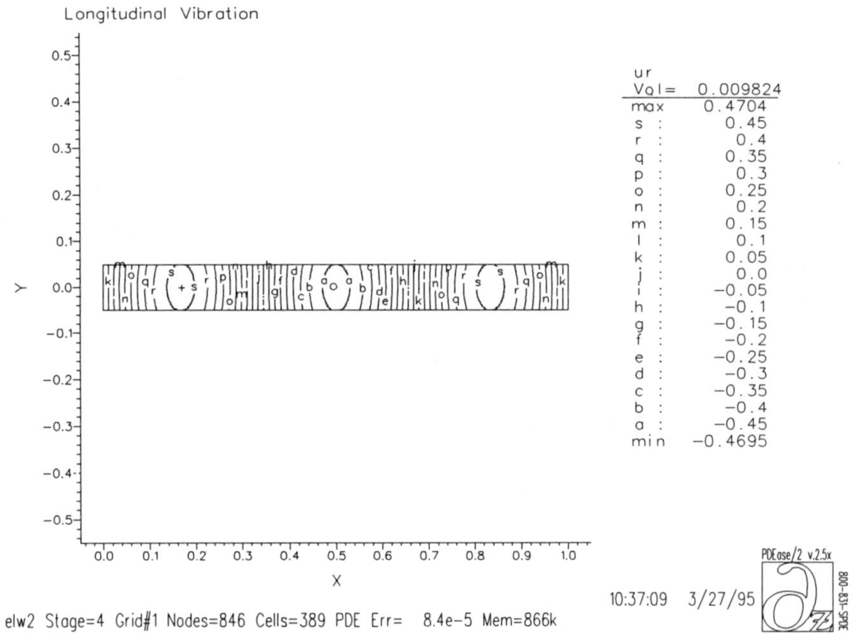

Bending Vibration

We shall now study the deformation of a steel bar as it is subjected to a time-sinusoidal, transverse displacement at the left end, the right end being completely free. The following descriptor gives a rough scan over the region of a resonance. We limit ourselves to one *elevation* plot per stage, showing the transverse displacement of the bar.

```
title                                                           { elw3.pde }
  'Bending Vibration'
select
  stages= 20
  errlim= 3e-5          nodelimit= 500
  contours= 10          surfacegrid= 30         paint= off
variables
  ur         vr
definitions
  omega= 2660+ 100*stage
  L= 1               h= 0.1
  dens= 7.8e3           mu = 0.3            E = 200e9                {Steel}
  c = E/(1-mu**2)       G = E/2/(1+mu)
  exr= dx(ur)        eyr= dy(vr)                  { x-, y- and xy-strains }
  exyr= dx(vr)+ dy(ur)
  sxr= c*(exr+ mu*eyr)   syr= c*(mu*exr+ eyr)                 { Stresses }
  sxyr= G*exyr
initial values
equations
  dx( sxr)+ dy( sxyr)+ dens*omega**2*ur= 0
  dx( sxyr)+ dy( syr)+ dens*omega**2*vr= 0
boundaries
          region 1
  start (0,-h/2)
  load(ur)= 0      load(vr)= 0        line to (L,-h/2)
  load(ur)= 0      load(vr)= 0        line to (L,h/2)             { Free end }
  load(ur)= 0      load(vr)= 0        line to (0,h/2)
  value(ur)= 0     value(vr)= 1e-3    line to finish                { Driven }
monitors
  elevation(vr) from (0,0) to (L,0)
plots
  elevation(vr) from (0,0) to (L,0)
end
```

We note that at the lower frequencies the left half of the bar swings in the same direction as the left end. In the region of the resonance the left half begins to swing in the opposite direction, and continues to do so. This phase shift of π radians is typical of resonance.

Even the first plot appearing on the screen exhibits higher displacement in certain regions than that applied at the left end. This amplification will increase and eventually become quite dramatic (150-200) close to resonance, as illustrated by the figure below.

The solution for the real parts of *u* and *v* gives us a good impression of the motion, although it does not seem to tell the whole story. If we wish, we could solve separately for the imaginary parts, but since the *value* as well as the *natural* boundary conditions are zero in this case, the imaginary parts would also be zero.

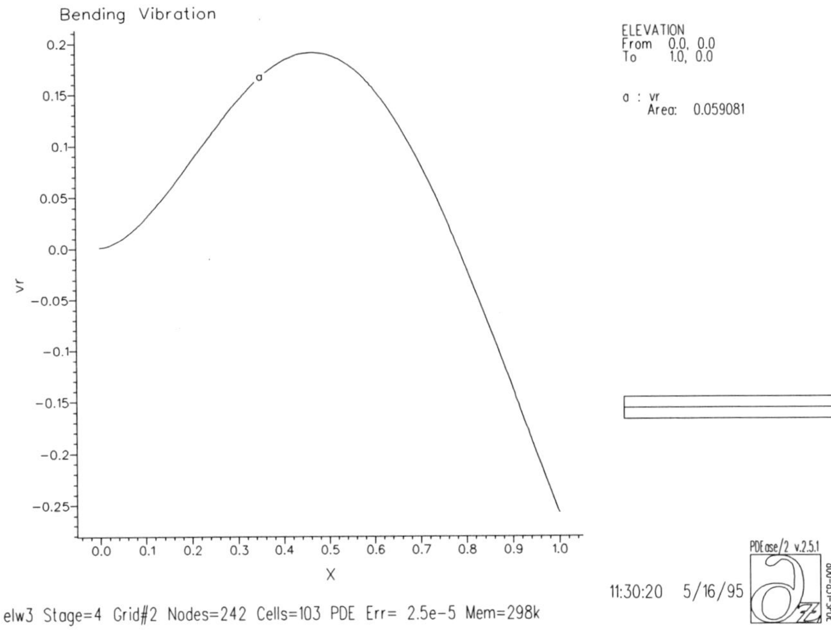

We shall now modify the boundary conditions such that we fix the *y*-displacement at the right end, leaving both ends free to wobble. We also include a *grid* plot of the deformation. The descriptor file required is similar to the preceding ones and only requires the following modifications.

```
                                                    { elw4.pde }
...
omega= 5400+ 5*stage
...
```

```
region 1
  start (0,-h/2)
  load(ur)= 0    load(vr)= 0    line to (L,-h/2)
  load(ur)= 0   value(vr)= 0    line to (L,h/2)                        { Fixed }
  load(ur)= 0    load(vr)= 0    line to (0,h/2)
  load(ur)= 0   value(vr)= 1e-3  line to finish                       { Driven }
  ...
plots
  elevation(vr) from (0,0) to (L,0)
  grid(x+0.5*ur, y+0.5*vr)
end
```

The figure below gives a snapshot of the deformation at a frequency close to resonance. The end faces of the bar are at an angle to the *y*-axis, in accord with our boundary conditions.

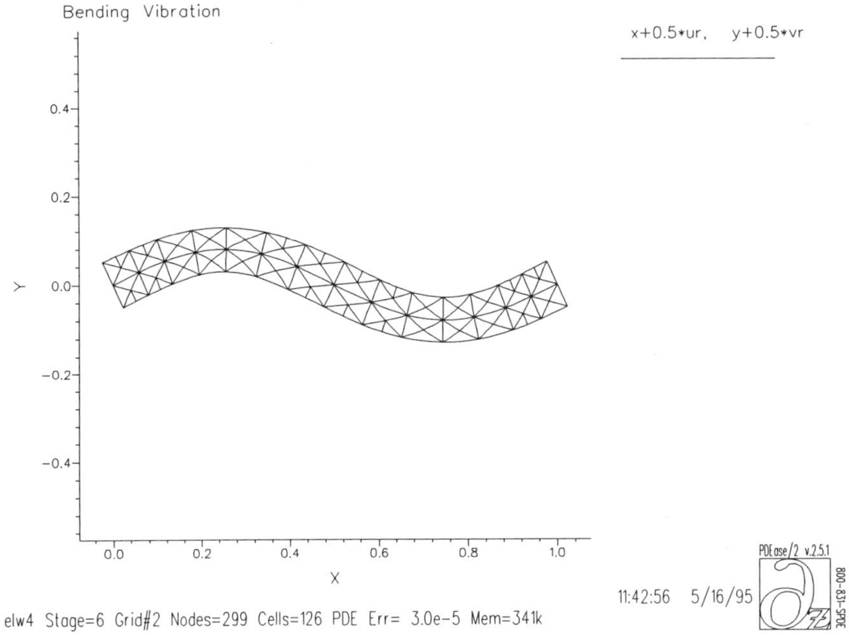

Exploring several resonances we see that forced bending vibration of a bar exhibits resonances at non-uniform frequency intervals. Shear and longitudinal vibration, on the other hand, have easily predictable resonant frequencies f in view of the relation $f\lambda = c$, where c is the appropriate velocity of sound.

Exercises

❑ Find the lowest-frequency resonance for shear vibration (ELW1.PDE). Add *contour* and *surface* plots of *vr*.
❑ Modify ELW2.PDE to find all longitudinal resonant modes at lower frequencies.
❑ Change ELW2.PDE to let the right end move freely in the *x* direction. Find the simplest resonant mode and add a *vector* plot of (*ur*,*vr*) as well as any elevation plots you need to understand the mode of oscillation.
❑ Find one lower and one higher resonance by modifying ELW3.PDE.
❑ Investigate whether the solution to ELW1.PDE is changed when using *load*(*ur*)=0 and *load*(*vr*)=0 on both horizontal faces of the bar.

Eigenstates in 2D

In the two preceding chapters, we noted that mechanical oscillations may reach orders-of-magnitude larger amplitudes inside a domain than that imposed at the boundary. This phenomenon, known as a resonance, occurs at certain discrete frequencies. In the absence of energy losses, such resonant oscillation may persists even if the excitation at the boundary is suppressed. The various modes of free oscillation under zero boundary conditions may also be called *eigenstates* (intrinsic states). In this chapter we shall study this type of vibration in plates under plane stress, and by two different methods.

Frequency Scan

We shall now study a rectangular steel plate, fixed at the boundaries except over a short segment where we apply a sinusoidal force. By the use of *stages* we scan the frequency *omega* over the range of the simplest resonance. The definition section includes an expression for the elastic energy density *energy_d*.

```
title
  'Vibrating Rectangular Plate'                              { elw11.pde }
select
  stages= 20
  errlim= 1e-5          nodelimit= 1000
  contours= 10          surfacegrid= 30         paint= off
variables
  ur        vr
definitions
  omega= 20200+ 10*stage
  L= 1    h= 0.8
  dens= 7.8e3    mu = 0.3       E = 200e9                       {Steel}
  c = E/(1-mu**2)     G = E/2/(1+mu)
```

```
  dabs= sqrt( ur**2+ vr**2)
  exr= dx(ur)          eyr= dy(vr)                         { x-, y- and xy-strains }
  exyr= dx(vr)+ dy(ur)
  sxr= c*(exr+ mu*eyr)   syr= c*(mu*exr+ eyr)                          { Stresses }
  sxyr= G*exyr
  energy_d= 0.5*(sxr*exr+ syr*eyr+ sxyr*exyr)                    { Energy density }
initial values
equations
  dx( sxr)+ dy( sxyr)+ dens*omega**2*ur= 0
  dx( sxyr)+ dy( syr)+ dens*omega**2*vr= 0
boundaries
          region 1
  start (0,-h/2)
  value(ur)= 0    value(vr)= 0                                            { Fixed }
  line to (L,-h/2)   to (L,h/2)  to (0,h/2)  to (0,-h/4)
  value(ur)= 1e-3    value(vr)= 0  line to finish                         { Driven }
plots
  contour( ur)      contour( vr)
  vector( ur, vr)
  contour( energy_d)
end
```

The following figures show the real parts of the *x*- and *y*-displacements for a frequency close to resonance. It may be surprising at first to find that the two patterns are so different. Inspecting the *vector* plot, also shown below, we realize that the motion is a simple left-right oscillation, entirely compatible with the *contour* plots. The *vector* plot presents a snapshot of the displacement, taken at the moment where the driving oscillation passes through its maximum. Half a period later, the displacement vectors will simply be reversed.

The energy density plot (not shown) will also prove useful to distinguish the various modes of vibration. Here, the energy density evidently is highest close to the left and right midpoints of the boundary. The maximum at the left is caused mostly by stretching, the one at the right by compression.

By analogy we expect a corresponding mode of vibration with displacements mostly in the *y* direction. Since the height is smaller than the width, this resonance should occur at a higher frequency.

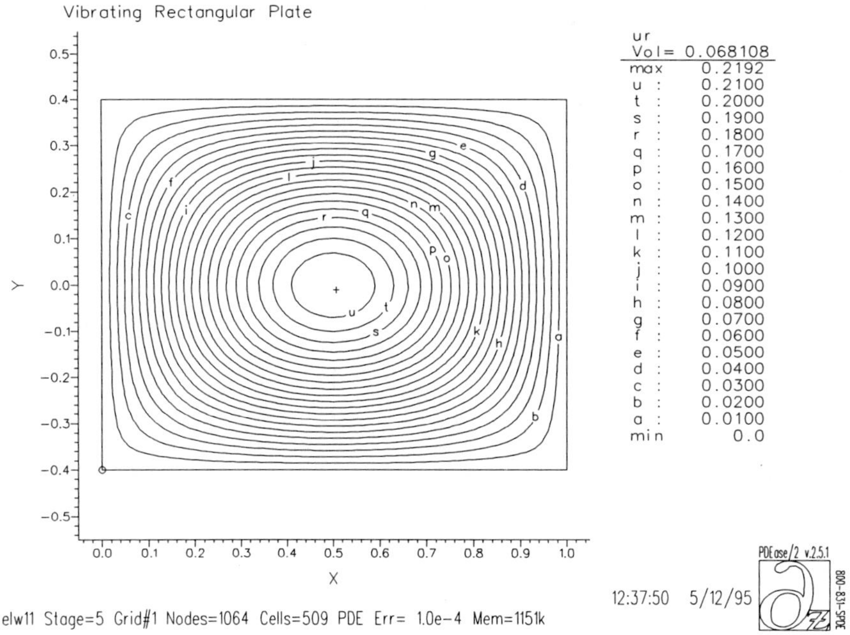
Vibrating Rectangular Plate
ur
Vol= 0.068108
max 0.2192
u : 0.2100
t : 0.2000
s : 0.1900
r : 0.1800
q : 0.1700
p : 0.1600
o : 0.1500
n : 0.1400
m : 0.1300
l : 0.1200
k : 0.1100
j : 0.1000
i : 0.0900
h : 0.0800
g : 0.0700
f : 0.0600
e : 0.0500
d : 0.0400
c : 0.0300
b : 0.0200
a : 0.0100
min 0.0
X
Y
12:37:50 5/12/95
PDEase/2 v.2.5.1
800-831-SPDE
elw11 Stage=5 Grid#1 Nodes=1064 Cells=509 PDE Err= 1.0e-4 Mem=1151k

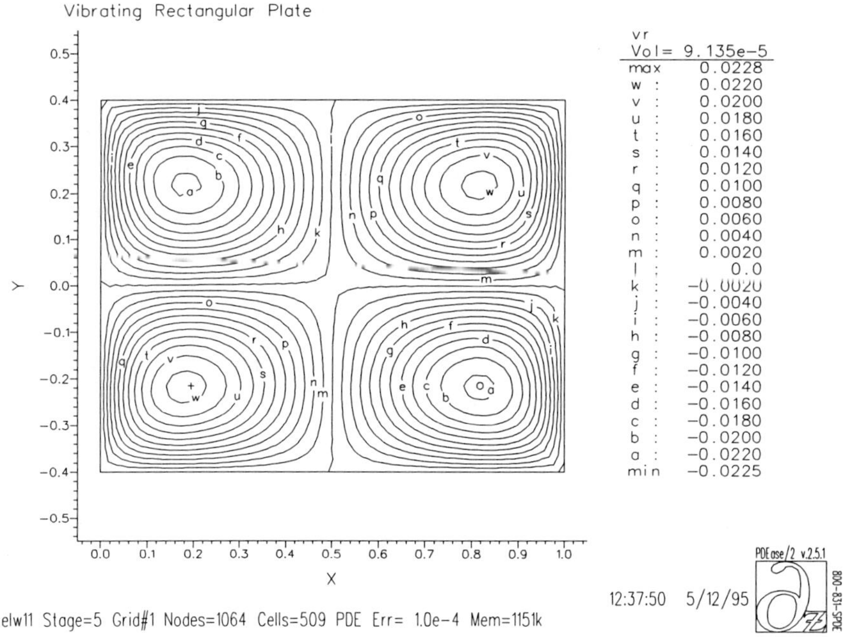
Vibrating Rectangular Plate
vr
Vol= 9.135e-5
max 0.0228
w : 0.0220
v : 0.0200
u : 0.0180
t : 0.0160
s : 0.0140
r : 0.0120
q : 0.0100
p : 0.0080
o : 0.0060
n : 0.0040
m : 0.0020
l : 0.0
k : -0.0020
j : -0.0040
i : -0.0060
h : -0.0080
g : -0.0100
f : -0.0120
e : -0.0140
d : -0.0160
c : -0.0180
b : -0.0200
a : -0.0220
min -0.0225
X
Y
12:37:50 5/12/95
PDEase/2 v.2.5.1
800-831-SPDE
elw11 Stage=5 Grid#1 Nodes=1064 Cells=509 PDE Err= 1.0e-4 Mem=1151k

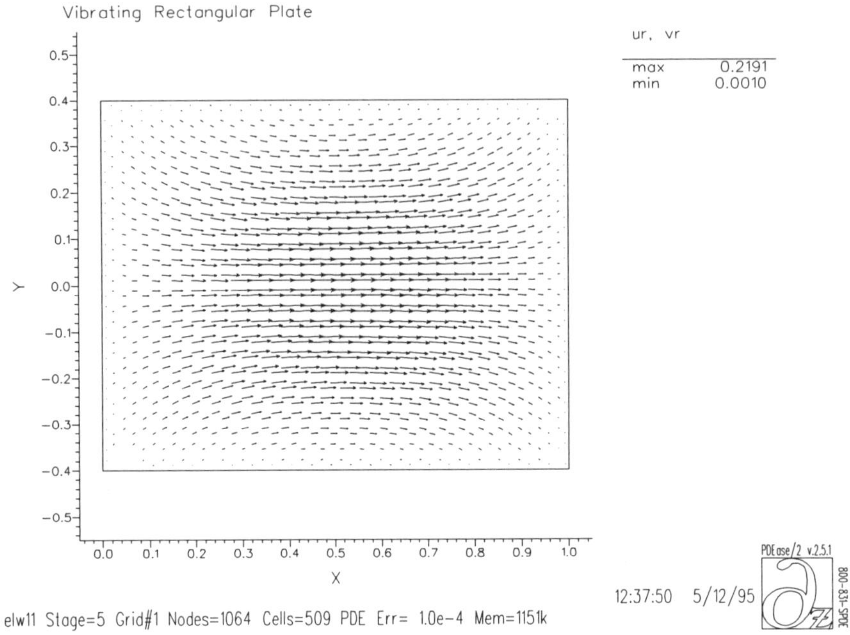

Finding Eigenstates

From the above example and those in the preceding chapter we know how to detect and analyze various modes of vibration. Frequency scanning is, however, rather inconvenient and time-consuming. Fortunately, PDEase allows us to calculate a given number of low-frequency eigenstates in a single run, after a slight modification of the descriptor, illustrated below.

First we specify the number of eigenstates (*modes*) required in the *select* segment. Then we modify the PDEs by substituting the reserved variable *lambda* for $\rho\omega^2$. We finally specify zero values on the entire boundary curve. The program responds by displaying plots for each of the modes. In addition, it creates a text file with the extension .EIG, containing a list of the eigenvalues *lambda*. From these values we may then calculate the resonance frequencies.

```
title                                                        { eig11.pde }
  'Vibrating Rectangular Plate'                              { Eigenstates }
select
  MODES= 8
  errlim= 1e-2          nodelimit= 500
  contours= 10          surfacegrid= 30         paint= off
variables
  ur  vr
definitions
  L= 1    h= 0.8
  dens= 7.8e3           mu = 0.3            E = 200e9              {Steel}
  c = E/(1-mu**2)       G = E/2/(1+mu)
  dabs= sqrt( ur**2+ vr**2)
  exr= dx(ur)           eyr= dy(vr)                 { x-, y- and xy-strains }
  exyr= dx(vr)+ dy(ur)
  sxr= c*(exr+ mu*eyr)        syr= c*(mu*exr+ eyr)              { Stresses }
  sxyr= G*exyr
  energy_d= 0.5*(sxr*exr+ syr*eyr+ sxyr*exyr)           { Energy density }
initial values
equations                                          { Eigenvalue = LAMBDA }
  dx( sxr)+ dy( sxyr)+ LAMBDA*ur= 0
  dx( sxyr)+ dy( syr)+ LAMBDA*vr= 0
boundaries
          region 1
  start (0,-h/2)
  value(ur)= 0   value(vr)= 0
  line to (L,-h/2) to (L,h/2) to (0,h/2)  to finish
plots
  contour(ur)       contour(vr)
  vector( ur, vr)
  contour( energy_d)
end
```

The following two figures show the modes of displacement for eigenstates 3 and 8. The first one suggests a rotational oscillation around the center of the plate. The second one is analogous but involves four twisting centers.

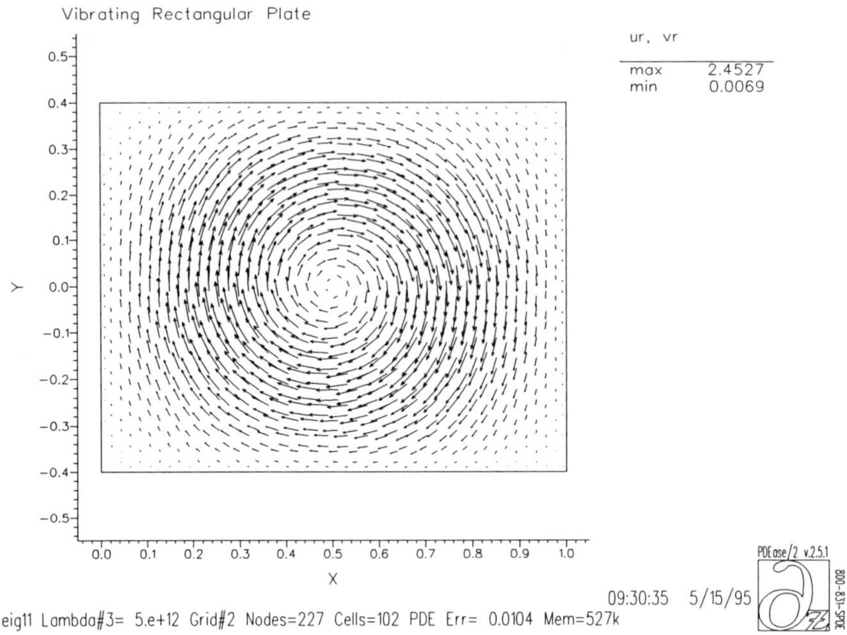

Vibrating Rectangular Plate
ur, vr
max 2.4527
min 0.0069
Y
X
0.5
0.4
0.3
0.2
0.1
0.0
−0.1
−0.2
−0.3
−0.4
−0.5
0.0 0.1 0.2 0.3 0.4 0.5 0.6 0.7 0.8 0.9 1.0
PDEase/2 v.2.5.1
800-831-SPDE
09:30:35 5/15/95
eig11 Lambda#3= 5.e+12 Grid#2 Nodes=227 Cells=102 PDE Err= 0.0104 Mem=527k

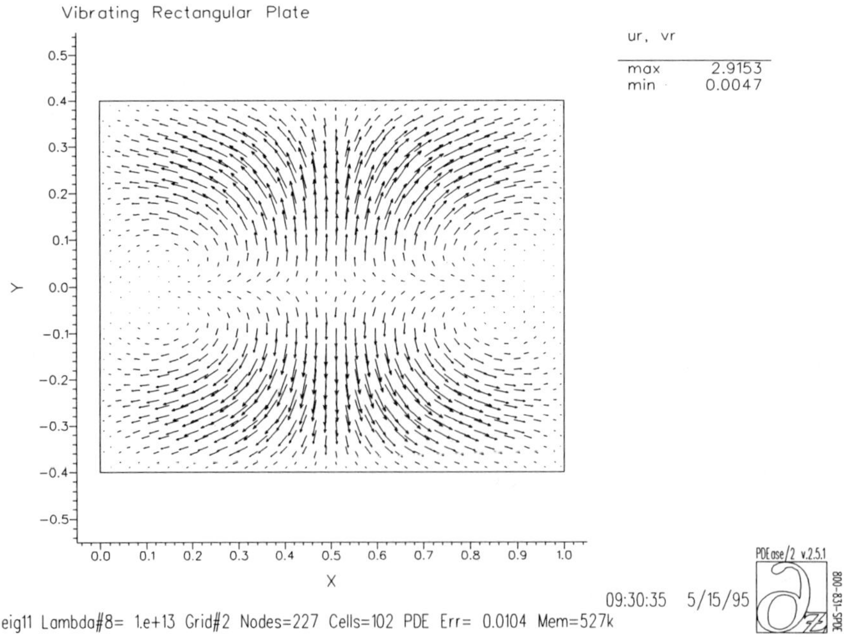

Vibrating Rectangular Plate
ur, vr
max 2.9153
min 0.0047
Y
X
0.5
0.4
0.3
0.2
0.1
0.0
−0.1
−0.2
−0.3
−0.4
−0.5
0.0 0.1 0.2 0.3 0.4 0.5 0.6 0.7 0.8 0.9 1.0
PDEase/2 v.2.5.1
800-831-SPDE
09:30:35 5/15/95
eig11 Lambda#8= 1.e+13 Grid#2 Nodes=227 Cells=102 PDE Err= 0.0104 Mem=527k

Plate With Free Boundaries

Stationary oscillations may also occur in cases of one or more free boundaries. To convert the preceding descriptor EIG11.PDE to such conditions we only need to change the boundary conditions as suggested below. Since even the boundaries oscillate in this case, we add a *grid* plot of the displacement.

```
          region 1                                        { eig12.pde }
  start (0,-h/2)
  load( ur)= 0      load( vr)= 0  line to (L,-h/2)            { Free }
  value( ur)= 0     value( vr)= 0  line to (L,h/2)           { Fixed }
  load( ur)= 0      load( vr)= 0  line to (0,h/2)
  value( ur)= 0     value( vr)= 0  line to finish
plots
  contour( ur)      contour( vr)
  vector( ur, vr)
  grid( x+0.05*ur, y+0.05*vr)
  contour( energy_d)
end
```

The following two figures illustrate the mode of vibration in the 4th eigenstate. The *vector* plot indicates that the motion may be described in terms of rotations around centers close to the corners, the sense of rotation changing sign twice per period. The net result for the horizontal boundaries is a constriction of the plate, followed by swelling during the next half-period. Since the magnitude of the displacement must approach zero at the left and right boundaries, the corners remain right-angled.

Transforming the above eigenvalue descriptor according to ELW11.PDE, we may easily verify that forced vibration over part of the boundary leads to very large amplitudes at the frequencies computed from the eigenvalues in the text file EIG12.EIG

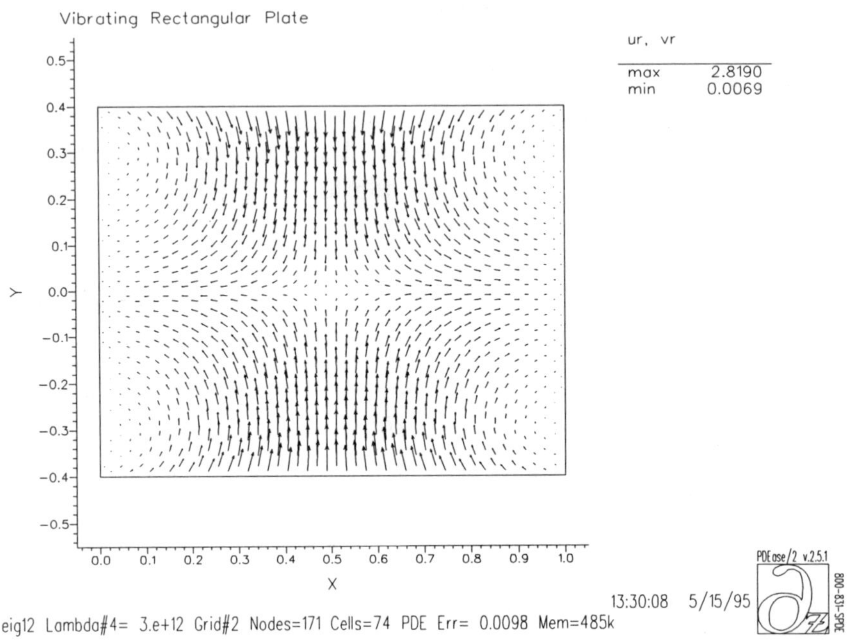
Vibrating Rectangular Plate
ur, vr
max 2.8190
min 0.0069
X
Y
13:30:08 5/15/95
eig12 Lambda#4= 3.e+12 Grid#2 Nodes=171 Cells=74 PDE Err= 0.0098 Mem=485k
PDEase/2 v.2.5.1
800-831-SPDE

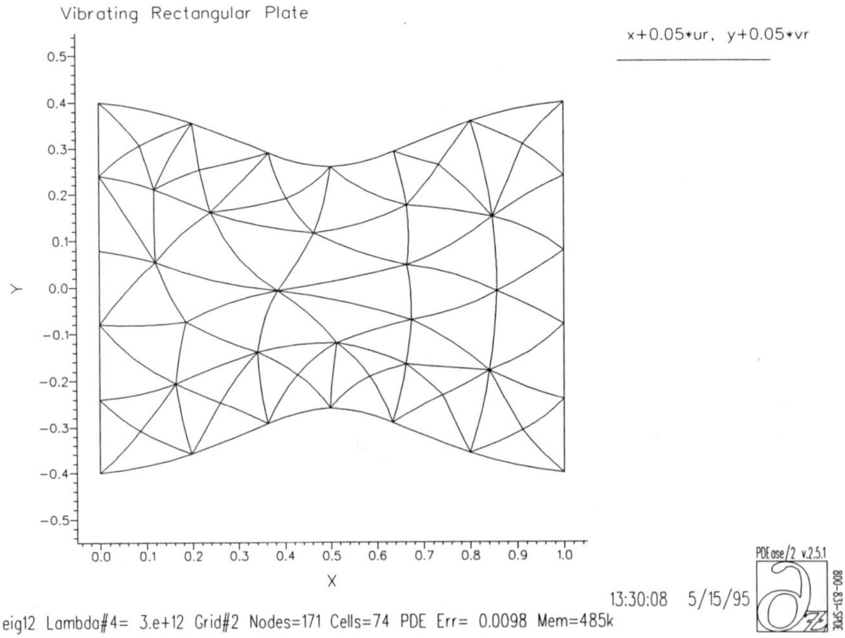
Vibrating Rectangular Plate
x+0.05*ur, y+0.05*vr
X
Y
13:30:08 5/15/95
eig12 Lambda#4= 3.e+12 Grid#2 Nodes=171 Cells=74 PDE Err= 0.0098 Mem=485k
PDEase/2 v.2.5.1
800-831-SPDE

Exercises

❑ Scan the frequency to find a vibrational mode with displacements perpendicular to those on p.314.

❑ Show that the 3rd eigenstate given in EIG11.EIG in fact leads to resonance under forced vibration over part of the boundary.

❑ Excite the 4th eigenstate in EIG12.EIG by scanning and notice what happens as the frequency passes over the amplitude peak.

❑ Modify the preceding problem, replacing the displacement on the boundary by an oscillating force.

❑ Study the eight first eigenstates of in-plane vibration for the plate shown below. The left boundary is completely fixed and all the others are free.

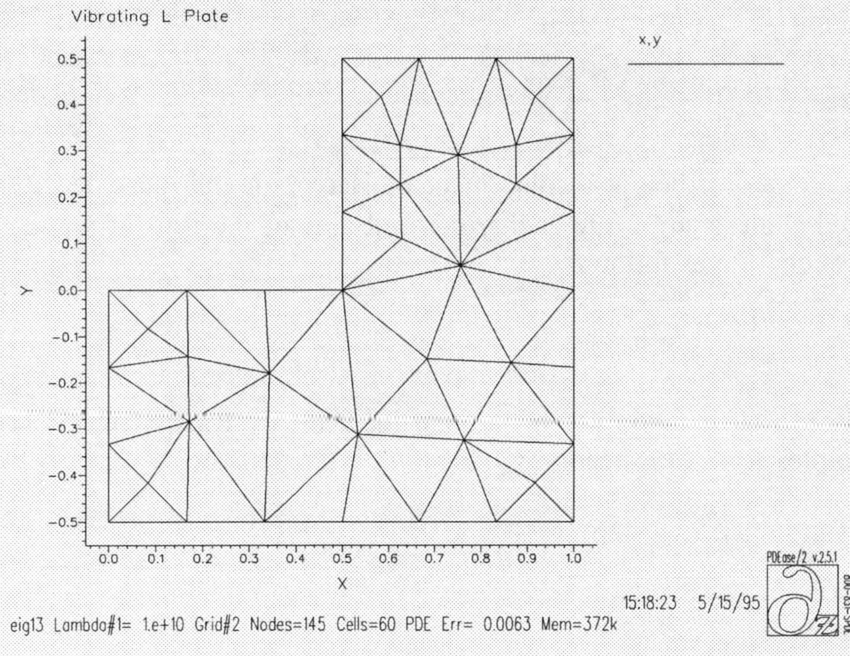

Eigenstates of a Membrane

In previous chapters we dealt only with oscillations in the *xy* plane. It is possible, however, to treat vibrations perpendicular to that plane, provided that the governing PDEs do not contain derivatives with respect to *z*. Such is the case of a liquid film, attached to a frame in the *xy* plane. The surface energy is accurately proportional to the area of the film, which means that the surface tension *T* lies in *xy* plane and has the same magnitude everywhere. This remains true for any fictitious cut made across the film.

For a vibrating liquid film the restoring forces in the *z* direction are proportional to the curvatures[1p438], which leads to the equation of motion

$$\rho d\frac{\partial^2 w}{\partial t^2} = T\left(\frac{\partial^2 w}{\partial x^2} + \frac{\partial^2 w}{\partial y^2}\right)$$

for the perpendicular displacement w. Here ρ is the mass density, d the thickness and T the energy per unit area of the film. If we recast this equation as follows

$$\nabla^2 w - \frac{\rho d}{T}\frac{\partial^2 w}{\partial t^2} = 0$$

we see that it is similar to the one for acoustic waves (p.285). The complex form thus becomes

$$\begin{cases} \nabla^2 w_r + \dfrac{\rho d\omega^2}{T} w_r = 0 \\ \nabla^2 w_i + \dfrac{\rho d\omega^2}{T} w_i = 0 \end{cases}$$

and the eigenvalue form

$$\nabla^2 w_r + \lambda w_r = 0$$ •

The above PDE also describes the vibration of a drumhead, although only approximately. In that case, the membrane may have a thickness depending on the coordinates x and y, and the elastic properties may even be anisotropic.

Drumheads have the advantage that node curves may be verified by a simple experiment. If you sprinkle a suitable non-sticking powder over the membrane and drive it by a small loudspeaker, the powder will come to rest at loci of vanishing amplitude.

Circular Drumhead

Let us first find the eigenstates of a circular membrane. With the error limit recommended below the run may take an hour or two on a slow PC, but *errlim*=0.1 is sufficient to render the main features of the modes. The eigenvalues *lambda* are given on the plots but may also be listed by printing the file DRUM1.EIG. Taking the square roots of these values we obtain relative values of the frequency.

```
title
  'Eigenstates of a Drumhead'                                    { drum1.pde }
select
  modes= 16
  errlim=0.01             nodelimit= 500
  contours= 10            surfacegrid= 30          paint= off
variables
  w
definitions
  r0= 0.1
equations
  del2( w)+ lambda*w= 0
boundaries
            region 1
  value(w) = 0
  start(r0,0) arc to (0,r0) to (-r0,0) to (0,-r0) to finish
monitors
  contour( w)
plots
  contour( w)         surface( w)
end
```

The resulting plots contain some duplicates associated with a given frequency. The node patterns of such pairs are identical but have different angular orientation. The figures below show the lowest states, in the order of increasing frequency.

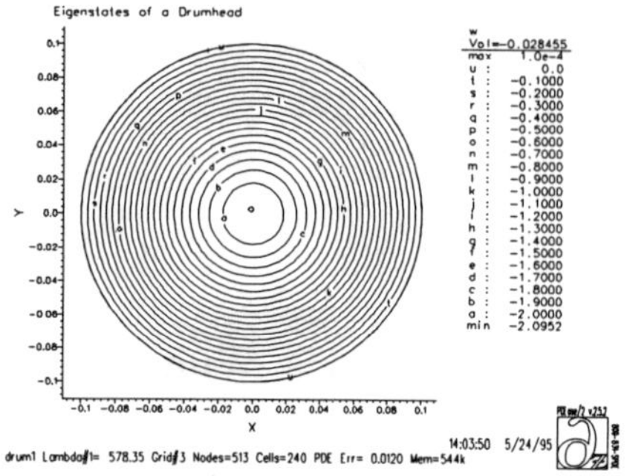

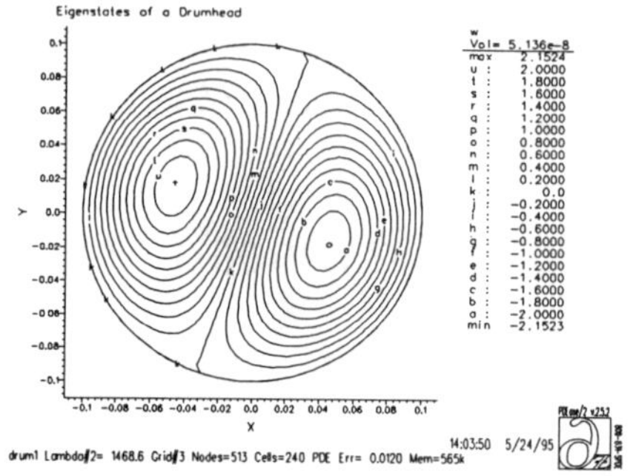

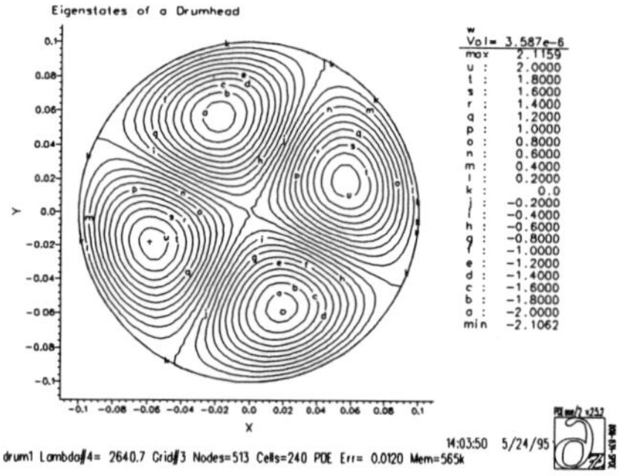

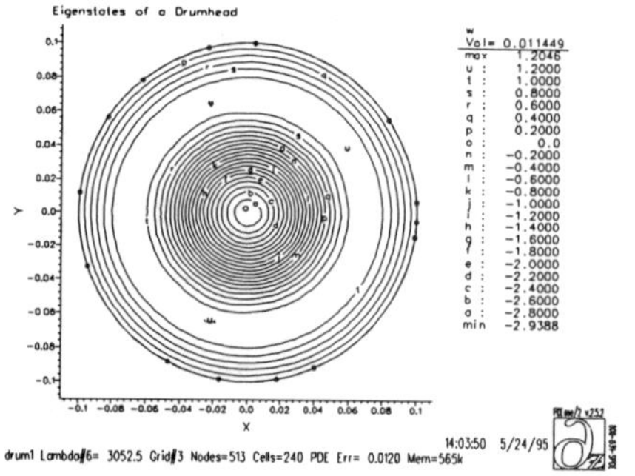

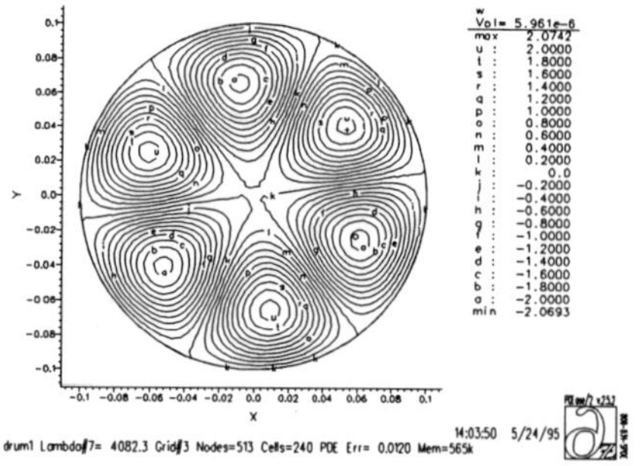

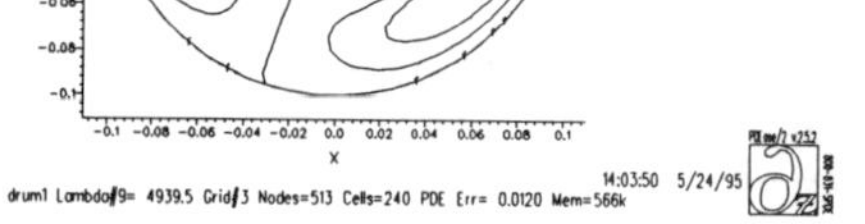

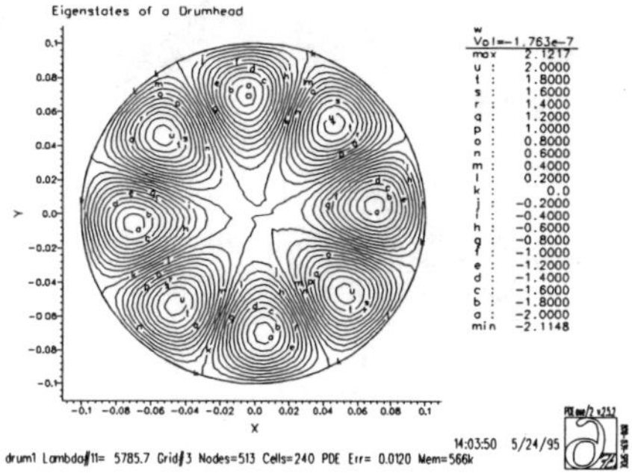

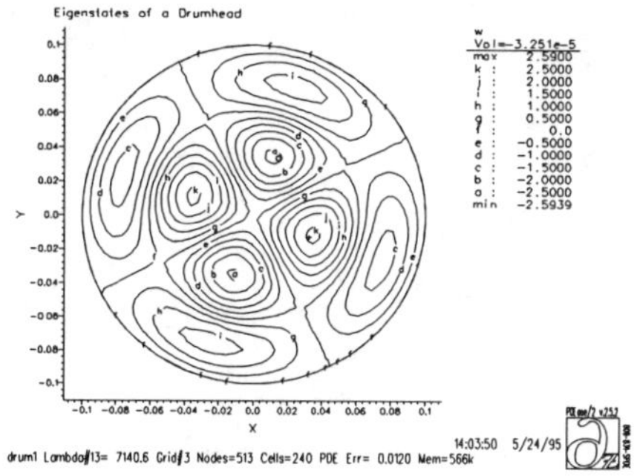

Triangular Drumhead

This example concerns a drumhead of unusual shape, i.e. an isosceles triangle. Only a few modifications of the preceding descriptor are needed, as shown by the list below.

```
...                                                  { drum2.pde }
modes= 8
...
L= 0.3
...
            region 1
value(w) = 0
start(0,0) line to (L,0) to (0,L) to finish
...
```

Four of the resulting plots are displayed below.

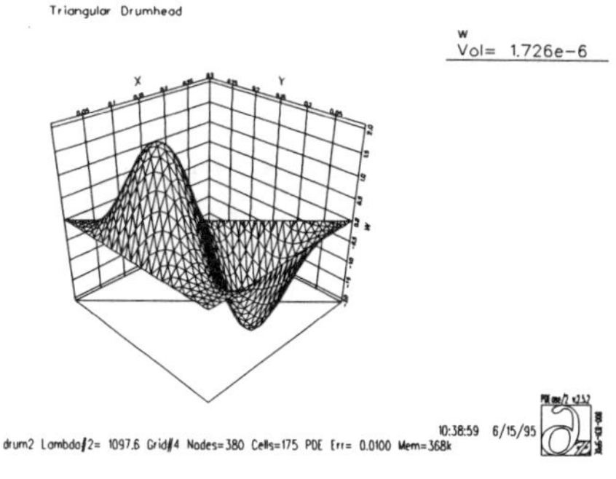

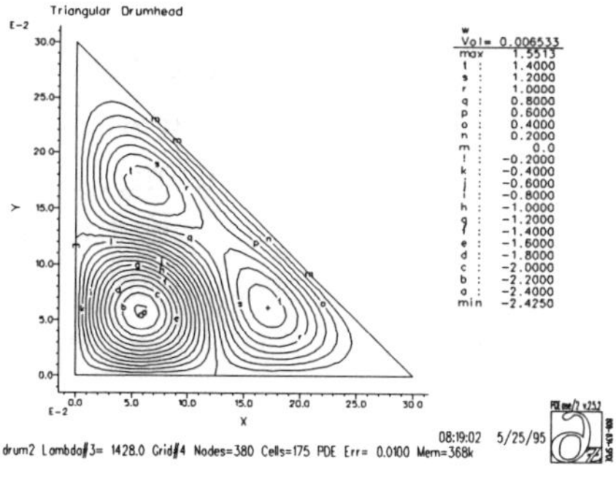

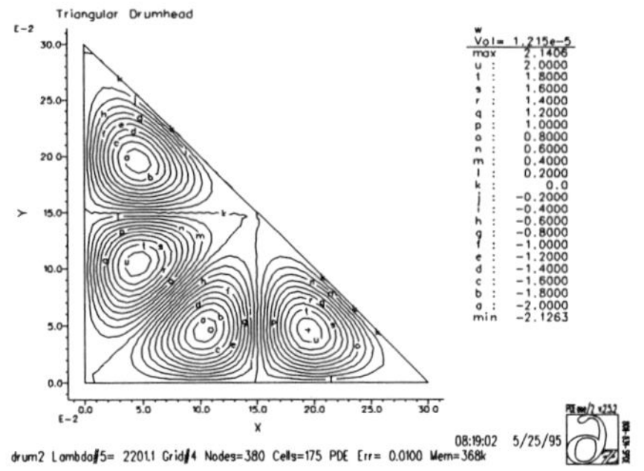

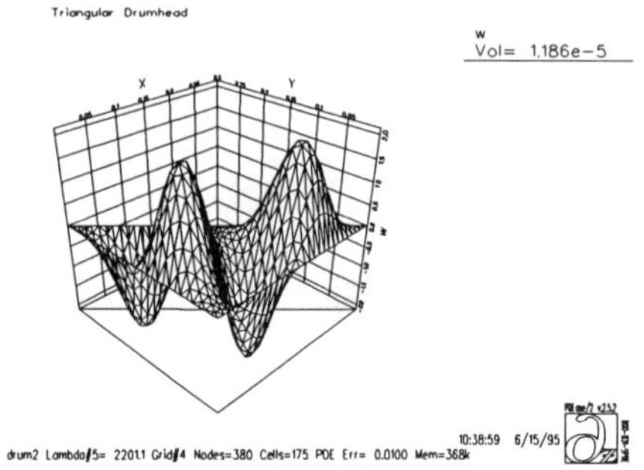

The symmetry of this example does not produce pairs of plots corresponding to the same frequency, as is evident from the file DRUM2.EIG. The existing mirror symmetry, however, is clearly reflected in the mode patterns.

Fancy Drumhead

Our next example involves a drum of curious cross-sectional shape, the boundary consisting of two arcs meeting at acute angles. The changes required are few, as seen below.

```
...                                                { drum4.pde }
        region 1
value(w) = 0
start(-L,0)  arc to (0,-L/2) to (L,0)
        arc to (0,L/2) to finish
...
```

The two figures on the next page give examples of the vibrational modes encountered in the solution of this problem. It is interesting to note, when browsing through the other plots, that two *y*-directed node lines (#3) appear before the first *x*-directed node line (#4). In addition, another state with y-directed nodes (#5) occurs before we observe a state with one horizontal and one vertical node line (#6).

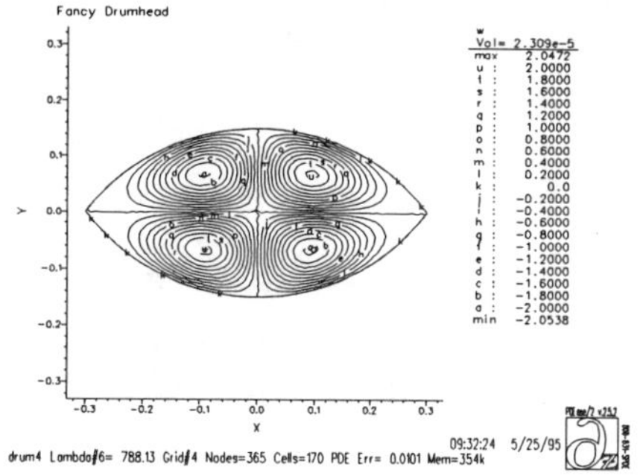

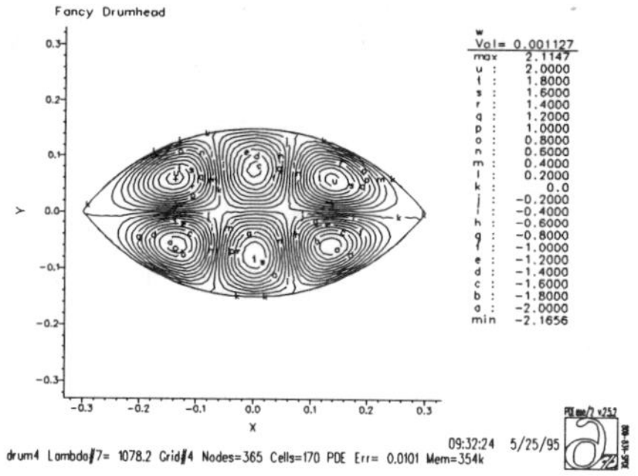

In the above examples the eigenvalue parameter was defined by the constant expression $\lambda = \rho d\omega^2 / T$. With a drumhead it is conceivable, however, that the density ρ, the thickness d or the surface tension T could be functions of (x,y). In such a case the eigenvalue equation should be written

$$\nabla^2 w_r + \lambda g(x,y) w_r = 0,$$

with a function $g(x,y)$ containing the variable factors. PDEase is able to handle this situation as well.

Exercises

❑ Change the triangular drumhead into an asymmetric one by letting the horizontal side extend to $1.5L$.

❑ Find the eigenstates of a membrane shaped as a regular pentagon. Compare the solutions to those of a circular membrane.

❑ Let $\rho d / T$ in the problem DRUM4.PDE be variable and equal to $1 + 3x$. Find the 8 lowest eigenvalues and their eigenfunctions.

Stress Propagation

We have studied some cases of sinusoidal oscillation in elastic objects. In the present chapter we shall consider transient phenomena, such as the propagation of a single stress pulse along a beam.

PDEs for Transient Stress

We are already familiar with the equations of motion required for the new task (p.300), i.e.

$$\begin{cases} \dfrac{\partial \sigma_x}{\partial x} + \dfrac{\partial \tau_{xy}}{\partial y} + F_x = \rho \dfrac{\partial^2 u}{\partial^2 t} \\ \dfrac{\partial \tau_{xy}}{\partial x} + \dfrac{\partial \sigma_y}{\partial y} + F_y = \rho \dfrac{\partial^2 v}{\partial^2 t} \end{cases}$$

We still neglect the terms F_x, F_y, corresponding to gravity forces, etc.

These PDEs are essentially different from those we used for transient heat flow (p.139), in the sense that the derivatives with respect to time are of 2nd order. PDEase does not solve such equations directly, but we may introduce auxiliary variables to make them 1st order. The inconvenience is to have twice as many unknowns.

Having four dependent variables will of course make the runs longer, but we should notice that we also had four equations in the case of sinusoidal vibrations, although we solved only for the real parts. We still obtained a reasonably good picture of the modes of oscillation, but to explore the time dependence fully we would have needed the imaginary parts as well.

Using the new dependent variables

$$u_t \equiv \frac{\partial u}{\partial t} \text{ and } v_t \equiv \frac{\partial v}{\partial t}$$

we thus modify the above system to read

$$
\begin{cases}
u_t = \dfrac{\partial u}{\partial t} \\
\dfrac{\partial \sigma_x}{\partial x} + \dfrac{\partial \tau_{xy}}{\partial y} + F_x - \rho \dfrac{\partial u_t}{\partial t} = 0 \\
v_t = \dfrac{\partial v}{\partial t} \\
\dfrac{\partial \tau_{xy}}{\partial x} + \dfrac{\partial \sigma_y}{\partial y} + F_y - \rho \dfrac{\partial v_t}{\partial t} = 0
\end{cases}
$$

There are now as many dependent variables as there are equations, and we have to supply boundary conditions for each of them.

Shear Pulse Propagation Along a Bar

We first consider one of the simplest problems available, i.e. the transmission of a pulse along a beam of rectangular cross-section. The following descriptor is required.

```
title                                                              { eltr3.pde }
  'Propagation of a Shear Pulse'
select
  ngrid= 30              regrid= off
  errlim= 0.05
  contours= 10           surfacegrid= 30   paint= off
variables
  u(range= 1e-3)         ut(range= 1e2)
  v(range= 1e-3)         vt(range= 1e2)
definitions
  L= 1      h= 0.05
  dens= 7.8e3    mu = 0.3      E = 200e9                              {Steel}
  c = E/(1-mu**2)     G = E/2/(1+mu)
  ex= dx(u)          ey= dy(v)                       { x-, y- and xy-strains }
  exy= dx(v)+ dy(u)
  sx= c*(ex+ mu*ey)   sy= c*(mu*ex+ ey)                           { Stresses }
  sxy= G*exy
  v_pulse= if t<2e-5 then 1e-3 else 0
initial values
```

```
  u= 0       ut= 0         v= 0          vt= 0
equations
  ut= dt( u)
  dx( sx)+ dy( sxy)- dens*dt( ut)= 0
  vt= dt( v)
  dx( sxy)+ dy( sy)- dens*dt( vt)= 0
boundaries
                region 1
  start (0,-h/2)
  value(u)= 0    load(v)= 0    line to (L,-h/2)
  load(u)= 0   value(v)= 0   line to (L,h/2)                    { Fixed }
  value(u)= 0    load(v)= 0    line to (0,h/2)
  value(u)= 0   value(v)= v_pulse  line to finish               { Sheared }
monitors
plots
  for t= 2e-5 by 2e-5 to 1e-3
    elevation( v) from (0,0) to (L,0)
end
```

In order to increase the speed of execution we specify a fixed grid density under *select*, thereby inhibiting adaptive gridding, and choose such a small beam height (h) that there will be only one row of finite elements. In the definitions segment we define a shear displacement (*v_pulse*) of 20 μs duration, to be applied to the left face of the beam. The opposite end is kept immobile in the y direction.

This descriptor generates a large number of solutions, and it may take many hours to run on a slow PC system.

The following sequence of figures show that the pulse propagates to the right until the end it reached. There it changes sign and starts to travel to the left. Coming back to the origin it is again reflected and its direction of displacement is reversed.

The shear pulse imposed on the left end of the bar would be strictly rectangular if presented in the figures below. The pulse transmitted along the bar, on the other hand, has a rounded-off shape because of the finite size of the volume elements. Evidently there is also a certain amount of “ringing” before and after the pulse, and these anomalous features become increasingly visible at larger times. Such oscillations may in part be caused by the strictly periodic subdivision of the domain.

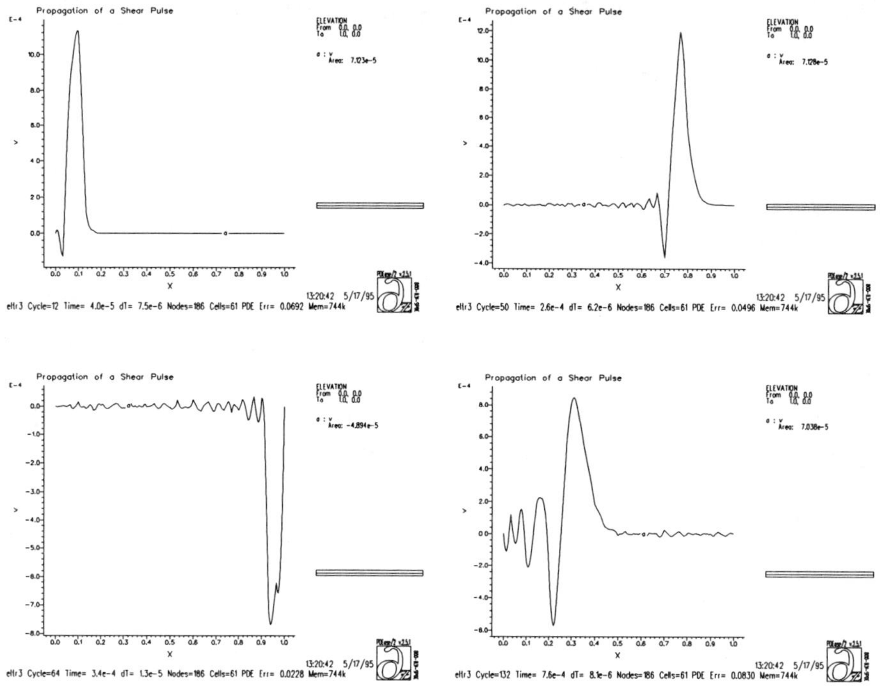

Shear Pulse Along a Tapered Bar

In the next example we shall let the pulse travel along a tapered beam. We shall also evaluate the total elastic energy of the pulse by displaying *contour* plots of the energy density *energy_d.* This problem requires a few modifications of the preceding descriptor, as shown by the list below.

```
                                                        { eltr5.pde }
...
energy_d= 0.5*(sx*ex+ sy*ey+ sxy*exy)              { Energy density }
...
              region 1
start (0,-h/2)
value(u)= 0      load(v)= 0        line to (L,-h/5)
```

```
  load(u)= 0        load(v)= 0            line to (L,h/5)                  { Free }
  value(u)= 0       load(v)= 0            line to (0,h/2)
  value(u)= 0       value(v)= v_pulse  line to finish                 { Sheared }
monitors
plots
  for t= 2e-5 by 2e-5 to 1e-3
      elevation( v) from (0,0) to (L,0)
      contour( energy_d)
end
```

The following four plots present a selection of the results. Browsing through the whole sequence of figures you get a more complete impression of the mode of propagation. The most striking feature is the gradual increase of the peak height as the pulse moves toward the right end.

The plots of the energy density (not shown) permit us to read off the integrals over the domain, i.e. the total elastic energy of the pulse. We notice that the energy remains essentially constant, although the values scatter considerably, corresponding to a standard deviation of a few percent.

Finally, it is evident that the sign of the displacement is unchanged as the pulse is reflected against the free end.

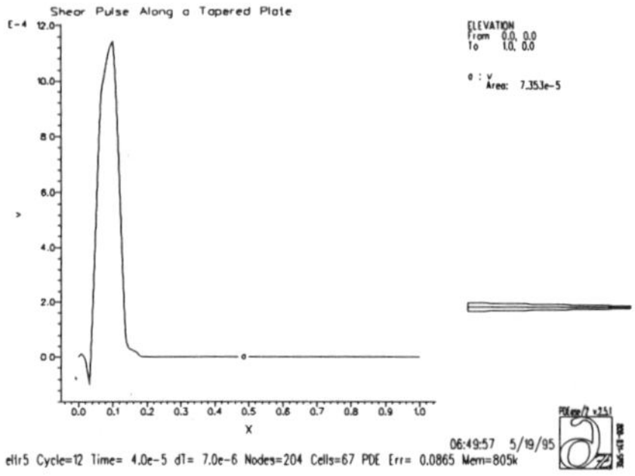

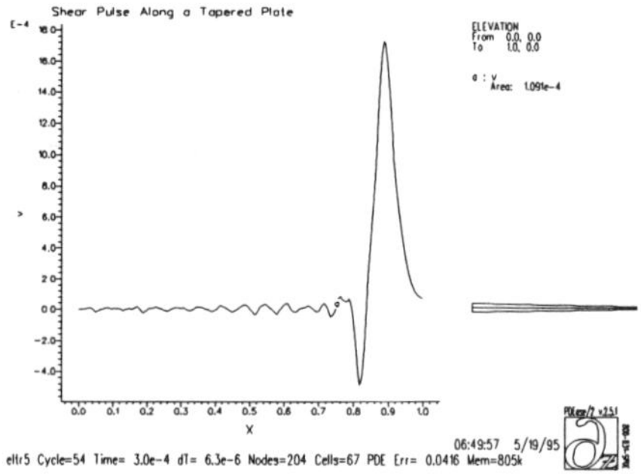

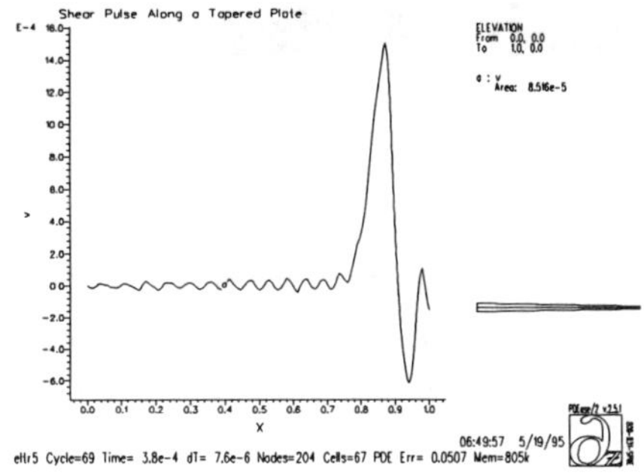

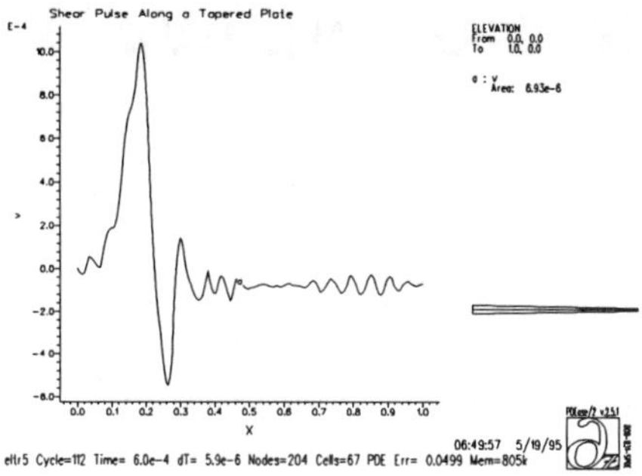

Exercises

❑ Convert ELTR3.PDE into a strictly one-dimensional descriptor, requiring only two PDEs. Assess the improvement in speed of execution and quality of pulse reproduction. Compare the speed of propagation with the theoretical expression for the shear wave velocity $(\sqrt{G/\rho})$.

❑ Modify ELTR3.PDE to treat longitudinal pulses. Compare the speed of propagation with the theoretical expression for the corresponding wave velocity $(\sqrt{E/\rho})$. Include a plot of the elastic energy density to check the constancy of the total energy.

❑ Modify ELTR3.PDE by applying a step function of shear displacement at the left end. Notice the effect of reflection against the right end.

Electromagnetic Waves in Matter

In the following we simplify by assuming that the *magnetic* properties are the same throughout the domain. The derivation of the PDE starts with the Maxwell equation[3p323]

$$\nabla \times \mathbf{E} = -\frac{\partial \mathbf{B}}{\partial t},$$ ●

on which we apply the curl operation once more to obtain

$$\nabla \times (\nabla \times \mathbf{E}) = -\nabla \times \frac{\partial \mathbf{B}}{\partial t} = -\frac{\partial}{\partial t}\nabla \times \mathbf{B} = -\mu \frac{\partial}{\partial t}\nabla \times \mathbf{H},$$

where we made use of the constancy of μ. According to another of Maxwell's equations we have

$$\nabla \times \mathbf{H} = \mathbf{j} + \frac{\partial \mathbf{D}}{\partial t},$$ ●

which permits us to write

$$\nabla \times (\nabla \times \mathbf{E}) = -\mu \frac{\partial}{\partial t}\nabla \times \mathbf{H} = -\mu \frac{\partial}{\partial t}\left(\mathbf{j} + \frac{\partial \mathbf{D}}{\partial t}\right) = -\mu \frac{\partial \mathbf{j}}{\partial t} - \mu \frac{\partial^2 \mathbf{D}}{\partial t^2},$$

or, with $\mathbf{j} = \sigma \mathbf{E}$ and $\mathbf{D} = \varepsilon \mathbf{E}$,

$$\nabla \times (\nabla \times \mathbf{E}) = -\mu\sigma \frac{\partial \mathbf{E}}{\partial t} - \mu\varepsilon \frac{\partial^2 \mathbf{E}}{\partial t^2}$$ ●

Using a transformation of the left hand side, well-known from vector analysis, we may now write

$$\nabla(\nabla \cdot \mathbf{E}) - \nabla^2 \mathbf{E} = -\mu\sigma \frac{\partial \mathbf{E}}{\partial t} - \mu\varepsilon \frac{\partial^2 \mathbf{E}}{\partial t^2}$$

If ε is constant and the charge density vanishes, the first term on the left side of this equation may be recast as $(1/\varepsilon)\nabla(\nabla \cdot \varepsilon \mathbf{E}) = (1/\varepsilon)\nabla(\nabla \cdot \mathbf{D}) = 0$, leaving us with

$$\nabla^2\mathbf{E} - \mu\varepsilon\frac{\partial^2\mathbf{E}}{\partial t^2} - \mu\sigma\frac{\partial\mathbf{E}}{\partial t} = 0 \qquad \bullet$$

We could well use the above PDE to calculate the general evolution of the electric field in space and time, but that would mean mixing transient and periodic contributions to the solution. If we restrict our interest to sinusoidal oscillations, we may effectively eliminate the time variable from the PDE. A stationary state of oscillation may be described by

$$\mathbf{E}(\mathbf{r}) = \mathbf{E}_0(\mathbf{r})\exp(i\omega t)$$

where ω is the angular frequency. Just as in the case of temperature waves (p.130) the real part of this solution is interpreted as the instantaneous intensity of the field. If we substitute this expression into the PDE we obtain

$$\exp(i\omega t)\nabla^2\mathbf{E}_0 + \exp(i\omega t)\mu\varepsilon\omega^2\mathbf{E}_0 - \exp(i\omega t)i\mu\sigma\omega\mathbf{E}_0 = 0$$

or

$$\nabla^2\mathbf{E}_0 + \mu\varepsilon\omega^2\mathbf{E}_0 - i\mu\sigma\omega\mathbf{E}_0 = 0 \qquad \bullet$$

which means that we have similar second-order PDEs for each of the components E_{0x}, E_{0y} and E_{0z}. These components are in general complex quantities, to be multiplied by the exponential time factor.

In the present version of PDEase we can only obtain solutions for **E** as a function of two space variables. The direction of **E**, however, is not limited to the *xy*-plane.

Plane Wave in a Conducting Material

Let us now consider a very simple case: an electromagnetic wave driven by an oscillating field, **E** parallel to the *z*-axis, the resulting wave somehow terminated at the other end of the domain. For this case we have the PDE

$$\nabla^2 E_z + \mu\varepsilon\omega^2 E_z - i\mu\sigma\omega E_z = 0$$

where we have dropped the subscript 0 for the amplitude. Remembering that $E_z = E_{zr} + iE_{zi}$ is complex, we separate the equation into its real and imaginary parts

$$\begin{cases} \nabla^2 E_{zr} + \mu\varepsilon\omega^2 E_{zr} + \mu\sigma\omega E_{zi} = 0 \\ \nabla^2 E_{zi} + \mu\varepsilon\omega^2 E_{zi} - \mu\sigma\omega E_{zr} = 0 \end{cases}$$

This system of PDEs may now be typed into a descriptor and solved by PDEase.

Plane Wave in a Homogeneous Medium

Let us start with the simplest possible example. We drive the wave by an oscillating field **E** of constant amplitude and directed along the z axis. We choose the boundary conditions such that the field becomes independent of y. Since the domain has to be finite, we terminate it by a superconducting plane, perpendicular to the x-axis. The quantities involved thus depend only on x, although the solution domain extends in the y direction as well.

```
select                                                          { emw1.pde }
  errlim= 1e-4          nodelimit= 500
  converge= 1e-6        subconverge= 1e-6
  contours= 10          surfacegrid= 30              paint= off
variables
  Ezr       Ezi                                 { Real and imaginary parts }
definitions
  Lx= 1.0               Ly= 0.2
  c0= 3.00e8                                             { Speed of light }
  eps0= 8.85e-12        eps= eps0
  mu0= 4*pi*1e-7        mu= mu0
  sigma= 1e-1                                    { Conductivity of medium }
  omega= 5e9                                        { Angular frequency }
  Ein= 1.0                                              { Input field Ez }
  Eabs= sqrt(Ezr**2+ Ezi**2)
  phase= sign(Ezi)* arccos(Ezr/Eabs)/pi*180
equations
  del2( Ezr)+ mu*eps*omega**2*Ezr+ mu*sigma*omega*Ezi= 0
  del2( Ezi)+ mu*eps*omega**2*Ezi- mu*sigma*omega*Ezr= 0
```

```
boundaries
          region 1
  start (0,0)
  natural(Ezr)= 0   natural(Ezi)= 0 line to (Lx,0)
  value(Ezr)= 0   value(Ezi)= 0 line to (Lx,Ly)             { Conducting wall }
  natural(Ezr)= 0  natural(Ezi)= 0 line to (0,Ly)
  value(Ezr)= Ein    value(Ezi)= 0   line to finish             { Input E field }
monitors
  elevation( Ezr) from (0,Ly/2) to (Lx,Ly/2)
plots
  elevation( Ezr, Ezi, Eabs) from (0,Ly/2) to (Lx,Ly/2)
  elevation( phase) from (0,Ly/2) to (Lx,Ly/2)
  contour( Ezr)       contour( Ezi)
  end
```

In the above descriptor we apply a field of 1+0*i, in complex notation and SI units. This implies that the phase angle is zero on the left boundary plane. As is evident from the following plot, the curves start at the expected values at the left and tend to zero over the first half of the range. This rapid vanishing of the wave illustrates the skin effect[3p370], and permits us to estimate the skin depth.

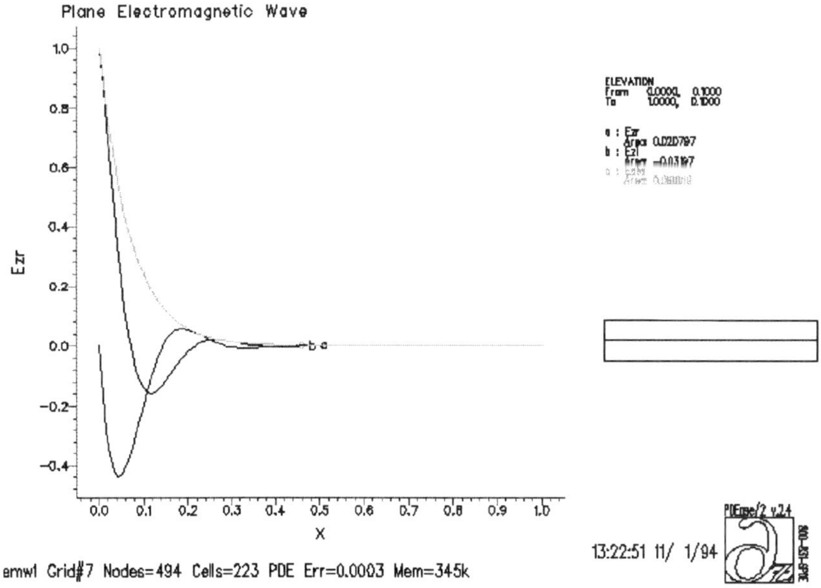

The plot of the phase angle against x (not shown) reminds us of the corresponding figure for a temperature wave (p.134). In fact, the theory of thermal conduction is similar to that of electrical conduction in a pure metal.

After having seen the rapid attenuation of the input wave we now decrease the conductivity *sigma* to 1e-2. The amplitude then still decreases as we go to the right, but it remains large quite close to the end.

If we take one step further, putting *sigma*= 1e-3, completely new features appear. We note that the magnitude of **E** now becomes larger than the input value, and the plot of the phase angle (below) shows unexpected shapes.

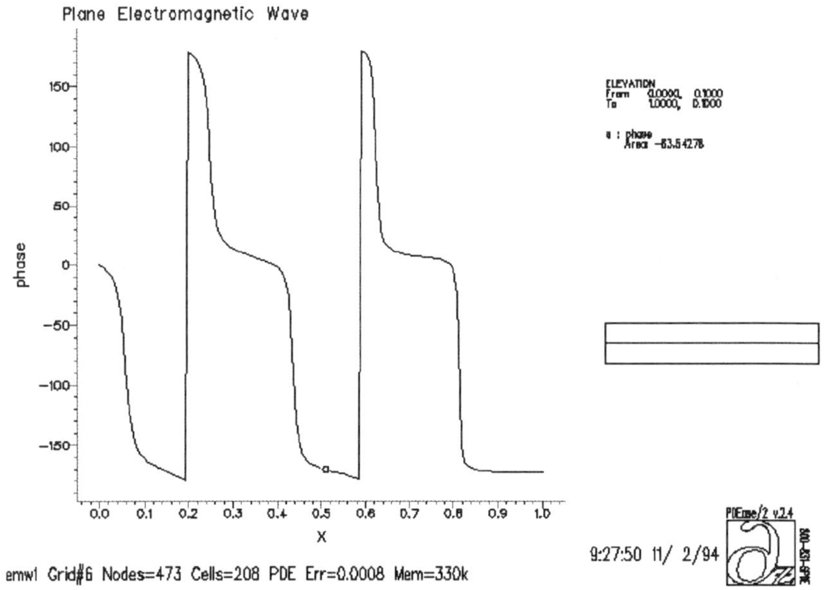

Here we may begin to doubt the numerical solutions. Fortunately, there is an exact solution, which we may use for comparison. We assume a solution of the form

$$E_z = E\exp(-\alpha x - ikx)$$

and substitute it into the complex PDE. When separated into its real and imaginary part the PDE yields a system of equations for α and k. Eliminating k we obtain a 2nd-degree algebraic equation in α^2, which we easily solve. Then we immediately obtain α and k. This leads to two solutions, which may be superimposed as follows

$$E_z = E_{1z} \exp(-\alpha_1 x - ik_1 x) + E_{2z} \exp(-\alpha_2 x - ik_2 x)$$

It now remains to adapt this general solution to the boundary conditions, which results in four linear equations for the complex coefficients E_{1z} and E_{2z}. Solving these is an exercise in easy but tedious algebra. In order to test our numerical solution to the PDE we only need to include the following extra lines in the next descriptor file.

```
...
alpha1= omega* sqrt(mu*eps)*
       sqrt(-0.5+ 0.5*sqrt(1+sigma**2/(omega**2*eps**2)))
k1= mu*sigma*omega/(2*alpha1)
alpha2= -alpha1
k2= mu*sigma*omega/(2*alpha2)
e1= exp(-alpha1)              e2= exp(-alpha2)
c1= cos(k1*Lx)                c2= cos(k2*Lx)
s1= sin(k1*Lx)                s2= sin(k2*Lx)
E1r= -Ein*[e2*c2/(e1*s1-e2*s2)+e2*s2/(e1*c1-e2*c2)] /
       [(e1*c1-e2*c2)/(e1*s1-e2*s2)-(-e1*s1+e2*s2)/(e1*c1-e2*c2)]
E1i= -E1r*(e1*c1-e2*c2)/(e1*s1-e2*s2)-Ein*e2*c2/(e1*s1-e2*s2)
E2r= Ein- E1r
E2i= -E1i
Ezr_ex= E1r*exp(-alpha1*x)*cos(k1*x)+ E1i*exp(-alpha1*x)*sin(k1*x)
  +E2r*exp(-alpha2*x)*cos(k2*x)+ E2i*exp(-alpha2*x)*sin(k2*x)
Ezi_ex= -E1r*exp(-alpha1*x)*sin(k1*x)+ E1i*exp(-alpha1*x)*cos(k1*x)
  -E2r*exp(-alpha2*x)*sin(k2*x)+ E2i*exp(-alpha2*x)*cos(k2*x)
Eabs_ex= sqrt(Ezr_ex**2+ Ezi_ex**2)
phase_ex= sign(Ezi_ex)* arccos(Ezr_ex/Eabs_ex)/pi*180
...
elevation( Ezr, Ezr_ex) from (0,Ly/2) to (Lx,Ly/2)
elevation( Ezi, Ezi_ex) from (0,Ly/2) to (Lx,Ly/2)
elevation( phase, phase_ex) from (0,Ly/2) to (Lx,Ly/2)
end
```

Running the new file we find that the PDEase solution virtually coincides with the exact one, as shown below for the imaginary part. You will also find that the exact solution confirms our strange-looking phase curve.

It is easy to modify this descriptor by substituting different values of *sigma* and *omega*, which sometimes results in dramatically changed solutions. For instance, putting *sigma*= 1e-4 and *omega*= 4.1**pi***c*0/*Lx* produces a wave of much larger amplitude than before, as seen in the next figure. In fact, *omega*= 4**pi***c*0/*Lx* would correspond to a resonant condition, where *Lx* equals two full wavelengths.

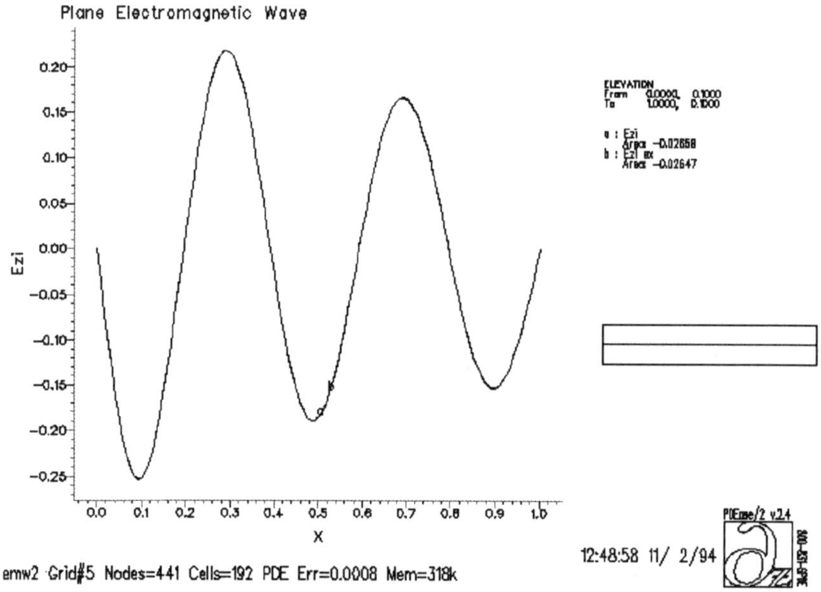

It is equally easy to obtain solutions for frequencies closer to resonance. If the amplification is smaller than ten, the agreement with the exact expression is still good. When this factor reaches several orders of magnitude, however, the discrepancy gradually becomes appreciable. This is caused by round-off errors in PDEase as well as in the evaluation of the exact expression near resonance.

How should we interpret the above curves for E_{zr} and E_{zi}? Any electric field component must of course have a real value at all times. According to the principles of phasor (" $j\omega$ ") analysis of oscillations the quantity of interest is the real part of the time-dependent expression. In our case we have

$$E_z = \mathrm{Re}\left[(E_{zr} + iE_{zi})\exp(i\omega t)\right] = E_{zr}\cos(\omega t) - E_{zi}\sin(\omega t)$$

which gives us the field as a function of space and time. This expression is periodic, and it should be noted that whenever $\omega t = \pi n$, with n integral, the second term vanishes and E_z becomes equal to the real part of the solution. The plot of E_{zr} may be regarded as a "snapshot" of the field at certain values of t.

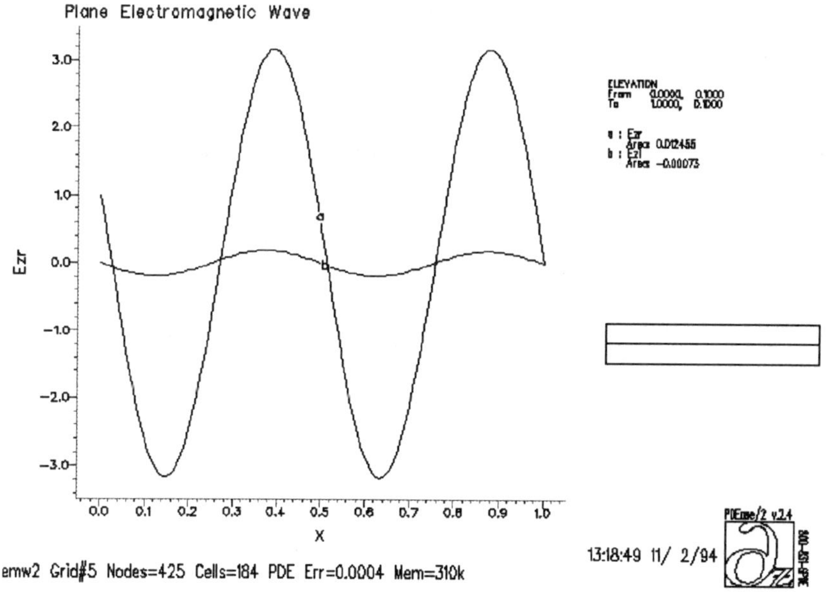

Standing Wave in a Non-Homogeneous Medium

When deriving the PDEs in the beginning of this chapter (p.332) we had to assume the magnetic permeability μ to be constant throughout the solution domain, although σ could vary in space. Assuming the field to be directed along the z axis we expand the first term of the PDE (p.332),

$$\nabla \times (\nabla \times \mathbf{E}) + \mu\sigma \frac{\partial \mathbf{E}}{\partial t} + \mu\varepsilon \frac{\partial^2 \mathbf{E}}{\partial t^2} = 0 .$$

Here $\nabla \times \mathbf{E}$ simplifies to become

$$\nabla \times \mathbf{E} = \begin{vmatrix} \mathbf{i} & \mathbf{j} & \mathbf{k} \\ \dfrac{\partial}{\partial x} & \dfrac{\partial}{\partial y} & \dfrac{\partial}{\partial z} \\ 0 & 0 & E_z \end{vmatrix} = \begin{Bmatrix} \dfrac{\partial E_z}{\partial y} \\ -\dfrac{\partial E_z}{\partial x} \\ 0 \end{Bmatrix}$$

and using this vector in the first curl operation we obtain

$$\left[\nabla \times (\nabla \times \mathbf{E})\right]_z = \frac{\partial}{\partial x}\left(-\frac{\partial E_z}{\partial x}\right) - \frac{\partial}{\partial y}\frac{\partial E_z}{\partial y} = -\frac{\partial^2 E_z}{\partial x^2} - \frac{\partial^2 E_z}{\partial y^2}$$

Thus, when only the z component of $\mathbf{E}$ is different from zero, this expression takes the Laplacian form that we used earlier. This expression remains valid even if ε varies in space.

Let us consider the case of a plane wave in a domain consisting of empty space over the left half and solid matter over the right half. The solid chosen is like quartz but is fictitious in the sense that is has an appreciable electrical conductivity, caused by impurities. A perfect conductor at the end reflects the wave as before. As is evident from the descriptor below we specify the electrical properties for the two regions under *boundaries*.

```
select                                              { emw11.pde }
  errlim= 1e-4            nodelimit= 500
```

```
  converge= 1e-6       subconverge= 1e-6
variables
  Ezr       Ezi
definitions
  Lx= 1.0              Ly= 0.2
  c0= 3.00e8           eps0= 8.85e-12        eps
  mu0= 4*pi*1e-7       mu= mu0               sigma
  omega= 3e9
  Ein= 1.0                                              { Input E field }
  Eabs= sqrt(Ezr**2+ Ezi**2)
  phase= sign(Ezi)* arccos(Ezr/Eabs)/pi*180
equations
  del2( Ezr)+ mu*eps*omega**2*Ezr+ mu*sigma*omega*Ezi= 0
  del2( Ezi)+ mu*eps*omega**2*Ezi- mu*sigma*omega*Ezr= 0
boundaries
          region 1          eps= eps0             sigma= 0
  start(0,0)
  natural(Ezr)= 0  natural(Ezi)= 0 line to (Lx,0)
  value(Ezr)= 0  value(Ezi)= 0 line to (Lx,Ly)
  natural(Ezr)= 0  natural(Ezi)= 0 line to (0,Ly)
  value(Ezr)= Ein    value(Ezi)= 0   line to finish
          region 2          eps= 4*eps0           sigma= 1e-2
  start(Lx/2,0)  line to (Lx,0) to (Lx,Ly) to (Lx/2,Ly) to finish
monitors
  elevation( Ezr, Ezi) from (0,Ly/2) to (Lx,Ly/2)
plots
  elevation( Ezr, Ezi) from (0,Ly/2) to (Lx,Ly/2)
end
```

The following plot shows the real and imaginary parts of the solution. As expected from elementary theory, the wavelength is smaller in the region where the permittivity is higher, but there is no sudden change in the curves at the vacuum-solid interface. In this case the vector **E** is perpendicular to the plane of the figure, hence parallel to the interface. Electromagnetic theory[3p330] requires that the tangential component of **E** be continuous across the interface. This is confirmed by the figure, since E_z is a linear combination of E_{zr} and E_{zi} at all times.

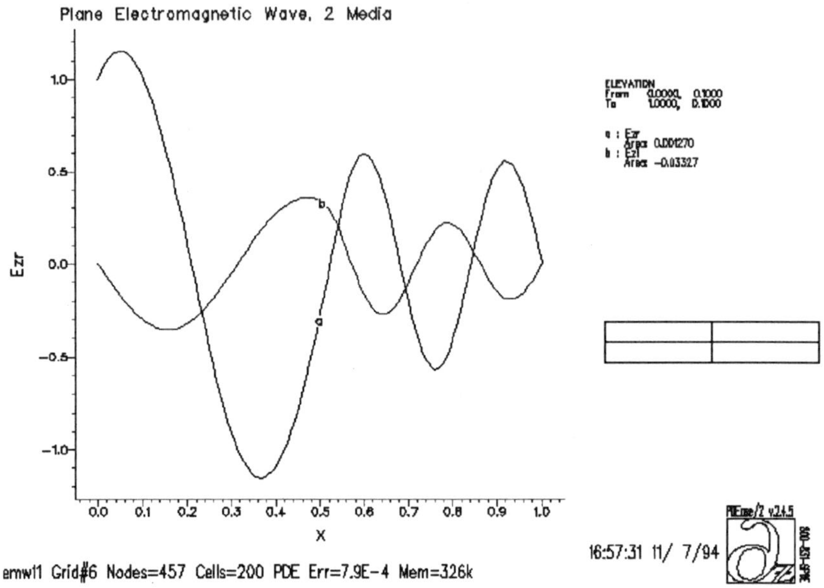

There is more to be explored using this configuration. Let us modify the above descriptor slightly, as follows.

```
...                                                         { emw12.pde }
omega= 4*pi*c0/ Lx
...
    region 2            eps= 4*eps0             sigma= 0
...
```

As shown by the next figure, the solution now becomes dramatically different, in the sense that the amplitude becomes several hundred times larger than the input field. This is not completely unexpected, however, since we have chosen the frequency to correspond to a resonance.

Even in this case, the problem can be solved exactly. The basic solution is of the same form as with a single medium, but in addition we have to impose boundary conditions at the interface. Thus there will be even more linear equations, and in this context we shall not indulge in the drudgery of handling the complicated expressions involved.

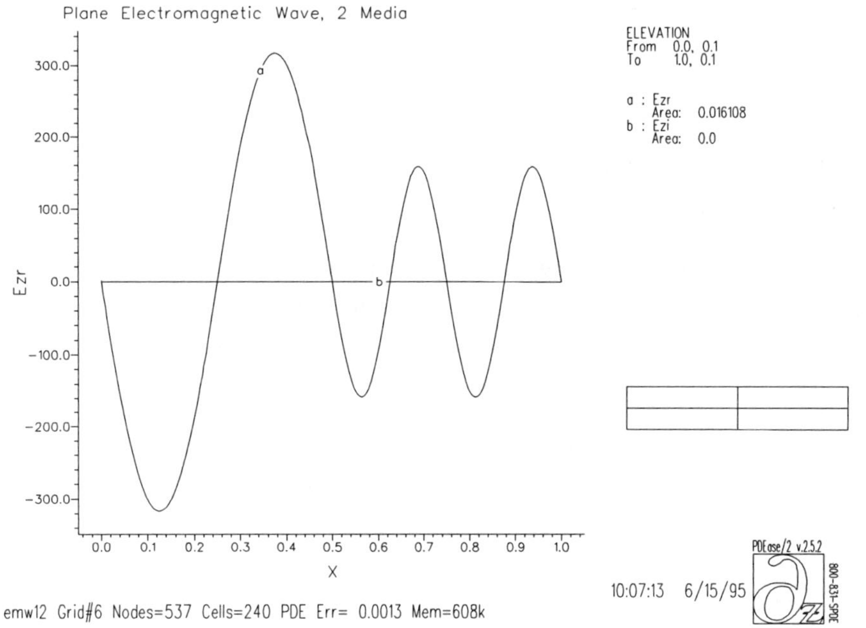

The Magnetic Wave Field

In the earlier part of this chapter we solved problems with respect to the electric field **E**. According to Maxwell's equations, however, a periodic electric field cannot exist without an accompanying magnetic field. Having obtained a solution $\mathbf{E}_0(x,y)$, we directly obtain the corresponding vector field $\mathbf{B}_0(x,y)$ from the first Maxwell relation (p.332).

$$\frac{\partial \mathbf{B}}{\partial t} = -\nabla \times \mathbf{E} = -\begin{vmatrix} \mathbf{i} & \mathbf{j} & \mathbf{k} \\ \frac{\partial}{\partial x} & \frac{\partial}{\partial y} & \frac{\partial}{\partial z} \\ E_x & E_y & E_z \end{vmatrix} = -\begin{Bmatrix} \frac{\partial E_{0z}}{\partial y} - \frac{\partial E_{0y}}{\partial z} \\ \frac{\partial E_{0x}}{\partial z} - \frac{\partial E_{0z}}{\partial x} \\ \frac{\partial E_{0y}}{\partial x} - \frac{\partial E_{0x}}{\partial y} \end{Bmatrix} \exp(i\omega t)$$

Let us write the expression for the magnetic field similarly as

$\mathbf{B}(\mathbf{r}) = \mathbf{B}_0(\mathbf{r})\exp(i\omega t)$,

where $\mathbf{B}_0(\mathbf{r})$ is a vector having complex components. When we substitute its first component into the corresponding row of the vector equation it becomes

$$i\omega B_{0x} \exp(i\omega t) = -\left[\frac{\partial E_{0z}}{\partial y} - \frac{\partial E_{0y}}{\partial z}\right] \exp(i\omega t)$$

and finally

$$B_{0x} = \frac{i}{\omega}\left[\frac{\partial E_{0z}}{\partial y} - \frac{\partial E_{0y}}{\partial z}\right]$$

with similar relations for the other two components, easily obtained by cyclic permutation.

In the particular case of a plane wave with an electric field in the z direction, B_z will vanish and for the other components we find

$$B_{0x} = \frac{i}{\omega}\left[\frac{\partial E_{0z}}{\partial y} - \frac{\partial E_{0y}}{\partial z}\right] = \frac{i}{\omega}\frac{\partial(E_{zr} + iE_{zi})}{\partial y} = -\frac{1}{\omega}\frac{\partial E_{zi}}{\partial y} + \frac{i}{\omega}\frac{\partial E_{zr}}{\partial y} \quad \bullet$$

$$B_{0y} = \frac{i}{\omega}\left[\frac{\partial E_{0x}}{\partial z} - \frac{\partial E_{0z}}{\partial x}\right] = -\frac{i}{\omega}\frac{\partial(E_{zr} + iE_{zi})}{\partial x} = \frac{1}{\omega}\frac{\partial E_{zi}}{\partial x} - \frac{i}{\omega}\frac{\partial E_{zr}}{\partial x} \quad \bullet$$

We first apply these relations to the earlier example involving two media (p.340). The previous descriptor should be modified as below.

```
select                                          { emw21.pde }
  errlim= 1e-8              nodelimit= 1000
```

```
  converge= 1e-6              subconverge= 1e-6
variables
  Ezr       Ezi
definitions
  Lx= 1.0                  Ly= 0.2
  c0= 3.00e8               eps0= 8.85e-12          eps
  mu0= 4*pi*1e-7           mu= mu0                 sigma
  omega= 3e9
  Ein= 1.0                                                { Input E field }
  Eabs= sqrt(Ezr**2+ Ezi**2)
  phase= sign(Ezi)* arccos(Ezr/Eabs)/pi*180
                                                        { Magnetic field ... }
  Bxr= -(1/omega)*dy(Ezi)          Bxi=  (1/omega)*dy(Ezr)
  Bxabs= sqrt(Bxr**2+ Bxi**2)
  Byr=  (1/omega)*dx(Ezi)          Byi= -(1/omega)*dx(Ezr)
  Byabs= sqrt(Byr**2+ Byi**2)
equations
  del2( Ezr)+ mu*eps*omega**2*Ezr+ mu*sigma*omega*Ezi= 0
  del2( Ezi)+ mu*eps*omega**2*Ezi- mu*sigma*omega*Ezr= 0
boundaries
      region 1             eps= eps0          sigma= 0
  start(0,0)
  natural(Ezr)= 0          natural(Ezi)= 0 line to (Lx,0)
  value(Ezr)= 0            value(Ezi)= 0 line to (Lx,Ly)
  natural(Ezr)= 0          natural(Ezi)= 0 line to (0,Ly)
  value(Ezr)= Ein          value(Ezi)= 0   line to finish
      region 2             eps= 4*eps0        sigma= 1e-2
  start(Lx/2,0)  line to (Lx,0) to (Lx,Ly) to (Lx/2,Ly) to finish
monitors
  elevation( Ezr, Ezi) from (0,Ly/2) to (Lx,Ly/2)
plots
  elevation( Ezr, Ezi, Eabs) from (0,Ly/2) to (Lx,Ly/2)
  elevation( Bxr, Bxi, Bxabs) from (0,Ly/2) to (Lx,Ly/2)
  elevation( Byr, Byi, Byabs) from (0,Ly/2) to (Lx,Ly/2)
end
```

This descriptor gives rise to the following figure for the y component of **B**. Since these curves are produced by differentiation of a numeric function, the computational errors become much larger than for **E**. The magnetic field does not vanish at the right end, nor is this required by Maxwell's equations, since **B** is parallel to the conducting surface.

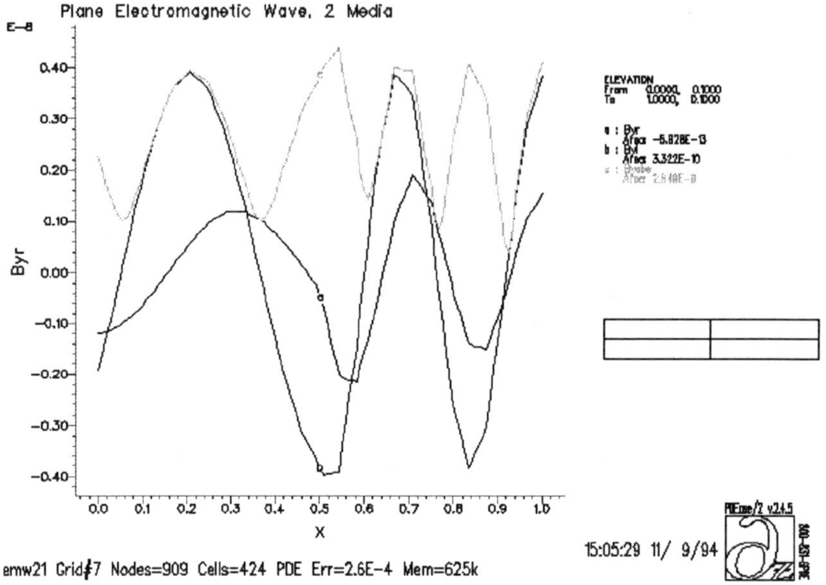

The other component, B_x, appears to be smaller by a factor of about a hundred, and the curve (not shown) is highly irregular. This confirms the theoretical prediction that this component must vanish.

Exercises

❑ Would you qualify the material in the first example (p.334) as a good conductor? Modify the parameters of the descriptor to estimate the skin depth of copper at 50 Hz or 60 Hz.
❑ Explore standing waves in a single medium using a sequence of closely-spaced frequencies. Select *stages*=9, choose *sigma*=0 and *omega*=(3.9+*stage**0.02)**pi***c*0/*Lx*. Keep only one plot per stage, the one for *Ezr* and *Ezi*.
❑ Solve for the same values of *omega* in the case of two media (p.341) with *eps*=4**eps*0 and *sigma*=0 in region 2.
❑ Repeat the preceding investigation with *sigma*=1e-2 in region 2.
❑ Find your own resonance frequency in the case of two media and scan over that peak.

Two-Dimensional Resonators

As we have seen in previous chapters, the maximum amplitude of a plane wave may be larger, at certain frequencies, than at the input end. We shall now turn to a more realistic situation, where three walls of a resonator box are perfect conductors and the input end is open.

Loss-Free Rectangular Cavity

First we assume the medium in the box to be a vacuum, so that there is no loss by conduction. The descriptor required is rather similar to that for the simplest plane wave. We just change the boundary conditions on the walls parallel to the *x*-axis and introduce 10 stages for scanning over a frequency range.

```
select                                                          { emw41.pde }
  stages= 10
  errlim= 1e-6            nodelimit= 500
  converge= 1e-6          subconverge= 1e-6
  contours= 10            surfacegrid= 30         paint= off
variables
  Ezr                Ezi
definitions
  Lx= 1.0                 Ly= 1.0
  eps0= 8.85e-12          eps= eps0
  mu0= 4*pi*1e-7          mu= mu0
  sigma= 0                omega= (3+0.1*stage)*1e9
  Ein= 1.0                                                 { Input E field }
  Eabs= sqrt(Ezr**2+ Ezi**2)
equations
  del2( Ezr)+ mu*eps*omega**2*Ezr+ mu*sigma*omega*Ezi= 0
  del2( Ezi)+ mu*eps*omega**2*Ezi- mu*sigma*omega*Ezr= 0
boundaries
            region 1
  start (0,0)
  value(Ezr)= 0  value(Ezi)= 0     line to (Lx,0) to (Lx,Ly) to (0,Ly)
```

```
  value(Ezr)= Ein    value(Ezi)= 0   line to finish
monitors
  contour( Ezr)
plots
  contour( Ezr)
end
```

Browsing through the plots we find high amplitudes and symmetric contours at *omega*=3.4e9, as shown below, and a completely different resonant pattern (not shown) at *omega*=3.9e9.

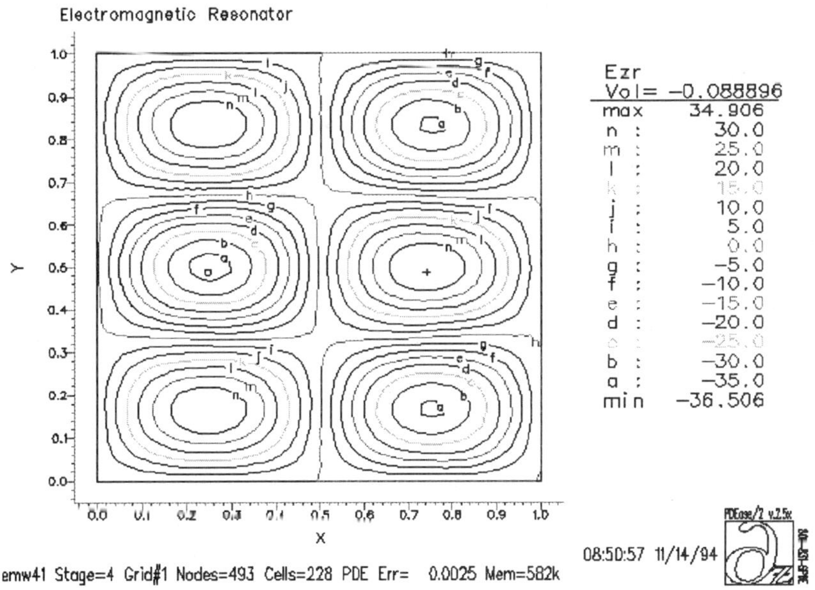

Resonance Model

Instead of scanning over all possible frequencies, we shall now derive an exact solution to a closely similar problem. Already in the examples with plane waves we saw that the maximum amplitude at a resonant frequency may be hundreds of times larger than the input amplitude. As a model for resonance we hence take a wave having zero amplitude on all boundaries, leaving aside the problem of how to create such a secluded wave in practice. We still assume that the medium has zero conductivity.

The exact solution for a resonant state in 1D (p.337) should be the real part of

$$\{E_{1z}\exp(-ik_1x)+E_{2z}\exp(-ik_2x)\}\exp(i\omega t)$$

where the coefficients E_{1z} and E_{2z} generally are complex. None of the two terms in curly brackets can vanish at both $x=0$ and $x=L_x$ unless the magnitude of the coefficient is zero. If the coefficients are equal and $k_2=-k_1\equiv k_x$, however, the total expression degenerates into

$$2i\sin(k_x x)\exp(i\omega t).$$

The sine factor always vanishes at the origin and generally at $k_x x=m\pi$, where m is any integer. The condition for zero field at the vertical boundaries thus is

$$k_x=\frac{m\pi}{L_x}.$$

An amplitude function of the form

$$E_z=A\sin\left(\frac{m\pi}{L_x}x\right)\sin\left(\frac{n\pi}{L_y}y\right)$$

vanishes at all of the four boundaries for any integral values m and n. We shall show that it also satisfies the PDE relevant to zero conductivity (p.334)

$$\nabla^2E_z+\mu\varepsilon\omega^2E_z=0$$

Substituting the function into the PDE we find that A and the sine factors cancel, leaving us with

$$-\left(\frac{m\pi}{L_x}\right)^2-\left(\frac{n\pi}{L_y}\right)^2+\mu\varepsilon\omega^2=0$$

This means that our trial function satisfies the PDE, provided that

$$\omega = \frac{\pi}{\sqrt{\mu\varepsilon}}\sqrt{\left(\frac{m}{L_x}\right)^2 + \left(\frac{n}{L_y}\right)^2} = \pi c\sqrt{\left(\frac{m}{L_x}\right)^2 + \left(\frac{n}{L_y}\right)^2}$$

where c is the wave velocity in the medium.

Resonant Modes

We shall now test whether a cavity really exhibits resonances according to the above prediction. For this purpose we may use the preceding descriptor with the following amendments. We eliminate *stages*, increase the number of nodes, type in the exact expression for *omega* at resonance and plot the magnetic vector components as well.

```
select                                                          { emw42.pde }
  errlim= 1e-8            nodelimit= 1000
  ...
  m=2                     n=3
  omega= pi/sqrt(mu*eps)*sqrt[(m/Lx)**2+ (n/Ly)**2]
  ...
  Bxr= -(1/omega)*dy(Ezi)          Bxi= (1/omega)*dy(Ezr)
  Byr= (1/omega)*dx(Ezi)           Byi= -(1/omega)*dx(Ezr)
  ...
  contour( Ezr)           contour( Ezi)
  contour( Bxr)           contour( Bxi)
  contour( Byr)           contour( Byi)
  ...
```

The plot of E_{zr} is the same as above, while the imaginary part E_{zi} vanishes. On the other hand, the real parts B_{xr} and B_{yr} both vanish, whereas the imaginary parts are non-zero.

The figure below shows B_{xi}. Here it is interesting to note the behavior at the boundaries. While the electric field vanishes at all four boundaries, B_{xi} only vanishes on the vertical lines. This is because the Maxwell boundary condition for the magnetic field at a perfect conductor is $B_n = 0$. Theory imposes no condition, however, on a magnetic field parallel to the conductor. For the plot of B_{yi} (not shown) the situation is of course reversed.

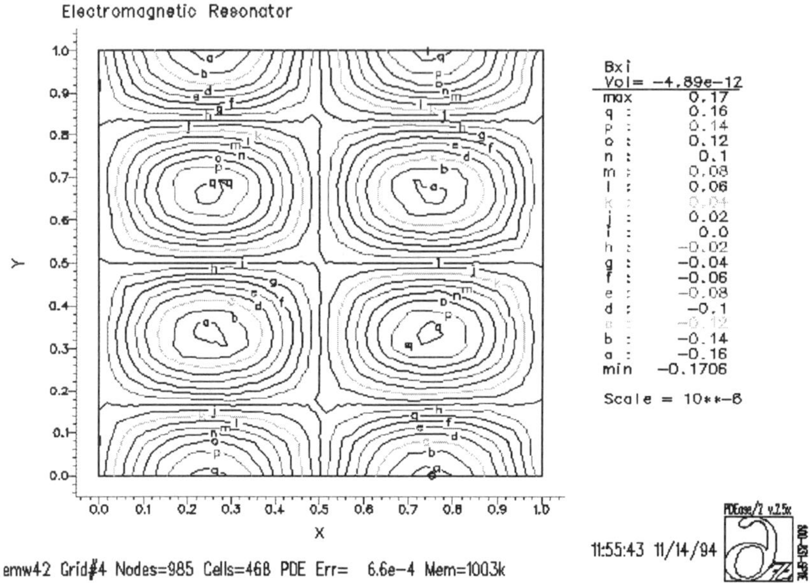

We now repeat the run with the parameters

```
m=3                n=2
```

At first you may be surprised to find that there is no change in the resonant pattern, although we have exchanged the values of *m* and *n*. The value of *omega* is the same for both configurations, and our procedure only yields one of the alternatives.

The problem seems to be that the driving effect of the input field vanishes because of symmetry. As a remedy, we perturb the field at the input in a less symmetric manner and also make the domain less symmetric than before.

```
                                                        { emw43.pde }
 ...
 Lx= 1.0            Ly= 1.234
 ...
                    region 1
 start (0,0)
```

```
  value(Ezr)= 0  value(Ezi)= 0
  line to (Lx,0)  to (Lx,Ly)  to (0,Ly)    to (0,Ly/2)
  value(Ezr)= Ein    value(Ezi)= 0   line to finish
  ...
  surface( Ezr)
  vector( Bxi, Byi)
end
```

This produces the expected new type of pattern for *Ezr*, which we display below as a *surface* plot. Now there are indeed three half-waves in the *x* direction. When investigating the result for other values of *m* and *n*, we should keep this device in mind.

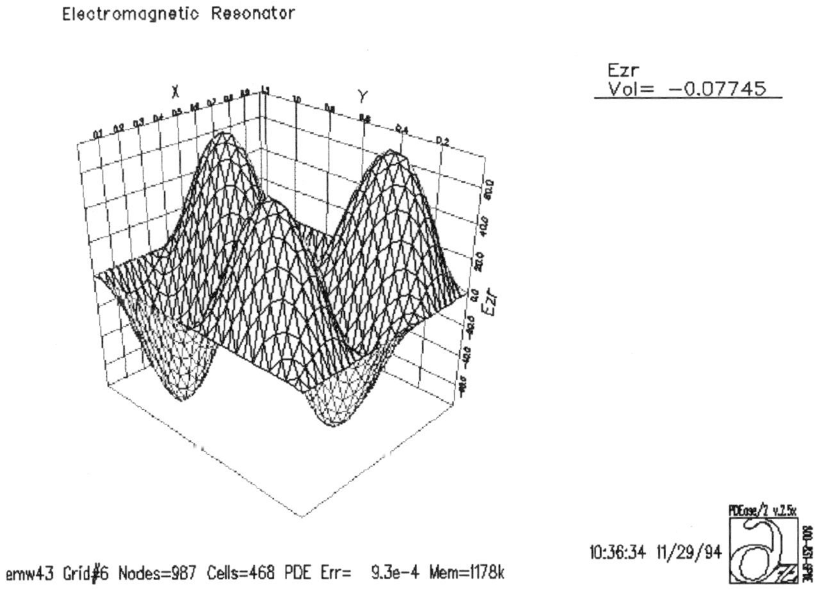

Since we have included the magnetic field in the descriptor, we may also obtain a *vector* plot of the imaginary *x* and *y* components of **B**, i.e. a snapshot of the field lines. In the plot below we first note that the magnetic field lines run along the conducting walls, as required by Maxwell's relations. In addition we find that **B** curls around crests in the electric field, which is perpendicular to the figure. Snapshots at other times would look just the same, except that the maximum magnitude of **B** would differ and that the sense of the circulation would reverse twice per period.

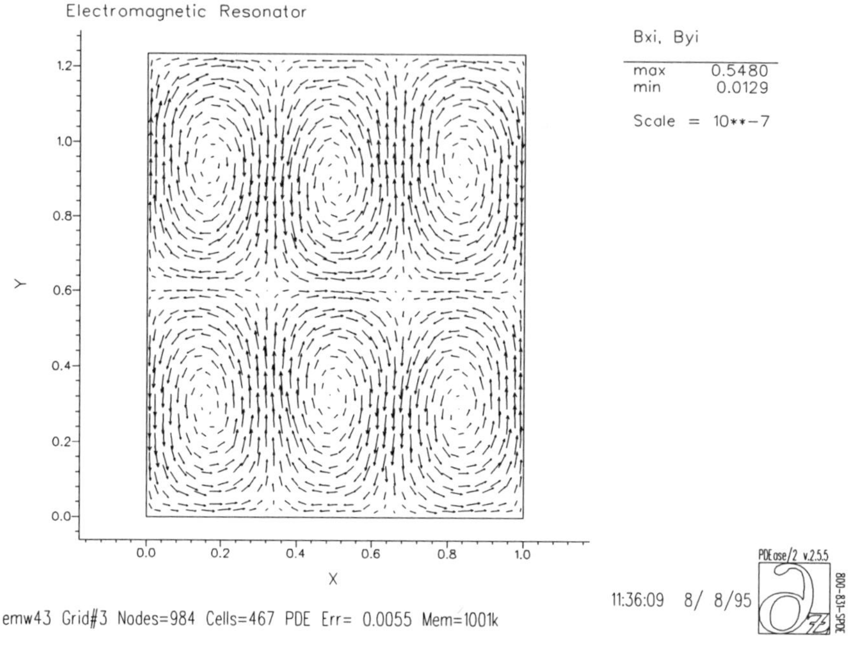

Eigenvalues and Eigenstates

Resonances without losses are stationary states of oscillation which persist even without external excitation, once the wave has been initiated. We came across similar phenomena in the chapters on elastic waves. PDEase permits us to detect and display such *eigenstates* directly and more conveniently without scanning the frequency (p.321). The descriptor required differs little from previous forms.

```
select                                              { emw44.pde }
  modes= 10
  errlim= 1e-2          nodelimit= 1000
  converge= 1e-6           subconverge= 1e-6
```

```
  contours= 10              surfacegrid= 30          paint= off
variables
  Ezr       Ezi
definitions
  Lx= 1.5              Ly= 1.0
  eps0= 8.85e-12       eps= eps0
  mu0= 4*pi*1e-7       mu= mu0
equations
  del2( Ezr)+ lambda*Ezr= 0
  del2( Ezi)+ lambda*Ezi= 0
boundaries
        region 1
  start (0,0)
  value(Ezr)= 0  value(Ezi)= 0
  line to (Lx,0) to (Lx,Ly) to (0,Ly) to (0,Ly/2) to finish
monitors
  contour( Ezr)
plots
  contour( Ezr)          surface( Ezr)
end
```

On running this descriptor we find a sequence of eigenfunctions, two of which are shown below as examples. The file EMW44.EIG contains a list of the eigenvalues, $\lambda = \mu\varepsilon\omega^2$. It exhibits similar pairs corresponding to eigenfunctions that only differ by their signs. From the eigenvalues we may now calculate the frequencies and compare them with the exact expression (p.351).

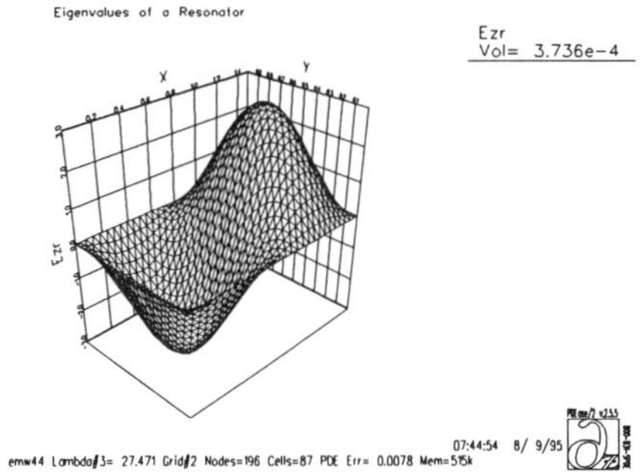

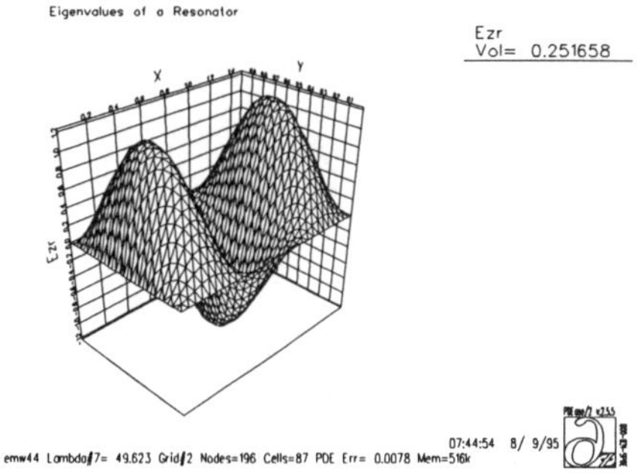

Cavity With Semi-Circular End

Next we explore a case for which there is no elementary solution. The geometry of the cavity is evident from the figure below. Since we have no guide as to the resonant frequencies we use PDEase in its *eigenvalue* mode. We use the preceding descriptor with the following amendments.

```
...                                                        { emw45.pde }
L= 1.0
...
            region 1
start (0,-L)
value(Ezr)= 0  value(Ezi)= 0
line to (L,-L) arc to (2*L,0) to (L,L)
line to (0,L) to finish
```

The next two figures are examples of the eigenfunctions obtained with this descriptor.

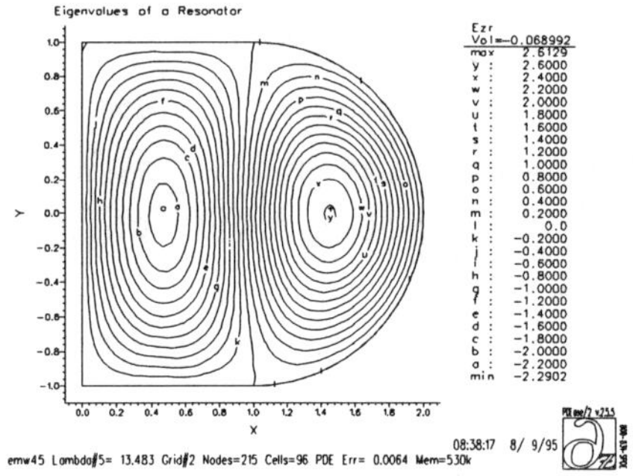

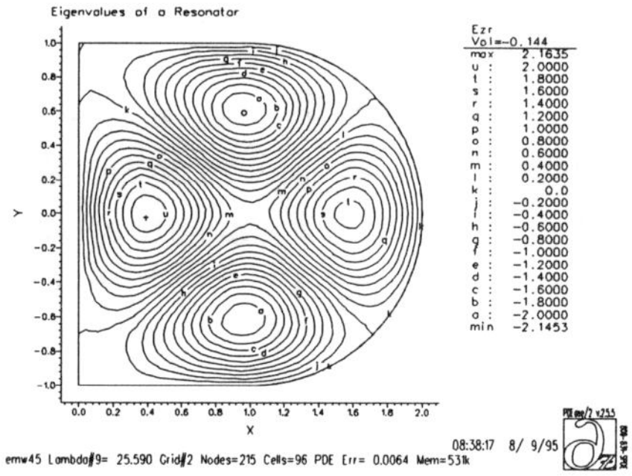

Exercises

❑ Use EMW41.PDE to explore the resonant states corresponding to several different combinations of m and n. What pattern do you obtain for the lowest frequency?
❑ To how high frequencies can you trace resonant modes? What circumstances put the upper limit on that frequency?
❑ Could you produce a snapshot of the magnetic field lines at a different time? (Hint: start the wave at a different phase.)
❑ Use the *eigenvalue* method for finding the 10 lowest stationary states in a resonator defined by an equilateral triangle.
❑ Modify EMW41.PDE to make the medium conductive. Scan over a resonance for different values of *sigma*.

Wave-Guides

In previous chapters we assumed that all field quantities were independent of z. As we have already seen (p.350), a solution for the complex amplitude of the z-component $E_{z0}(x,y,z)$ may sometimes be expressed as a product, such as $E_{z0}^{xy}(x,y)\,E_{z0}^{z}(z)$ (where the superscripts refer to the variables). This means that if we are able to guess a partial solution $E_{z0}^{z}(z)$, we may then solve for $E_{z0}^{xy}(x,y)$ by means of PDEase.

A waveguide[3p522] may be modeled as an infinitely long cavity, where a wave is guided through a dielectric material in the direction of the axis. The E_{z0}^{z} factor of the wave field may be described by

$$E_{z0}^{z}(z) = \exp(-\gamma z) \equiv \exp(-\alpha z)\exp(-i\beta z), \qquad \bullet$$

where $\gamma \equiv \alpha + i\beta$. The PDE for the amplitude of a time-sinusoidal E_z in a loss-free dielectric material is (p.333)

$$\frac{\partial^2 E_{z0}}{\partial x^2} + \frac{\partial^2 E_{z0}}{\partial y^2} + \frac{\partial^2 E_{z0}}{\partial z^2} + \mu\varepsilon\omega^2 E_{z0} = 0, \qquad \bullet$$

and if we substitute $E_{z0} = E_{z0}^{xy}(x,y)\,E_{z0}^{z}(z)$ into the above equation, it becomes

$$E_{z0}^{z}\frac{\partial^2 E_{z0}^{xy}}{\partial x^2} + E_{z0}^{z}\frac{\partial^2 E_{z0}^{xy}}{\partial y^2} + E_{z0}^{xy}E_{z0}^{z}\gamma^2 + \mu\varepsilon\omega^2 E_{z0}^{xy}E_{z0}^{z} = 0, \quad \text{or}$$

$$\frac{\partial^2 E_{z0}^{xy}}{\partial x^2} + \frac{\partial^2 E_{z0}^{xy}}{\partial y^2} + h^2 E_{z0}^{xy} = 0, \qquad \bullet$$

where $h^2 \equiv \gamma^2 + \mu\varepsilon\omega^2$. PDEase easily solves this equation, but if we are to obtain a periodic wave through the waveguide the exponent $\gamma \equiv \sqrt{h^2 - \mu\varepsilon\omega^2}$ must be purely imaginary, i.e. $i\beta \equiv \sqrt{h^2 - \mu\varepsilon\omega^2}$ in $E_{z0}^{z}(z) = \exp(-i\beta z)$. If the real part α were not zero, the wave would be exponentially attenuated. The above condition may also be written $\omega > h / \sqrt{\mu\varepsilon}$. ●

There is an additional condition involved, however. In the preceding chapter we obtained solutions for virtually any value of ω, but we had to supply excitation by means of an external field. It this input field were removed, the wave in the cavity would die out. Only at the resonance frequencies would the oscillating field be self-sustained. We observed many resonant states, with the characteristic feature that the amplitude inside the cavity was higher than the input amplitude. Finally, the resonance frequencies would depend on the cross-sectional shape and size.

In summary, if we solve the PDE for a given cross-section we obtain a series of resonance parameter values, ${h_k}^2$. To each of these critical values corresponds a relation

$$\omega = \frac{\sqrt{h_k^2 + \beta^2}}{\sqrt{\mu\varepsilon}}, \quad \text{or} \quad \beta_k \equiv \frac{2\pi}{\lambda_z} = \sqrt{\mu\varepsilon\omega^2 - h_k^2}$$ ●

connecting the allowed frequencies and the wave numbers β_k.

In practice, the origin of the wave may be a microwave source or a laser, usually operated at a fixed frequency ω. As we shall see, we obtain a sequence of resonance parameters h_k^2, and we may then use the above relation to calculate the corresponding longitudinal wave numbers β_k (and hence the axial wavelengths). For a given value of the input frequency ω there may thus be more than one possible mode of wave transmission, each characterized by a different node pattern (set of zero lines) in the *xy*-plane.

Waveguide of Circular Cross-Section

We shall illustrate the theory by a circular *xy*-domain. In order to introduce a field we need to make an axial slit in the tube. We choose the width of the slit so small that it will not be noticeable at resonance. We assume the tube to be a perfect conductor and the internal dielectric to have vanishing conductivity. The following descriptor shows how this can be implemented.

We scan over a suitable range of h^2 around the critical value, much as we did before with the frequency in the case of a resonator. We generally do not make use of all the twenty stages, since we can interrupt the calculation as soon as we have reached the peak. Since we excite the wave by an external field, we need PDEs for the imaginary as well as the real part of E_{z0}^{xy}. In the descriptor we choose the symbol *Ezr* for the real part.

```
select                                                           { emw51.pde }
  stages= 20
  errlim= 1e-6                    nodelimit= 300
  converge= 1e-6                  subconverge= 1e-6
  contours= 16                    surfacegrid= 30        paint= off
variables
  Ezr        Ezi
definitions
  r0= 1e-2
  phi= pi/50                        { Half-angle of slit, seen from the axis }
  eps0= 8.85e-12                  eps= eps0
  mu0= 4*pi*1e-7                  mu= mu0
  h2= (5.7800+ stage*0.0005)*1e4                                 { h2= h**2 }
  Ein= 1.0                                                   { Input E field }
equations
  del2( Ezr)+ h2*Ezr= 0
  del2( Ezi)+ h2*Ezi= 0
boundaries
        region 1
  start ( -r0*cos(phi), -r0*sin(phi) )
    value(Ezr)= 0          value(Ezi)= 0
    arc to (0,-r0) to (r0,0) to ( -r0*cos(phi), r0*sin(phi) )
                                    { Input field, real and imaginary part... }
  value(Ezr)= Ein          value(Ezi)= 0       line to finish
monitors
  contour( Ezr)
plots
  contour( Ezr)            surface( Ezr)
end
```

Running the above descriptor we find the first surface plot below, which illustrates the fundamental mode of oscillation. Increasing the parameter values to

```
h2= (1.4700+ stage*0.0005)*1e5
```

we obtain the second surface plot, which exhibits an azimuthal variation as well.

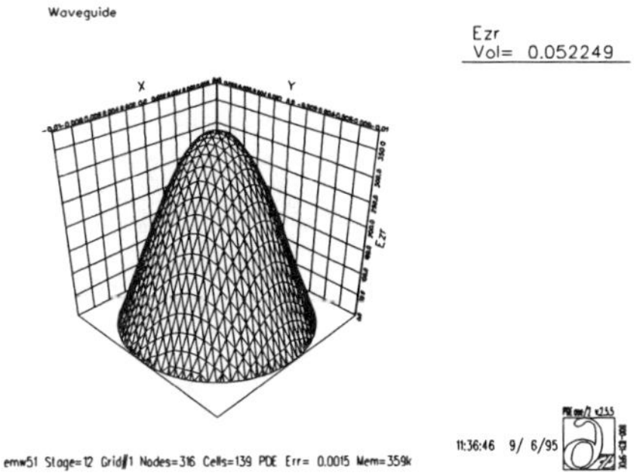

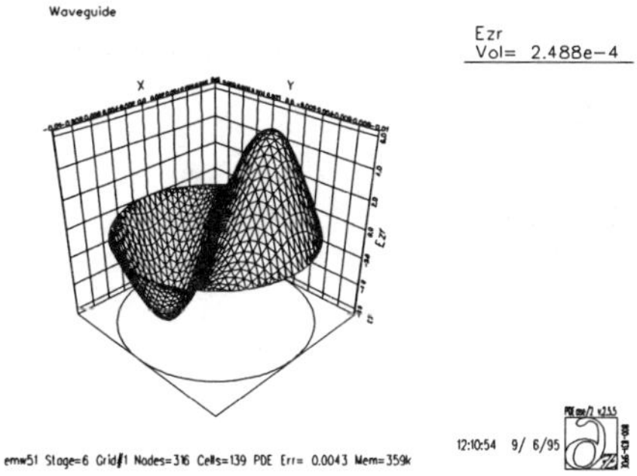

If these plots seem familiar, it is because the problem of the circular drumhead (p.321) is virtually identical. We will see this more clearly if we convert from forced oscillation through a slit to the eigenvalue mode. In other words, we close the slit and search for stationary, oscillating fields inside the conducting tube. Since there will be no phase shift with respect to a driving field, we only need a single equation

$$\frac{\partial^2 E_{z0}^{xy}}{\partial x^2} + \frac{\partial^2 E_{z0}^{xy}}{\partial y^2} + h^2 E_{z0}^{xy} = 0$$

where h^2 will be the eigenvalue to be calculated. The new descriptor will thus read

```
select                                                    { emw52.pde }
  modes= 10                  errlim= 1e-2
  converge= 1e-6             subconverge= 1e-6
  contours= 16               surfacegrid= 60       paint= off
```

```
variables
  Ez
definitions
  r0= 1e-2
equations
  del2( Ez)+ lambda*Ez= 0
boundaries
        region 1
  start(r0,0)
  value(Ez)= 0
  arc to (0,r0) to (-r0,0) to (0,-r0) to finish
monitors
  contour( Ez)
plots
  contour( Ez)            surface( Ez)
end
```

Depending on the PDEase version used you may accidentally obtain the error message “Eigenvalue near zero”. If this should occur, you should increase the number of initial nodes by selecting *ngrid*= 20 or higher, at the expense of a longer run. Alternatively, you may decrease the number of *modes* requested.

The above descriptor results in eigenfunctions of increasing complexity, as illustrated below by two examples.

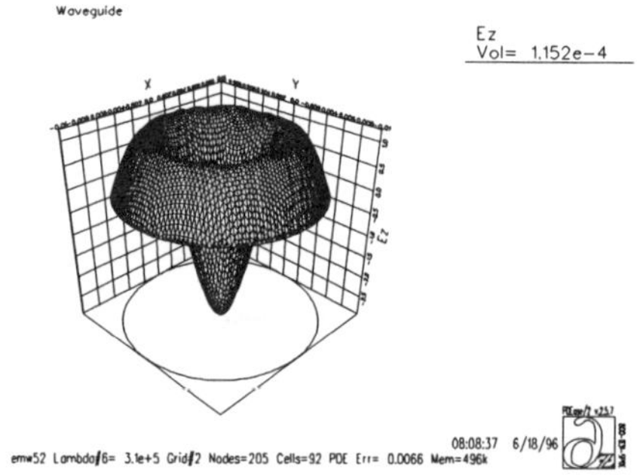

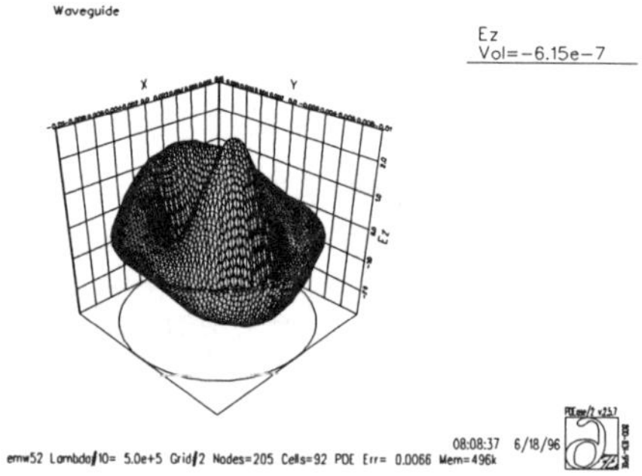

One family of nodes arises by successive addition of radial nodal (zero) lines. At higher eigenvalues we also come across a circular node, in addition to the usual one at the boundary. You may be astonished to find that the program reports pairs of eigenfunctions which are identical except for a rotation through 90^0. In fact we may obtain an infinite number of solutions by making linear combinations of such pairs. In a following chapter on the harmonic oscillator we shall illustrate that a function of *any* orientation may be generated by combining two basic states in the appropriate way.

A list of the eigenvalues *lambda* ($= h^2$) is available after the run in the file EMW52.EIG. All these eigenvalues also apply to a *resonator* of the same geometrical shape, but with no variation of the field with the *z*-coordinate. We obtain the critical frequencies for the resonator case by the relation $\mu\varepsilon\omega^2 = h^2$.

Expressions for Other Field Components

A waveguide is different from a resonator in the sense that a wave runs in the axial direction. So far we only have a solution for E_{z0}, but we may obtain the other two space components of the electric field, as well as those for the magnetic field, from the Maxwell equations.

The type of waveguide we have been studying in this chapter is called transverse-magnetic (TM) since the component of **B** in the axial direction vanishes. The electric components may all be different from zero, however, and in total we have to consider five field components. Fortunately, they all have similar *z*-dependence and may be calculated from E_{z0}^{xy}.

We shall now derive expressions for the remaining field components at resonance from a solution obtained by PDEase. Maxwell's curl equations yield the relations required. Starting with

$$\nabla \times \mathbf{E} = -\frac{\partial \mathbf{B}}{\partial t}$$

we may write, for the first component,

$$\frac{\partial E_z}{\partial y} - \frac{\partial E_y}{\partial z} = -\frac{\partial B_x}{\partial t} \quad .$$

After introducing factored expressions for the various field components, i.e.

$$E_{z0}(x,y,z) = E_{z0}^{xy}\exp[-i\beta z)]$$

and so on, we obtain

$$\exp[i\omega t - i\beta z)]\frac{\partial E_{z0}^{xy}}{\partial y} - (-i\beta)E_{y0}^{xy}\exp[i\omega t - i\beta z)] =$$

$$- i\omega B_{x0}^{xy}\exp[i\omega t - i\beta z)]$$

or

$$\frac{\partial E_{z0}^{xy}}{\partial y} + i\beta E_{y0}^{xy} = -i\omega B_{x0}^{xy}$$

After an analogous treatment of the other two components of **B** we arrive at

$$\begin{cases} \dfrac{\partial E_{z0}^{xy}}{\partial y} + i\beta E_{y0}^{xy} = -i\omega B_{x0}^{xy} \\ -i\beta E_{x0}^{xy} - \dfrac{\partial E_{z0}^{xy}}{\partial x} = -i\omega B_{y0}^{xy} \\ \dfrac{\partial E_{y0}^{xy}}{\partial x} - \dfrac{\partial E_{x0}^{xy}}{\partial y} = -i\omega B_{z0}^{xy} \end{cases} \bullet$$

Next we use the other curl equation

$$\frac{1}{\mu}\nabla \times \mathbf{B} = \frac{\partial \mathbf{D}}{\partial t}$$

in a similar manner to find three more equations connecting electric and magnetic field components.

$$\begin{cases} \dfrac{\partial B_{z0}^{xy}}{\partial y} + i\beta B_{y0}^{xy} = i\omega\mu\varepsilon E_{x0}^{xy} \\ -i\beta B_{x0}^{xy} - \dfrac{\partial B_{z0}^{xy}}{\partial x} = i\omega\mu\varepsilon E_{y0}^{xy} \\ \dfrac{\partial B_{y0}^{xy}}{\partial x} - \dfrac{\partial B_{x0}^{xy}}{\partial y} = i\omega\mu\varepsilon E_{z0}^{xy} \end{cases}$$

The above six equations for the six field components form a homogeneous system. Such systems may often be solved in terms of one of the unknowns. Here we assume $B_{z0}^{xy} = 0$ (transverse-magnetic mode), which permits us to solve for the other quantities in terms of E_{z0}^{zy}, the latter being known from PDEase. Eliminating between pairs of equations we arrive at the solution for the remaining field components

$$E_{x0}^{xy} = -\frac{i\beta}{h_k^2}\frac{\partial E_{z0}^{xy}}{\partial x}, \qquad E_{y0}^{xy} = -\frac{i\beta}{h_k^2}\frac{\partial E_{z0}^{xy}}{\partial y}$$

$$B_{x0}^{xy} = \frac{i\omega\mu\varepsilon}{h_k^2}\frac{\partial E_{z0}^{xy}}{\partial y}, \qquad B_{y0}^{xy} = -\frac{i\omega\mu\varepsilon}{h_k^2}\frac{\partial E_{z0}^{xy}}{\partial x}$$

In order to rephrase these results to fit PDEase we have to substitute the real and imaginary parts, i.e.

$$E_{z0}^{xy} \equiv E_{zr} + iE_{zi}$$

where we use the previous, simplified notation as in the descriptor. Our final results for the electric field are

$$E_{xr} = \frac{\beta}{h_k^2}\frac{\partial E_{zi}}{\partial x}, \qquad E_{xi} = -\frac{\beta}{h_k^2}\frac{\partial E_{zr}}{\partial x}$$

$$E_{yr} = \frac{\beta}{h_k^2}\frac{\partial E_{zi}}{\partial y}, \qquad E_{yi} = -\frac{\beta}{h_k^2}\frac{\partial E_{zr}}{\partial y}$$

and for the magnetic field

$$B_{xr} = -\frac{\omega\mu\varepsilon}{h_k^2}\frac{\partial E_{zi}}{\partial y}, \qquad B_{xi} = \frac{\omega\mu\varepsilon}{h_k^2}\frac{\partial E_{zr}}{\partial y}$$

$$B_{yr} = \frac{\omega\mu\varepsilon}{h_k^2}\frac{\partial E_{zi}}{\partial x}, \qquad B_{yi} = -\frac{\omega\mu\varepsilon}{h_k^2}\frac{\partial E_{zr}}{\partial x}$$

where $\beta_k \equiv \frac{2\pi}{\lambda_{zk}} = \sqrt{\mu\varepsilon\omega^2 - h_k^2}$. Multiplying by the factor $\exp[-i\beta z)]$ we could obtain the 3D form of the above solutions.

Other Field Components Using PDEase

We now modify EMW51.PDE to generate a specific eigenfunction. The new lines required in the *definitions* segment are as listed below. We must realize that the eigenvalues given by EMW52.EIG are approximate and may be in error by a few parts in a thousand. Scanning for state number 9 in the list we find the value for *h*2 cited below to yield very large amplitude. It remains to choose a suitable value for *omega* compatible with the inequality (p.359). From this we calculate *beta* and obtain all the field components. You could include plots of various kinds of any of these quantities *ad lib*.

```
...                                                    { emw61.pde }
errlim= 1e-3                    nodelimit= 1000
...
h2= 497380                      omega= 1.2*sqrt[ h2/(mu*eps)]
beta= sqrt(mu*eps*omega**2- h2)
Exr= beta/h2*dx(Ezi)            Exi= -beta/h2*dx(Ezr)
Eyr= beta/h2*dy(Ezi)            Eyi= -beta/h2*dy(Ezr)
Bxr= -omega*mu*eps/h2*dy(Ezi)   Bxi= omega*mu*eps/h2*dy(Ezr)
Byr= omega*mu*eps/h2*dx(Ezi)    Byi= -omega*mu*eps/h2*dx(Ezr)
...
```

The following two plots are examples of results that may emerge when applying this descriptor. The first one shows the directions of the electric field in the *xy* plane. Notice that the field meets the boundary at right angles, as must be, and that the magnitude is high at the boundary and near the center. These field components are thus non-zero, contrary to the situation in a resonator of the same geometry. If β_k becomes zero, however, the electric field vector in the solution plane vanishes, as may be seen from the above expressions.

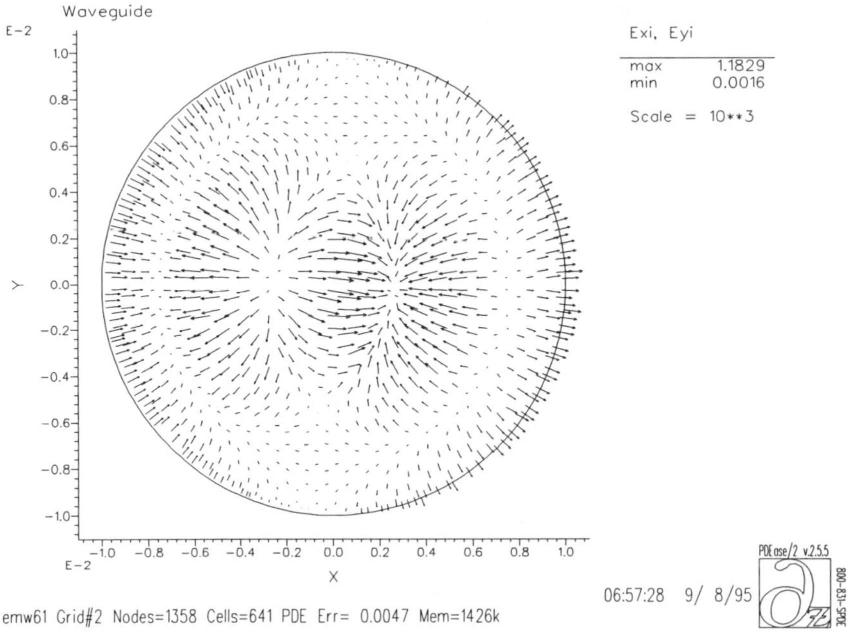

The second figure is a direction map of the magnetic field in the *xy* plane. First we see that the normal component vanishes at the metallic boundary, which is correct. Also we may notice that the magnetic field curls around the crests in E_z, as we would expect from Maxwell's relations.

It would be possible, as well, to solve for the field components in other planes parallel to the present one. The options are to multiply the solutions by $\exp(-i\beta z)$ or to excite the wave at another phase angle.

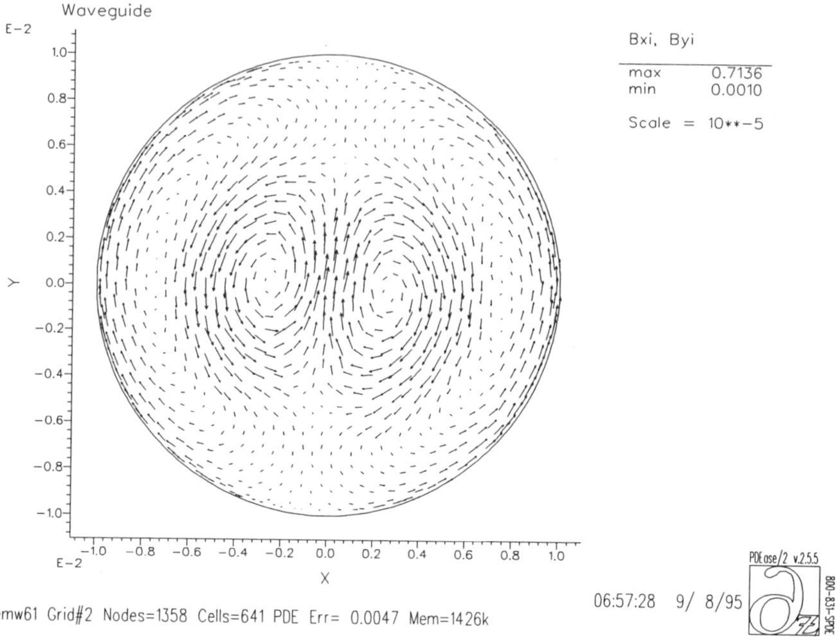

Exercises

❑ Study the *surface* plots corresponding to the four- and six-wedge patterns from EMW52.PDE. Would 3 radial node spokes be possible?
❑ Plot a resonant state for the circular waveguide, corresponding to a 90° shift of the input field.
❑ Make normalized *vector* plots (unit magnitudes) of the electric and magnetic fields in the *xy*-plane, corresponding to the last figures.
❑ Solve for a few allowed states of a rectangular waveguide. Find all allowed values of the wavelength, corresponding to a chosen value of *omega*.
❑ Explore some resonant modes of a waveguide having equilateral triangular cross-section. Include all electric and magnetic field components.

Harmonic Oscillators

We have seen that similar PDEs occur again and again in various fields of physics. Hence, it should not come as a surprise that wave mechanics also looks strongly familiar.

As explained in any book on quantum mechanics[10], the time-independent Schrödinger equation for a one-particle system may be written

$$\frac{h^2}{8\pi^2 m_e}\nabla^2\psi - V\psi + W\psi = 0, \qquad \bullet$$

where h is the Planck constant, m_e the mass of the particle, V the potential energy, W the total energy and ψ the wave function. The square of this real function, $\psi(x,y,z)^2$, is the probability per unit volume of finding the particle near the point (x,y,z). For an atomic system, V is a known function, and we may determine the allowed energy states W_i by eigenvalue methods. We just impose the condition that ψ must vanish on a distant boundary.

Oscillator in 1D

We start by considering one of the simplest systems in quantum mechanics: an elastically bound particle (e.g. an electron), free to move along the x-axis. Newton's classical equation of motion yields the stationary solution $x = a\sin(2\pi f_0 t)$ with $V = 2\pi^2 m_e f_0^2 x^2$ for the potential energy. Here, a is the amplitude, m_e the mass of the particle and f_0 the frequency of oscillation.

In order to solve the corresponding wave-mechanical problem using PDEase we write the equation in the form

$$\nabla^2\psi - 2\pi^2 m_e f_0^2 x^2 c\psi + Wc\psi = 0, \qquad \bullet$$

with the notation $c = 8\pi^2 m_e / h^2$. The descriptor is straightforward and should speak for itself. Recent versions of PDEase allow the use of a caret (^) for exponentiation, which we exploit here to improve readability.

As regards the PDE it should be noted that the term containing the eigenvalue λ must be written with that parameter as the *first* factor.

```
select                                                    { quant1.pde }
  modes= 10
  errlim= 1e-2    nodelimit= 500
  converge= 1e-6      subconverge= 1e-6
variables
  psi
definitions
  Lx= 20e-10                    Ly= Lx/100
  me= 9.1096e-31                h= 6.6262e-34
  c= (8* pi^2* me)/h^2          f0= 1e15
  V= 2*pi^2*me*f0^2*x^2
equations
  del2( psi)- V*c*psi+ lambda*c*psi= 0
boundaries
          region 1
  start(Lx,-Ly)
  value(psi)= 0   line to (Lx,Ly)
  natural(psi)= 0  line to (-Lx,Ly)                       { Enforce 1D }
  value(psi)= 0   line to (-Lx,-Ly)
  natural(psi)= 0  line to finish
          feature                                 { Enforce denser grid }
  start(-Lx/5,0) line to (Lx/5,0)
monitors
  elevation( psi) from (-Lx,0) to (Lx,0)
plots
  elevation( psi) from (-Lx,0) to (Lx,0)
  elevation( psi^2) from (-Lx,0) to (Lx,0)
end
```

The following plots show two samples of the solutions. We notice that these eigenfunctions are different from those arising in the mechanical and electromagnetic cases, in the sense that the wave function does not vanish to first order toward the boundary, but rather exponentially.

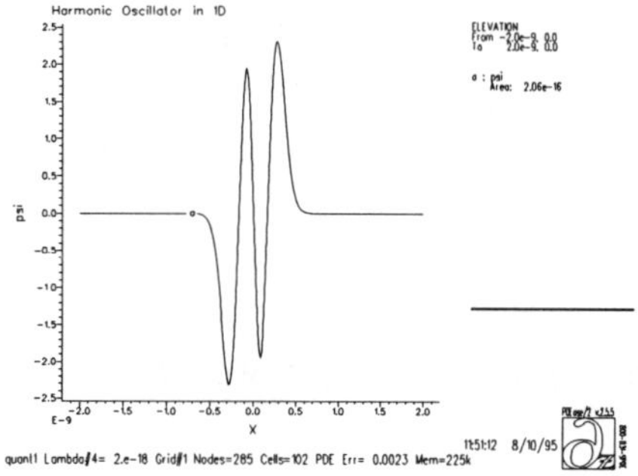

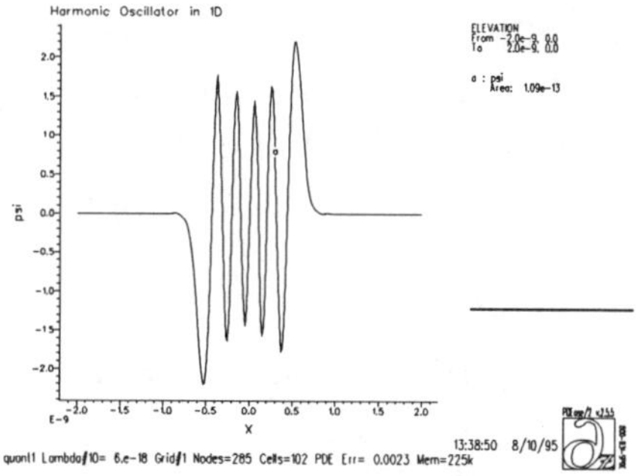

The file QUANT1.EIG contains the list of eigenvalues displayed to the right. The fact to notice is that the first value is one-half of the difference between the succeeding energies. This is in accord with the exact solution, which yields the energy values (traditionally with $n = 0,1,2,\ldots$)

$$W_i = (n + \tfrac{1}{2})hf_0 = (i - 1 + \tfrac{1}{2})hf_0$$

(here with $i = 1,2,3,\ldots$). It is a surprising and significant result of quantum mechanics that the energy of the lowest state of the oscillator does not vanish.

i	W_i
1	3.31e-19
2	9.95e-19
3	1.66e-18
4	2.32e-18
5	2.99e-18
6	3.66e-18
7	4.34e-18
8	5.01e-18
9	5.69e-18
10	6.37e-18

The plots of ψ^2 (not shown here) reflect the probability of existence of the particle. Since the latter must be present somewhere on the x-axis, we expect the integral over ψ^2 to be unity. PDEase does not automatically satisfy this condition, but we may easily take it into account when plotting the results. We only need to divide ψ^2 by the integral (with respect to x) of that function before plotting. PDEase provides a command for computing a *double* integral over the solution domain. Since our solution is independent of y, it is sufficient to divide this integral by the height of the domain, $2L_y$, to obtain the single integral.

It would also be of interest to compare the quantum-mechanical probability to that of the classical oscillator, say, for the highest state we calculate. We first extract the amplitude (a) of oscillation from the fact that the potential energy must be equal to the total energy of this highest state at the turning point.

$$2\pi^2 m_e f_0^2 a^2 = 9.5 h f_0$$

Inverting the classical solution to yield time t against x, we have

$$t = \frac{1}{2\pi f_0} \arcsin\left(\frac{x}{a}\right),$$

which has the derivative

$$\frac{dt}{dx} = \frac{1}{2\pi f_0} \frac{1}{\sqrt{a^2 - x^2}}.$$

We have now almost arrived at the probability, which is proportional to the time the particle spends in a given interval of x. It only remains to divide this derivative by the total range of time, i.e. one-half period, or $1/(2f_0)$.

The derivative goes steeply toward infinity close to the turning points, and we shall have to truncate the curve in order to plot it. The most convenient way is to reduce the function to zero outside the acceptable range of coordinate values. The logical statements of PDEase permit us to do this, as shown in the following amendment to the preceding descriptor.

```
  ...                                              { quant2.pde }
  int_x= integral( psi^2)/ (2*Ly)                  { Integral over x }
  a= sqrt(9.5*h/[2*pi^2*me*f0])                    { Amplitude }
  dPdx= 1/pi/sqrt(a^2-x^2)                         { Differential probability }
  prob= if abs(x)>0.99*a then 0   else dPdx
  ...
  elevation( psi) from (-Lx,0) to (Lx,0)
  elevation( psi^2/ int_x,  prob ) from (-Lx,0) to (Lx,0)
end
```

The following plot illustrates the relation between the classical and quantum-mechanical probability functions.

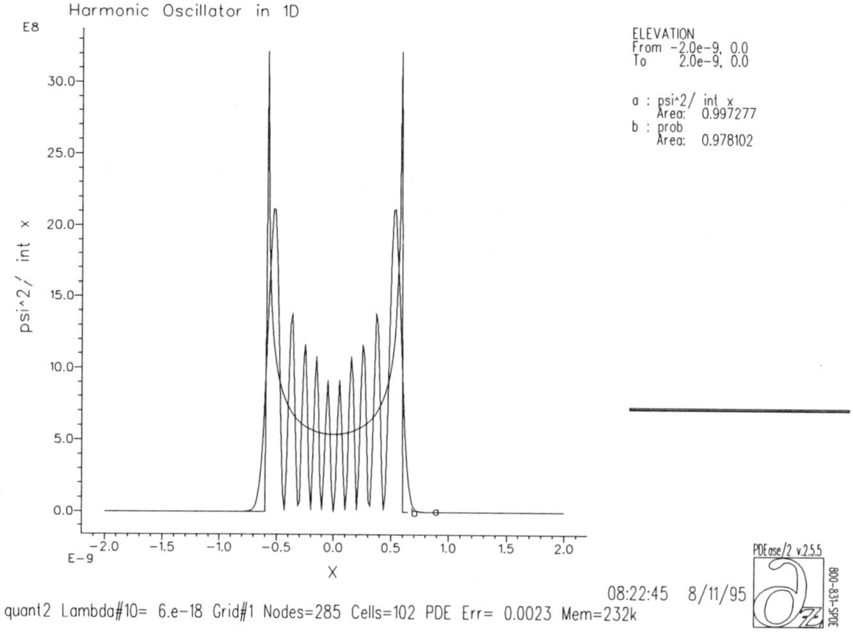

Oscillator in (x,y) Space

Our next example will be a particle elastically bound to a point, free to move in the *xy* plane. As is apparent from the descriptor below there are only a few changes with respect to QUANT1.PDE. The smaller value of *errlim* may make the run take an hour on a slow PC, but it is always possible to stop prematurely by <Esc>F. The expression for the potential energy *V* is similar but depends on the square of the radius, rather than *x*. In this case we require the probability integral over the 2D domain to be unity, which means that we can use *integral* directly to provide correct absolute values. The circular *feature* provides smaller triangular cells near the origin, where the wave function is likely to vary steeply.

```
select                                                          { quant11.pde }
  modes= 10
  errlim= 1e-3            nodelimit= 500
  converge= 1e-6          subconverge= 1e-6
  contours= 10            surfacegrid= 60         paint= off
variables
  psi
definitions
  r0= 2e-10                       r1= 10e-10
  me= 9.1096e-31                  h= 6.6262e-34
  c= (8* pi^2* me)/h^2            f0= 1e15
  V= 2*pi^2*me*f0^2* (x^2+y^2)
  prob= psi^2/ integral( psi^2)
equations
  del2(psi)- V*c*psi+ lambda*c*psi= 0
boundaries
            region 1
  start(r1,0)
  value(psi)= 0   arc to (0,r1) to (-r1,0) to (0,-r1) to finish
            feature
  start (r0,0) arc(center= 0,0) angle= 360
monitors
  contour( psi)
plots
  contour( psi)
  surface( prob) zoom(-r1/2,-r1/2, r1,r1)
end
```

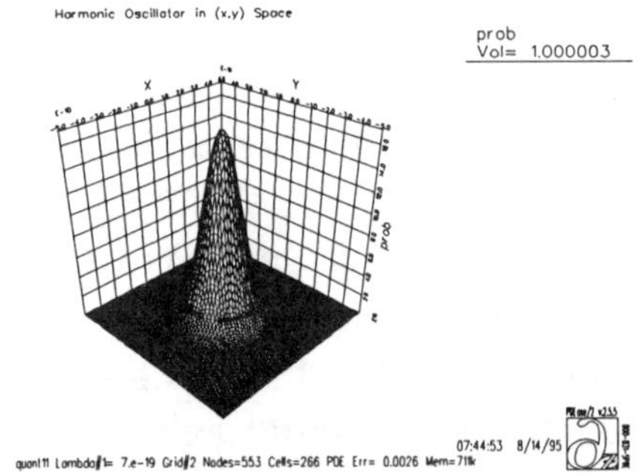

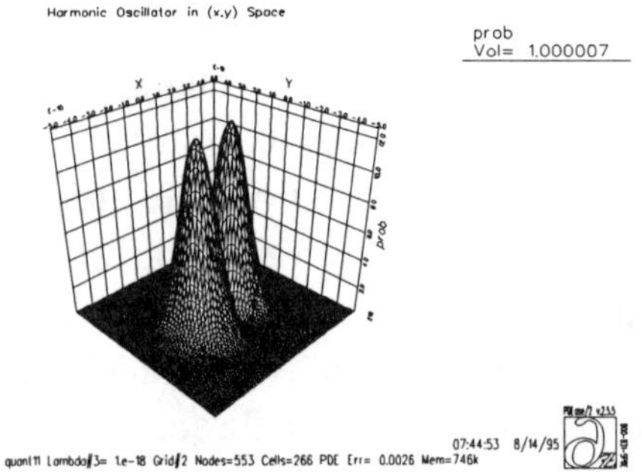

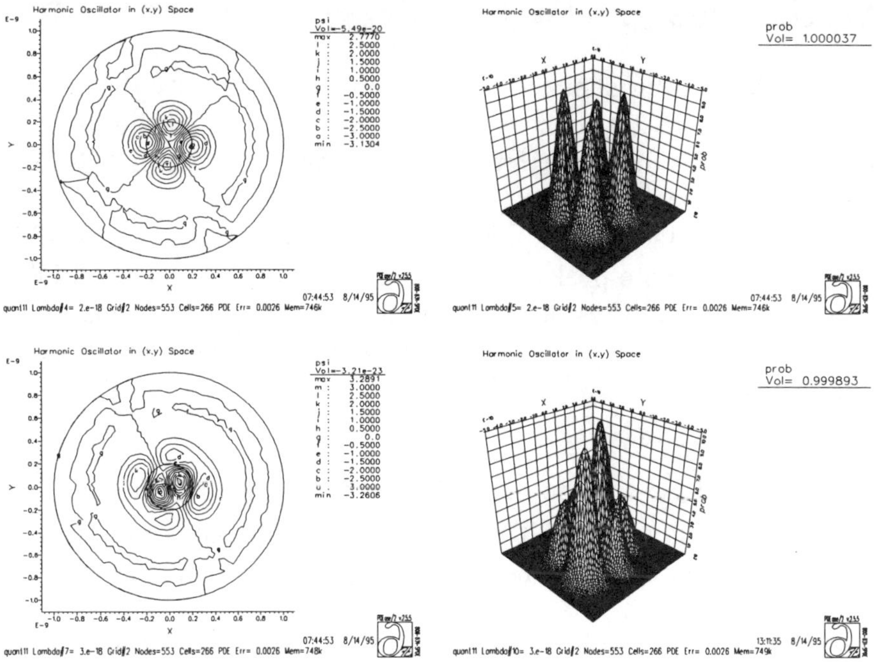

The above figures show a few samples of the plots produced by this descriptor, and the table to the right lists the energy eigenvalues. Evidently, the shapes of the eigenfunctions are more varied than in 1D. The probability distributions integrate (*Vol.*) to a value close to unity, as expected.

i	W_i
1	6.63e-19
2	1.33e-18
3	1.33e-18
4	2.00e-18
5	2.00e-18
6	2.00e-18
7	2.66e-18
8	2.66e-18
9	2.67e-18
10	2.67e-18

Exact solutions exist in this case as well, with the energy values

$$W_i = (n+1)hf_0 = ihf_0$$

$(i = 1,2,3,\ldots)$ as confirmed by the numerical results.

As the table shows, the 2nd and 3rd states have the same energy, hence known as *degenerate*. The next eigenvalue corresponds to three eigenfunctions, and the last one to four. We shall discuss degeneracy in more detail in the three-dimensional case.

Mixed States

You may have been shocked to see that the probability distribution for the 3rd state in 2D consists of two peaks (p.374), considering that the potential has no preferred direction. Why do we not find an axially symmetric probability?

The energy state corresponding to the two peaks is two-fold, however, so there are peaks at two different angles. Since the PDE is linear, we may always add the two solutions in various ways, producing a new wave function which would be equally valid. Let us take, for instance,

$$\psi = \sin(v)\,\psi_2 + \cos(v)\,\psi_3$$

where ψ_2 and ψ_3 are the second and third eigenfunctions of the harmonic oscillator in 2D. The trigonometric functions permit us to combine the functions in many different ways just by changing the angle v.

We shall now explore what happens when we make linear combinations of basic solutions. In order to gain access to two different solutions in the same descriptor, we have to pass by a data file, a procedure provided by PDEase. First, we must modify QUANT11.PDE as indicated below to produce this file.

```
 ...                                        { quant12.pde }
modes= 3
 ...
plots
  contour( psi)  print
end
```

This modification using *print* creates a data file named QUANT12.P01 and two others with extensions P02 and P03. The latter files contain our degenerate wave functions.

In order to study linear combinations of these functions we write a short descriptor as below. Here we change the angle v in steps by the *stages* device. Furthermore we read the files containing the solutions by the *table* statement.

```
  ...                                                            { quant15.pde }
select
  stages= 20
  contours= 10       surfacegrid= 60       paint= off
definitions
  r1= 10e-10
  psi2= table('quant12.p02')        psi3= table('quant12.p03')
  v= stage*pi/10                    psi= sin(v)*psi2+ cos(v)*psi3
  prob= psi^2/integral(psi^2)
boundaries
          region 1
  start(r1,0)
  arc to (0,r1) to (-r1,0) to (0,-r1) to finish
plots
  contour( psi)
end
```

Running this descriptor you will find that wave functions of any orientation in space may be generated by linear combination.

Oscillator in (ρ,z) Space

This problem is about a particle free to move in 3D and elastically bound to a point. To solve it completely we would need three space variables, but we shall reduce the number to two by separating the dependent variable into two factors. Using cylindrical coordinates we write

$$\psi = \psi_1(\varphi)\,\psi_2(\rho,z)$$

and modify the Laplacian term in the wave equation as follows.

$$\left\{\frac{1}{\rho}\frac{\partial}{\partial\rho}\left(\rho\frac{\partial\psi}{\partial\rho}\right)+\frac{1}{\rho^2}\frac{\partial^2\psi}{\partial^2\varphi}+\frac{\partial^2\psi}{\partial^2 z}\right\}-Vc\psi+Wc\psi=0$$

Substituting our factored function we obtain

$$\psi_1 \frac{1}{\rho}\frac{\partial}{\partial\rho}\left(\rho\frac{\partial\psi_2}{\partial\rho}\right)+\psi_2\frac{1}{\rho^2}\frac{\partial^2\psi_1}{\partial^2\varphi}+\psi_1\frac{\partial^2\psi_2}{\partial^2 z}$$
$$-Vc\psi_1\psi_2+Wc\psi_1\psi_2=0$$

We discover the convenience of the factored solution on multiplying all the terms in the PDE by $\rho^2/\psi_1\psi_2$. Then each of the terms either contains (ρ,z) or φ, never both of them. Hence we may rearrange the equation so that the (ρ,z) terms fall on the left side of the equals sign and the φ term on the right side.

$$\frac{1}{\psi_2}\rho\frac{\partial}{\partial\rho}\left(\rho\frac{\partial\psi_2}{\partial\rho}\right)+\frac{1}{\psi_2}\rho^2\frac{\partial^2\psi_2}{\partial^2 z}-Vc\rho^2+Wc\rho^2$$
$$=-\frac{1}{\psi_1}\frac{\partial^2\psi_1}{\partial^2\varphi}$$

Since the two sides are equal, they cannot depend on (ρ,z), nor on φ, but must be constant. Let us now solve the equation for ψ_1,

$$\frac{1}{\psi_1}\frac{\partial^2\psi_1}{\partial^2\varphi}=-m^2,$$

using the notation m^2 for the common constant. We obtain the simple solution

$$\psi_1 = A\sin(m\varphi)+B\cos(m\varphi)\equiv C\sin(m\varphi+\delta), \qquad \bullet$$

where A, B, C and δ are constants. If ψ_1 is to be a unique function of φ, adding 2π to the angle must yield the same value. This happens if and only if m is an integer $(0, \pm1, \pm2, \ldots)$. The value m=0 obviously yields a constant function ψ_1. The values of A and B may else be chosen to produce extrema (lobes) pointing at any angle with respect to the x-axis, while making the integral over a full turn equal to 2π.

If we multiply through by ψ_2 and introduce the constant m the PDE takes the form

$$\rho\frac{\partial}{\partial\rho}\left(\rho\frac{\partial\psi_2}{\partial\rho}\right)-m^2\psi_2+\rho^2\frac{\partial^2\psi_2}{\partial^2 z}-Vc\rho^2\psi_2+Wc\rho^2\psi_2=0 \qquad \bullet$$

If we also substitute an allowed value for *m*, PDEase solves this equation as easily as the previous ones.

```
select                                                    { quant21.pde }
  modes= 10
  errlim= 1e-3           nodelimit= 500
  converge= 1e-6         subconverge= 1e-6
  contours= 10           surfacegrid= 60          paint= off
  alias(x)= 'r'          alias(y)= 'z'
variables
  psi
definitions
  r0= 2e-10              r1= 20e-10
  m= 0
  me= 9.1096e-31         h= 6.6262e-34
  c= (8* pi^2* me)/h^2                  f0= 1e15
  V= 2*pi^2*me*f0^2*(x^2+y^2)
  int_rzphi= integral( 2*pi*x* psi^2)
  prob= psi^2/ int_rzphi
equations
  x*dx( x*dx( psi))- m^2*psi+ x^2*dyy( psi)
          - V*c*x^2*psi+ lambda*c*x^2*psi= 0
boundaries
          region 1
  start(0,-r1)       value(psi)= 0
  arc to (r1,0)  to (0,r1)
  natural(psi)= 0  line to finish
          feature
  start(0,-r0) arc(center= 0,0) to (r0,0) to (0,r0)
monitors
  contour( psi)
plots
  contour( prob)
  surface( prob)  zoom(0,-r1/2,  r1/2,r1)  viewpoint(r1,-r1/2, 30)
end
```

In the above descriptor we again integrate ψ^2 over the domain, which now effectively is three-dimensional. Since the integral of $\psi_1(\varphi)$ is set to be 2π, we need only multiply the integrand by $2\pi\rho$ to take the azimuthal variation into account.

The plots below show some of the probability functions for $m = 0$ in (ρ, z) space. In order to generate a mental picture of the 3D distribution one has to imagine these plots rotating about the z-axis.

The wave functions for $m = 1,2,3,\ldots$(not shown) gradually detach themselves from the symmetry axis as m increases. The general aspect of the probability distributions is a set of tori of different sizes, centered on the axis.

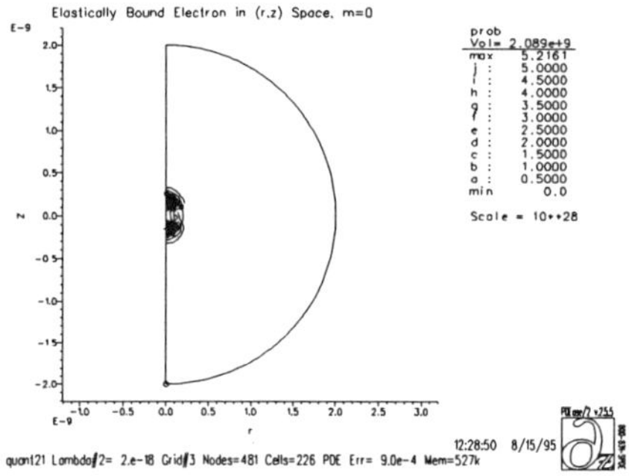

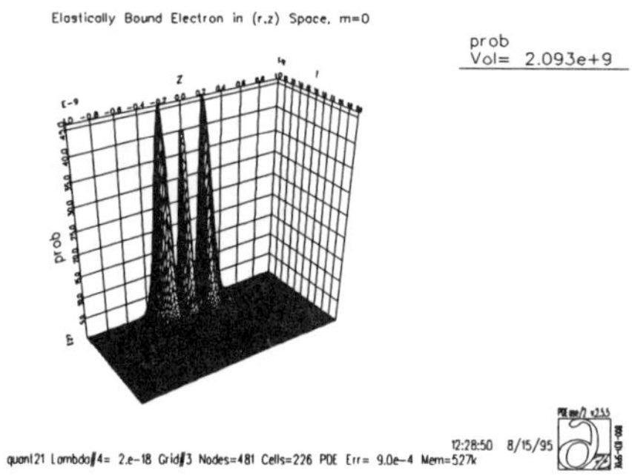

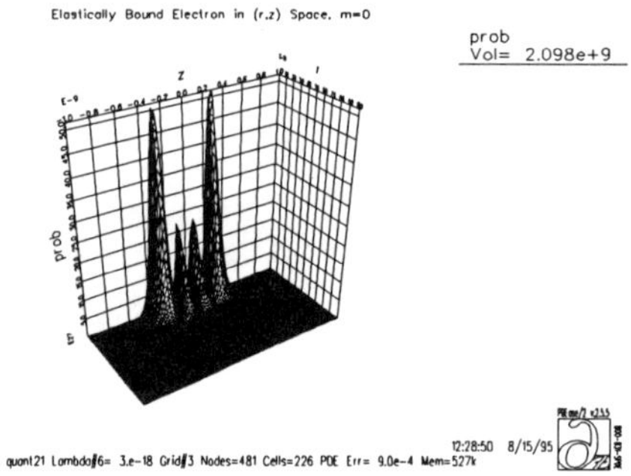

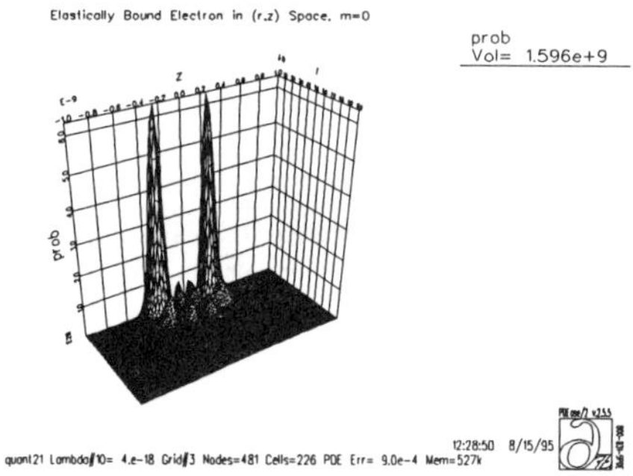

The following table shows the energies obtained for a few values of the quantum number m. Evidently, a given energy may occur repeatedly and in different columns. It is possible to solve the problem of the harmonic oscillator in 3D by exact means, and the result for the energies is

$$W_i = (n + 3/2)hf_0 = (i - 1 + 3/2)hf_0$$

$(i = 1,2,3,\ldots)$, which is well confirmed by the values in the table.

The exact solution also yields an expression for the number of states q corresponding to each energy eigenvalue:

$q = (n+1)(n+2)/2 = i(i+1)/2$

This result is also born out by the numerical results in the table, but when counting the values one must remember that there are states for $m = 0, \pm 1, \pm 2, \ldots$, which means that most of them occur twice with different signs.

i	*q*	*m*= 0	*m*= 1	*m*= 2	*m*= 3
1	1	9.92e-19			
2	3	1.66e-18	1.66e-18		
3	6	2.31e-18 2.32e-18	2.32e-18	2.32e-18	
4	10	2.98e-18 2.99e-18	2.99e-18 2.99e-18	2.99e-18	2.98e-18
5		3.64e-18 3.66e-18 3.66e-18	3.65e-18 3.66e-18	3.65e-18 3.66e-18	3.65e-18
6		4.33e-18	4.33e-18 4.33e-18 4.33e-18	4.33e-18 4.33e-18	4.32e-18 4.32e-18
7			5.00e-18	4.99e-18 5.00e-18 5.02e-18	4.99e-18 4.99e-18
8				5.67e-18	5.65e-18 5.66e-18 5.67e-18

Exercises

❑ Plot the solutions for the harmonic oscillator in 1D as $\ln(\psi^2)$ to explore how the function tends to zero toward the boundary.
❑ Find the eigenstates in (x,y) for a potential energy which is zero in the central region and 1e-17 at radii larger than 8e-10 (SI units). Define the potential energy using *ustep*.
❑ Modify the potential term in QUANT11.PDE to make the space-dependent factor take the form (abs(x)^2.1+ abs(y)^2.1).
❑ Modify QUANT11.PDE again to make the potential energy take the anisotropic form (abs(x)^2.1+ abs(y)^2).
❑ Replace the potential energy expression in QUANT21.PDE by V= 5e-9*sqrt(x^2+y^2) and solve for the new eigenstates.

Hydrogenic Systems

Although there are many systems in physics which are nearly-harmonic oscillators, a *single* particle in such a potential is mostly used as a theoretical model for illustrating the principles of quantum theory. There are two realistic systems, however, which we may treat by PDEase, i.e. the hydrogen atom and the hydrogen molecule ion. Both of these may be modeled as an electron in a fixed potential from atomic nuclei.

The Hydrogen Atom

In the simplest of all atoms, an electron moves under the electric attraction of a proton. We should treat this atom using an inertial system, which means that the proton will also move. The orbit of the nucleus becomes 1833 times smaller, however, so we shall completely neglect its influence on the eigenstates.

We thus consider the electron to have the spherically symmetric potential energy

$$V = -\frac{e^2}{4\pi\varepsilon_0 r} = -\frac{e^2}{4\pi\varepsilon_0\sqrt{\rho^2+z^2}}$$

where r is the distance to the nucleus, also expressed in (ρ, z) coordinates. Using the procedure in the preceding chapter we solve separately for the azimuthal (φ) variation of the wave function, which leaves us with the PDE (p.378)

$$\rho\frac{\partial}{\partial\rho}\left(\rho\frac{\partial\psi_2}{\partial\rho}\right) - m^2\psi_2 + \rho^2\frac{\partial^2\psi_2}{\partial^2 z} - Vc\rho^2\psi_2 + Wc\rho^2\psi_2 = 0$$

Apart from the new form for potential energy V, this problem is similar to that of the harmonic oscillator. The function V is peculiar, however, in the sense that it is negative and tends to infinity at the center. The latter fact makes it desirable to create tiny triangular cells close to the origin using *feature*.

The fact that V is negative also makes the eigenvalues negative, and we shall have to take special steps to avoid confusion. The eigenvalue solver gives us a specified number of modes, in the order of their absolute values. The energy values we shall obtain are thus all negative, there is an infinite number of them, and they crowd toward vanishing energy. In order to scan the energy eigenvalues from the negative end we must add a constant term V_0 to the potential energy as follows.

$$\rho\frac{\partial}{\partial\rho}\left(\rho\frac{\partial\psi_2}{\partial\rho}\right)-m^2\psi_2+\rho^2\frac{\partial^2\psi_2}{\partial^2 z}-(V+V_0)\,c\rho^2\psi_2+Wc\rho^2\psi_2=0 \quad \bullet$$

A glance at this PDE convinces us that the effect of this device is to increase all eigenvalues W by the same amount. After solving we shall thus obtain the true energy eigenvalues as $E=W-V_0$.

Another unusual feature of the hydrogen atom is that the size of the object will change from one energy eigenvalue to the next. In the case of the drumhead we did not encounter that complication, and the size of the harmonic oscillator changed only modestly. We shall thus solve for states of small and large dimensions by separate descriptors, where we allow for more space in the latter case. The extra term V_0 and the *modes* command permit us to select the appropriate ranges for the eigenvalues. The *gridlimit* command limits the number of re-grid operations.

As before (p.378), we take the integral of the azimuthal factor ψ_1^2 to be 2π; hence we simply multiply the integrand by $2\pi\rho$ when calculating the global probability (=1).

```
title                                                    { quant31.pde }
  'Hydrogen Atom, m=0'
select
  modes= 3              errlim= 1e-4            gridlimit= 2
  converge= 1e-6        subconverge= 1e-6
  contours= 16          surfacegrid= 40         paint= off
  alias(x)= 'r'         alias(y)= 'z'
```

```
variables
  psi
definitions
  r0= 0.2e-10     r1= 2e-10          r2= 10e-10          rz= 5e-10
  m= 0
  e= 1.60219e-19              me= 9.1096e-31           h= 6.6262e-34
  c= (8* pi^2* me)/h^2
  eps0= 8.854e-12
  V= -e^2/[(4*pi*eps0)*sqrt(x^2+y^2)]
  V0= 30e-19                                                    { E= W- V0 }
  int_rzphi= integral( 2*pi*x* psi^2)
  psi1= psi/ sqrt( int_rzphi)
  prob= psi1^2
equations
  x*dx(x*dx(psi))- m^2*psi+ x^2*dyy(psi)
       -(V+V0)*c*x^2*psi + lambda*c*x^2*psi= 0
boundaries
          region 1
  start(0,-r2)       value(psi)= 0
  arc to (r2,0)  to (0,r2)
  natural(psi)= 0 line to finish
          feature
  start(0,-r0) arc(center= 0,0) to (r0,0) to (0,r0)              { Small cells }
  start(0,-r1) arc(center= 0,0) to (r1,0) to (0,r1)
monitors
  contour( psi)
plots
  contour( psi1)
  contour( lambda*1- V0) as 'Energy '
  surface( prob)  zoom(0,-rz,  rz,2*rz)  viewpoint(1,-1,20)
end
```

The above descriptor yields three solutions. The energies E of the stationary states become -21.85e-19, -5.5e-19 and -5.4e-19 (joule). The two latter states, shown below, have closely the same energy, and according to the exact solution the values are equal. The exact energy values may be obtained from the expression

$$E_n = -\frac{2\pi^2 m_e e^4}{n^2 h^2 (4\pi\varepsilon_0)^2}.$$

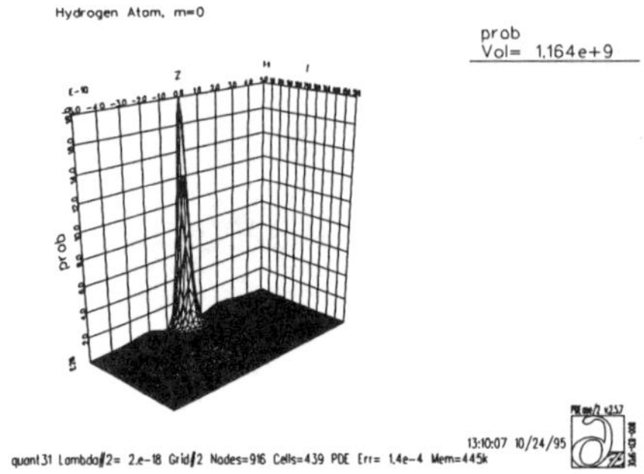

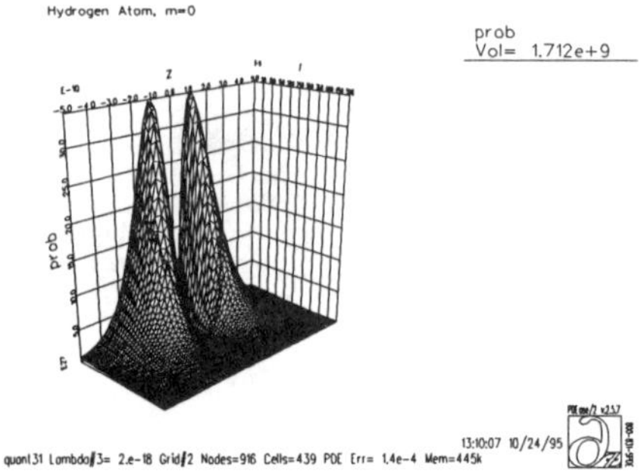

We next attempt to capture higher energy states, choosing the offset $V_0 = 3\mathrm{e}-19$ to exclude those we have already seen. In this example we shall need a larger number of *modes* and a larger size for the domain. Since this run would take a long time to complete under high precision, we make do with the first grid to obtain a quick solution. The following lines indicate the modifications to the preceding file.

```
  ...                                                          { quant32.pde }
select
  modes= 12               errlim= 1e-4             gridlimit= 1
  converge= 1e-6          subconverge= 1e-6
  contours= 16            surfacegrid= 40          paint= off
  alias(x)= 'r'           alias(y)= 'z'
variables
  psi
definitions
  r0= 1e-10       r1= 5e-10       r2= 30e-10
  ...
  V0= 3e-19
  ...
plots
  contour( psi1)
  contour( lambda*1- V0)  as 'Energy '
  surface( prob)  viewpoint(1,-1,20)
```

The *contour* plots below are examples of some states detected. The corresponding energy values identify the quantum number *n*, and the theory tells us that the number of radial node surfaces (spokes in the figure, cones or a plane in 3D) yields the remaining quantum number *l*. All of these states correspond to *m*=0. Using all information from this run we assign quantum numbers (*n,l,m*) to the following wave functions as (3,2,0), (3,1,0), (3,0,0) and (4,3,0).

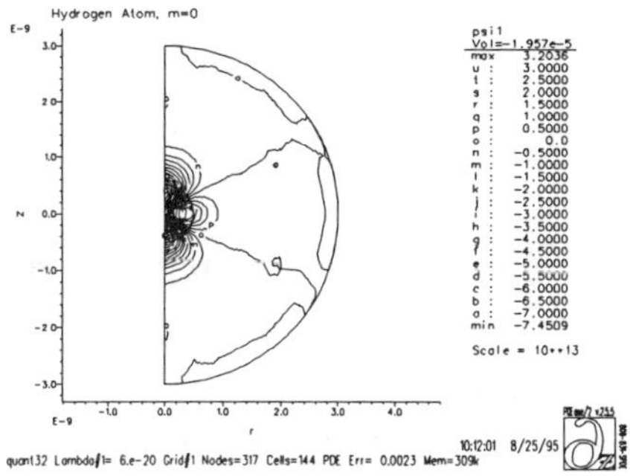

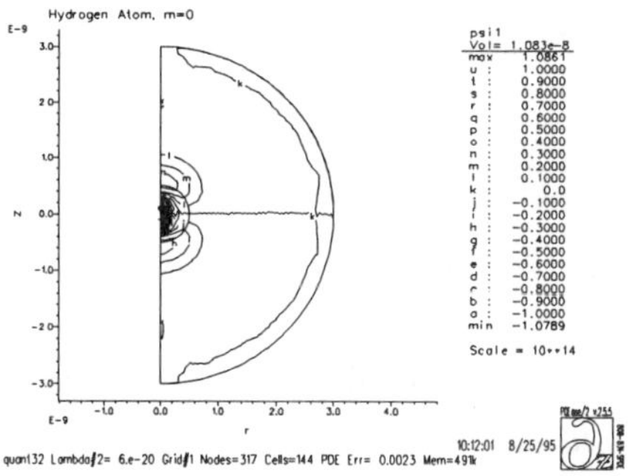

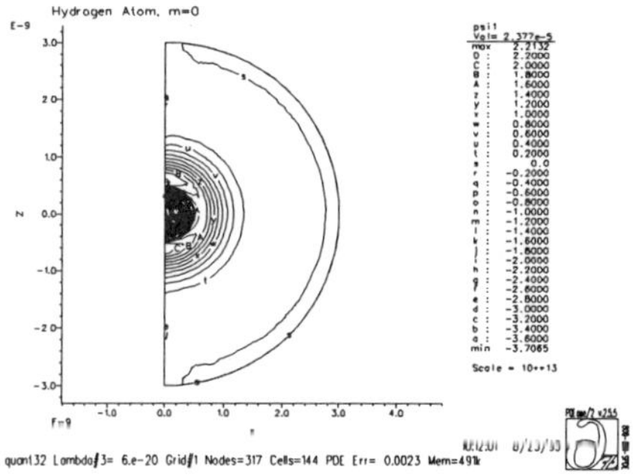

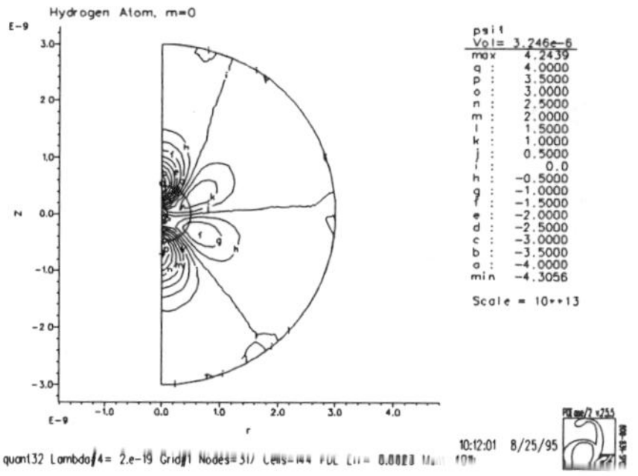

It now remains to explore higher values of the quantum number *m*. For this it is sufficient to modify *m* in the descriptor, and we may also decrease *modes* to 8.

The following plots were obtained for *m*= 2. According to the complete rules, the quantum number *l* equals the number of nodes going through the origin, plus *m*. Both of the following wave functions are without nodes and should thus have *l*= 2.

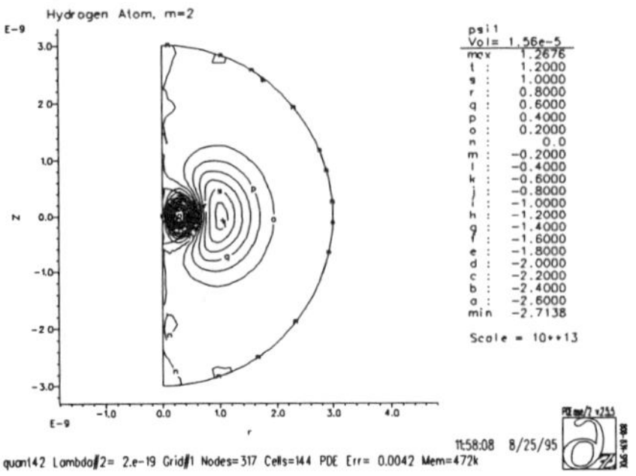

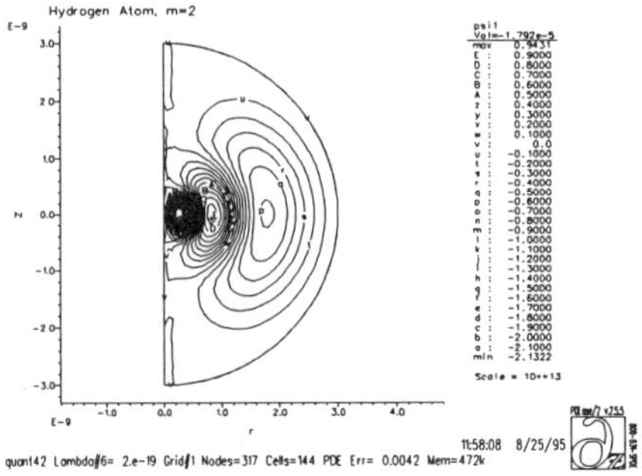

Expectation Values

We have already used wave functions to calculate the probability per unit volume of finding an electron at various points in space. There are also a number of quantities for which one may calculate the expectation (or average) value. The average value of a quantity Q is generally defined as

$$<Q>=\frac{\int Q(\mathbf{r})w(\mathbf{r})\,dv}{\int w(\mathbf{r})\,dv},$$

where the quantity Q and the weight function w depend on the space variables. In quantum mechanical calculations the weight function is taken to be $w(\mathbf{r})=\psi(\mathbf{r})^2$. To find the expectation value of the radial distance, r, we thus compute

$$<r>=\frac{\int \psi\, r\, \psi\, dv}{\int \psi^2\, dv},$$

where we write $\psi\, r\, \psi$ instead of $r\psi^2$ as a bridge to a more general notation we shall use later.

We could add lines to previous descriptors for calculating expectation values, but it is more accurate and instructive to use exact wave functions for such exercises. In any textbook on quantum mechanics[10] you will find the exact solutions for the hydrogen atom.

The hydrogen wavefunctions, as they appear in ρz plane, take the form

$$\psi(\rho,\theta) = P_{nl}(\rho_0)\ \exp(-\rho_0 / 2)\ F_{lm}(\theta)$$

where P is a polynomial in ρ_0, a reduced radius defined by

$$\rho_0 = 2\rho\sqrt{-\frac{8\pi^2 m_e W}{h^2}},$$

and F a polynomial in $\cos\theta$ and $\sin\theta$. The exponential factor is common to all states, and the two polynomials may be quoted from the following table. Strictly, we also require a constant factor which will make the integral of ψ^2 over space equal to unity, but we may easily compute this factor by direct integration.

n	l	$P_{nl}(\rho_0)$	l	m	$F_{lm}(\theta)$
1	0	1	0	0	1
2	0	$2-\rho_0$	1	0	$\cos\theta$
	1	ρ_0		1	$\sin\theta$
3	0	$6-6\rho_0+\rho_0^2$	2	0	$3\cos^2\theta-1$
	1	$(4-\rho_0)\rho_0$		1	$\sin\theta\cos\theta$
	2	${\rho_0}^2$		2	$\sin^2\theta$
4	0	$24-36\rho_0+12{\rho_0}^2-{\rho_0}^3$	3	0	$5\cos^3\theta/3-\cos\theta$
	1	$(20-10\rho_0+{\rho_0}^2)\rho_0$		1	$\sin\theta(5\cos^2\theta-1)$
	2	$(6-\rho_0){\rho_0}^2$		2	$\sin^2\theta\cos\theta$
	3	${\rho_0}^3$		3	$\sin^3\theta$

Let us now prepare a descriptor for plotting one of the exact wavefunctions of the hydrogen atom, and for calculating the expectation value of the radius at the same time. As an example we choose the state (3,2,0), which requires the following descriptor.

The expectation value for the radius, *r_exp*, can only be displayed by a figure, most simply by a *contour* plot.

```
title                                                    { atom320.pde }
  'Hydrogen Wave Function'
select
  contours= 16          surfacegrid= 60          paint= off
  alias(x)= 'R'         alias(y)= 'Z'
definitions
  r0= 2e-10             r1= 20e-10
  e= 1.60219e-19        me= 9.1096e-31           h= 6.6262e-34
  c= (8* pi^2* me)/h^2                eps0= 8.854e-12
  n= 3                                               { Quantum number }
  Wn= -2*pi^2*me*e^4/[(4*pi*eps0)^2*h^2* n^2]
  alpha= sqrt( -8*pi^2*me*Wn/h^2)
  rr= sqrt(x^2+y^2)       rho= 2*alpha*rr
  th= pi/2-arctan(y/x)                              { Polar angle theta }
                                           { Constant factors neglected }
  psi= rho^2*  exp( -rho/2)*  (3*cos(th)^2-1)
  int_rzphi= integral( 2*pi*x* psi^2)              { Integral over volume }
  psi1= psi/ sqrt( int_rzphi)
  prob= psi^2/ int_rzphi                        { Probability distribution }
  r_exp= integral( 2*pi*x* psi* sqrt(x^2+y^2) *psi)/ int_rzphi
boundaries
          region 1
  start(0,-r1)
  arc to (r1,0)  to (0,r1)  line to finish
          feature
  start(0,-r0) arc(center= 0,0) to (r0,0) to (0,r0)
plots
  contour( psi1)                                    { Normalized psi }
  surface( prob) viewpoint(20,-20,20)
  contour( r_exp)
end
```

The resulting plot of the wave function and the one indicating the average radius are shown below.

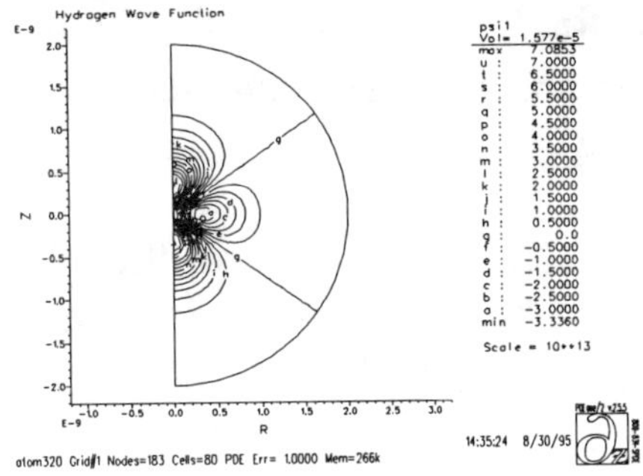

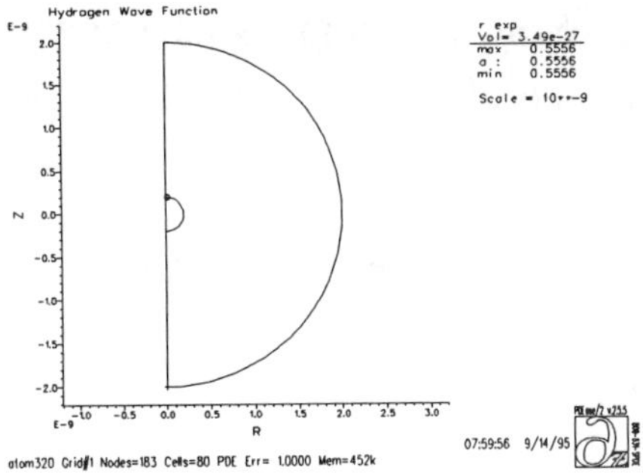

Another quantity of interest is the angular momentum **L**, for which the classical expression is

$$\mathbf{L} = \mathbf{r} \times \mathbf{p},$$

where **p** is the momentum. In quantum mechanics, momentum is defined as the differential operator

$$\mathbf{p} = -i\frac{h}{2\pi}\nabla$$

and hence the components of **L** may be written

$$L_x = -i\frac{h}{2\pi}\left(y\frac{\partial}{\partial z} - z\frac{\partial}{\partial y}\right) \equiv \frac{h}{2\pi}l_x$$

and similarly for the others by cyclic permutation. We shall only use the dimensionless number l_x, etc., given by

$$l_x = -i\left(y\frac{\partial}{\partial z} - z\frac{\partial}{\partial y}\right).$$

For the purposes of calculating the angular momentum of our solutions we need to express l_x, l_y and l_z in cylindrical coordinates. Since $x = \rho\cos(\varphi)$, etc., we may express the partial derivative as

$$\frac{\partial\psi}{\partial\rho} = \frac{\partial\psi}{\partial x}\frac{\partial x}{\partial\rho} + \frac{\partial\psi}{\partial y}\frac{\partial y}{\partial\rho} + \frac{\partial\psi}{\partial z}\frac{\partial z}{\partial\rho}$$

with similar equations for the other cylindrical coordinates. We may then solve the resulting system of linear equation to obtain $\partial\psi / \partial x$

and so on, which gives us expressions for l_x, etc.. The final operation is to combine these components into an absolute value squared $l^2 = l_x^2 + l_y^2 + l_z^2$.

Although the above transformations are straightforward, the labor required to carry them out makes it advisable to engage a program for symbolic manipulation. The author used Maple® to obtain the expression

$$l^2 = -z^2 \frac{\partial^2}{\partial \rho^2} + \left(\rho - \frac{z^2}{\rho} \right) \frac{\partial}{\partial \rho} + 2\rho z \frac{\partial^2}{\partial \rho \partial z} - \rho^2 \frac{\partial^2}{\partial z^2} + 2z \frac{\partial}{\partial z} - \left(1 + \frac{z^2}{\rho^2} \right) \frac{\partial^2}{\partial \varphi^2}$$

Here the operator is assumed to act on the total wave function, including the factor $\psi_1(\varphi)$ (p.378). To obtain expectation values we shall eventually integrate over the volume, and in this context the only effect of the factor ψ_1 is to give us an expression for the last derivative

$$\frac{\partial^2 \psi}{\partial \varphi^2} = -m^2 \psi .$$

In order to obtain the expectation value of l^2 we apply the rule for averages

$$< l^2 > = \frac{\int \psi \, l^2(\psi) \, dv}{\int \psi^2 \, dv},$$

but now the middle object of the integrand is not a function but the above differential operator. According to quantum mechanical theory the value of $< l^2 >$ should be equal to $l(l+1)$, and hence we solve to extract the quantum number l.

Let us now introduce the operator for l^2 into the preceding descriptor and calculate the expectation value for it. We need to add a few lines in the *definitions* section and preferably two plots.

```
...                                              { atom320.pde }
  n= 3                        m= 0
  ...
  L2=  integral(2*pi*x* psi*
         [ -y^2*dxx(psi)+ (x-y^2/x)*dx(psi)+ 2*x*y*dx(dy(psi))-
         x^2*dyy(psi)+ 2*y*dy(psi)+ m^2*(1+y^2/x^2)*psi] )/  int_rzphi
  L1= -0.5+ sqrt(0.25+ L2)
  ...
plots
  contour( L2)
  contour( L1)
```

Running this revised descriptor you will find $< l^2 >= 6$ and thus $l = 2$. In making this calculation we have integrated over the volume where the probability is appreciable, but confined the integration to the region inside the boundary. If we repeat the run with a smaller domain, say $r1$= 5e-10, you would expect to obtain a different value for $<l^2>$. Hence it will be a surprise to discover that it remains the same, even if you restrict the integration volume further. The reason is that $l(l+1)$ happens to be an eigenvalue of the operator l^2, which means that $l^2(\psi)$ may be replaced by $l(l+1)\psi$. The constant multiplier may be taken out of the upper integral, making the two integrals cancel.

We may verify this coincidence by a *contour* plot of $l^2(\psi)/\psi$, using *zoom* to exclude a tiny strip close to the *z* axis.

The Hydrogen Molecule-Ion

Our software can only deal with one electron in a fixed potential, but we may still study a system consisting of two hydrogen nuclei and one electron. For this case there is no exact solution. It is known from spectroscopic data for this molecular ion that the positive charges are 1.06e-10 m apart. If we assume that distance to be constant we may easily write a descriptor on the basis of a previous file. Locating the nuclei on the *z* axis, we may use the same cylindrical coordinates. The potential energy V will have two terms, one for each of the nuclei, as shown below.

```
title                                                   { quant61.pde }
  'Hydrogen Molecule Ion, m=0'
select
  modes= 5
  errlim= 1e-3          nodelimit= 1000
  converge= 1e-6        subconverge= 1e-6
  contours= 10          surfacegrid= 60    paint= off
  alias(x)= 'r'         alias(y)= 'z'
variables
  psi
definitions
  r0= 0.2e-10           r2= 5e-10          rz= 2e-10
  y0= 1.06e-10/2                    { Half-distance between protons }
  m= 0
  e= 1.60219e-19        me= 9.1096e-31         h= 6.6262e-34
  c= (8* pi^2* me)/h^2
  eps0= 8.854e-12
  V= -e^2/[(4*pi*eps0)*sqrt(x^2+[y-y0]^2)]
          -e^2/[(4*pi*eps0)*sqrt(x^2+[y+y0]^2)]
  V0= 50e-19                                              { E= W- V0 }
  int_rzphi= integral( 2*pi*x* psi^2)
  psi1= psi/ sqrt( int_rzphi)
  prob= psi1^2
  Energy = lambda*1- V0                                { For plotting }
equations
  x*dx(x*dx(psi))- m^2*psi+ x^2*dyy(psi)
          -(V+V0)*c*x^2*psi+ lambda*c*x^2*psi= 0
boundaries
          region 1
  start(0,-r2)      value(psi)= 0
  arc to (r2,0)  to (0,r2)
  natural(psi)= 0 line to finish
        feature
  start(0,-r0+y0) arc(center= 0,y0) to (r0,y0) to (0,r0+y0)
  start(0,-r0-y0) arc(center= 0,-y0) to (r0,-y0) to (0,r0-y0)
monitors
  contour( psi)
plots
  contour( Energy)
```

```
  contour( psi)
  surface( prob)  viewpoint(1,-1,20)
  surface( prob)  zoom(0,-rz, rz,2*rz)  viewpoint(1,-1,20)
end
```

The state of lowest energy, *W*= -48.14e-19, yields the wave function depicted below. The ground state of the hydrogen atom was found to be at about -22e-19, which implies that a double charge, if concentrated at one point, would lead to 4 times that energy, or -88e-19, thus much lower than for the same positive charge distributed over two protons.

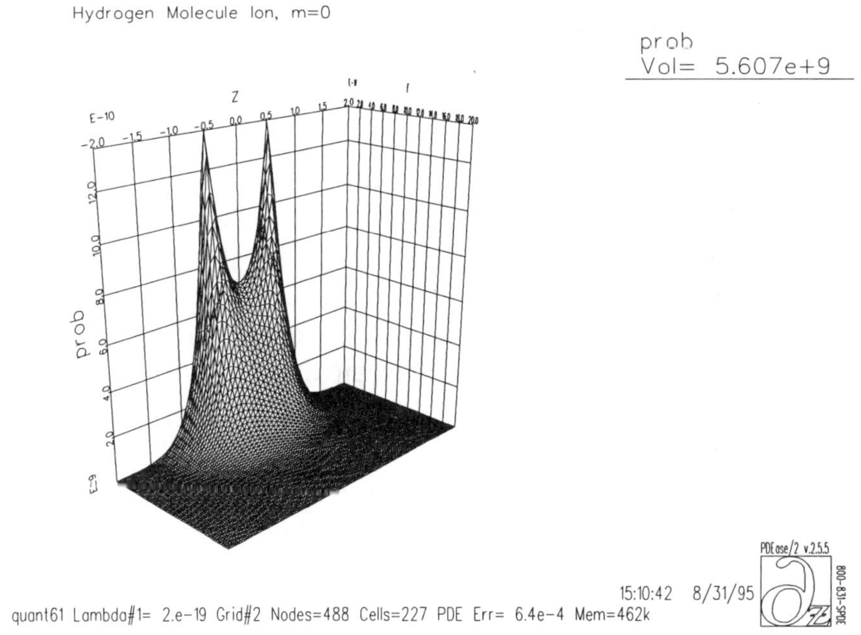

The molecular ion is strictly a three-particle system, but we treated it as if it consisted of two units, an electron and a fixed pair of positive charges. Zero energy for the atom referred to complete disruption of the system, the nucleus and the electron being so far apart as to have negligible energy of interaction. In the case of the molecule-ion, however, zero energy meant that the electron was far removed from the nuclei, the latter still being at the same distance from each other. In order to apply consistent definitions of zero energy for the molecule-ion and for the atom, we should add a term for the potential energy of the two nuclei, as indicated here.

```
...
V= -e^2/[(4*pi*eps0)*sqrt(x^2+[y-y0]^2)]
        -e^2/[(4*pi*eps0)*sqrt(x^2+[y+y0]^2)]
                +e^2/(4*pi*eps0* [2*y0])
...
```

Running the file again with this amendment we find a lowest energy of -26.37e-19, which is close to the empirical value for the energy of dissociation of the molecule-ion.

In the above calculations we accepted the equilibrium distance between the hydrogen nuclei on trust, but we could test if the literature value really corresponds to a minimum for the energy of the ground state. Running the same descriptor for $2y_0 = 1.16\text{e}-10$ and for 0.96e-10 as well we find that the lowest energy values are just above what we obtained for 1.06e-10. This state of the system is thus in stable equilibrium.

At the same time we discover that the energy of the next state varies appreciably with y_0, and in such a way as to decrease for increasing distance between the nuclei. That state is thus not in equilibrium at $2y_0 = 1.06\text{e}-10$, but the energy could still exhibit a minimum at a larger distance. In fact, further investigations show that the energy continues to decrease, and that there is no stable state of this type. There is another stable state at a larger inter-nuclear distance, however, which one of the exercises asks you to look for.

Exercises

❑ Plot a few exact wavefunctions of your choice and compare to solutions obtained by finite element analysis. Check some previous assignments of the quantum number *l*.

❑ Find and compare the values of $<r>$ for the exact states (1,0,0), (2,0,0) and (3,0,0).

❑ Compute $<l^2>$ for an atom in the state (3,2,2).

❑ Modify QUANT32.PDE to solve for a large value of *m*, say 10, expanding the domain as necessary. What is the general trend for large *m*?

❑ The potential from the hydrogen nucleus is spherically symmetric. Explore the effect of an anisotropy in *V*, replacing $1/\sqrt{\rho^2+z^2}$ by $1/\sqrt{1.2*\rho^2+z^2/1.2}$.

❑ Investigate if there is another stable state for the molecular ion, still with *m*= 0. Use inter-nuclear distances of about 5.5e-10.

Principles of Finite Element Analysis

The main idea of finite element analysis (FEA) is to divide the domain of interest into sub-regions of simple shapes and to solve over each one of those simultaneously. PDEase normally divides the domain into triangles, or rather prisms of triangular cross-section. The triangles we see on the screen are projections of the prisms on the viewing plane.

The program solves a PDE by determining values of the dependent variables at discrete points (nodes), i.e. at the corners of the triangles and at the midpoints between corners. The only record of the solution process is thus a list of these node values. In order to obtain function values and derivatives at other points of the triangular area, the program applies an algorithm for interpolation between nodes. This algorithm should be identical to the one used during the solution process.

Interpolation and Differentiation in 2D

Since the interpolation algorithm is of paramount importance in FEA, we begin by studying how it performs. At the same time we test the differentiating routines. *A priori* we do not know precisely how PDEase calculates derivatives, but we may easily check if the results are accurate.

Let us take a simple function of two variables, plot it as well as its derivatives and compare to the exact expressions for these. This is the objective of the following descriptor, which concerns the function sin(xy).

```
...
definitions
  Lx= pi/2                    Ly= pi/2
  f= sin(x*y)
```

```
  fx= dx(f)                 fx_ex= cos(x*y)*y
  fxy= dx(dy(f))            fxy_ex= -sin(x*y)*x*y+ cos(x*y)
boundaries
        region 1
  start(-Lx,-Ly)  line to (Lx,-Ly)  to (Lx,Ly)  to (-Lx,Ly) to finish
plots
  contour(f)                      surface(f)  interactive
  surface(fx)  interactive        surface(fxy) interactive
  contour(fx- fx_ex)              contour(fxy- fxy_ex)
end
```

The next figure is a *surface* plot of the mixed derivative. The *contour* plots of the errors (not shown) yield exactly zero, which must mean that the program internally uses symbolic calculation, employing the same exact procedures as any mathematician would.

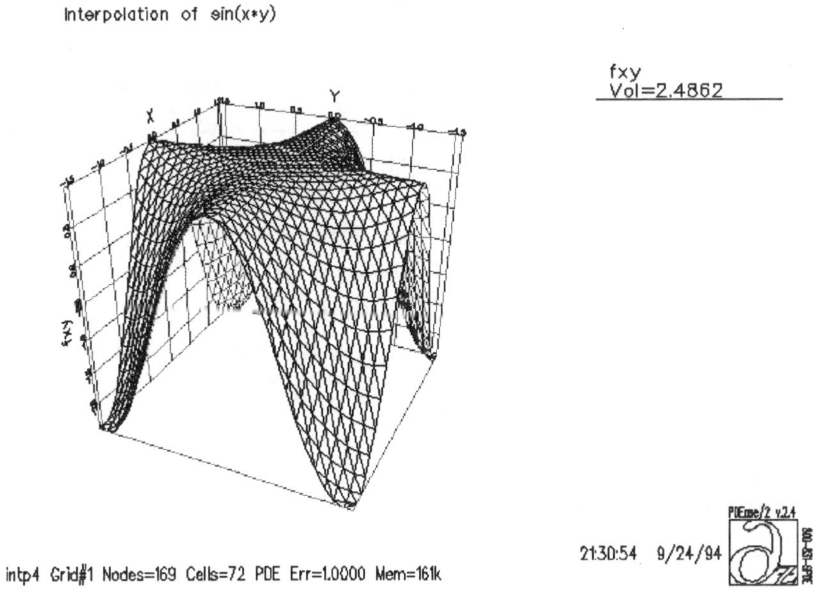

Our observations thus suggest that PDEase treats derivatives, or functions of derivatives, specified in the *definitions* segment by exact means. In other words, differentiation does not introduce additional errors. Whenever we request a plot, however, all expressions have to be computed numerically; hence there is a limit to the precision attainable. The source code of PDEase is written in Pascal, and the compiled programs usually work internally with a precision of about 16 digits, which certainly is good enough for our purposes.

Interpolating to Obtain a Solution

Once PDEase has produced a solution, it stores it in a table, where the values of the dependent variable (or variables) are given for each node point. The user would like to know the solution at any point in space, however, and to that end the program interpolates the table by a quadratic algorithm.

The procedure involved in this interpolation is simple. For each triangular region PDEase uses a polynomial of the following form

$$P(x,y) = a_0 + a_1 x + a_2 y + a_3 xy + a_4 x^2 + a_5 y^2.$$

Function values are known at three corners and three midpoints, a total of six points, corresponding exactly to the number of coefficients in $P(x,y)$. In order to determine the coefficients a_i the program just solves a system of six linear equations.

We shall now solve a simple PDE, the Poisson equation, and observe how PDEase interpolates to obtain a complete solution on the basis of node values only. Starting with an arbitrary function, we apply the Laplacian operation to it. We then use the result $r(x,y)$ of that operation to construct a PDE which has the function we selected as a solution. The equation $\nabla^2 U - r(x,y) = 0$ obviously meets our needs. The *value* boundary conditions we obtain directly from the initial function, and the descriptor becomes as listed below.

For clarity, we wish to have a very low number of nodes, hence the *nodelimit* and *gridlimit* statements. Furthermore, we request a small number of contours, since we are mostly interested in the zero curve. Finally we use the maximum grid density in the *surface* plots in order to see the error function in the finest possible detail.

```
  ...
select
  errlim= 1e-15       nodelimit= 5          regrid= off
  contours= 1         surfacegrid= 60       paint= on
variables
  u
definitions
  Lx= 1         Ly= 1
  u_ex= y**2* sin(x)
equations
  del2(u)+ (y**2-2)*sin(x)= 0
boundaries
            region 1
  start(-Lx,-Ly)            value(u)= u_ex
  line to (Lx,-Ly)  to (Lx,Ly)  to (-Lx,Ly)  to finish
plots
  surface(u- u_ex)
  grid(x,y)  zoom(0,0, Lx,Ly)
  contour(u- u_ex)  zoom(0,0, Lx,Ly)
  surface(u- u_ex)  zoom(0,0, Lx,Ly)  interactive
end
```

The error with respect to the exact solution is of particular interest. Inspecting the following *contour* plot we notice that curve *a* corresponds to zero error. Hence that curve should pass through all the nodes. Since the contour curves are generated by interpolation of the error surface, which is rather ragged, we cannot expect to find crossings exactly at the node points. If we compare the plot below with the zoomed *grid* plot (not shown) for the same region, however, we find that the zero curve mostly passes through the corners and the midpoints.

The *surface* plots (not shown) depict valleys going through the node points, where the error must be zero by definition. By navigating between the hills of the zoomed plot and by following the colored strip corresponding to the zero level you should be able to glean the zero-crossings connected with the nodes on the triangular grid.

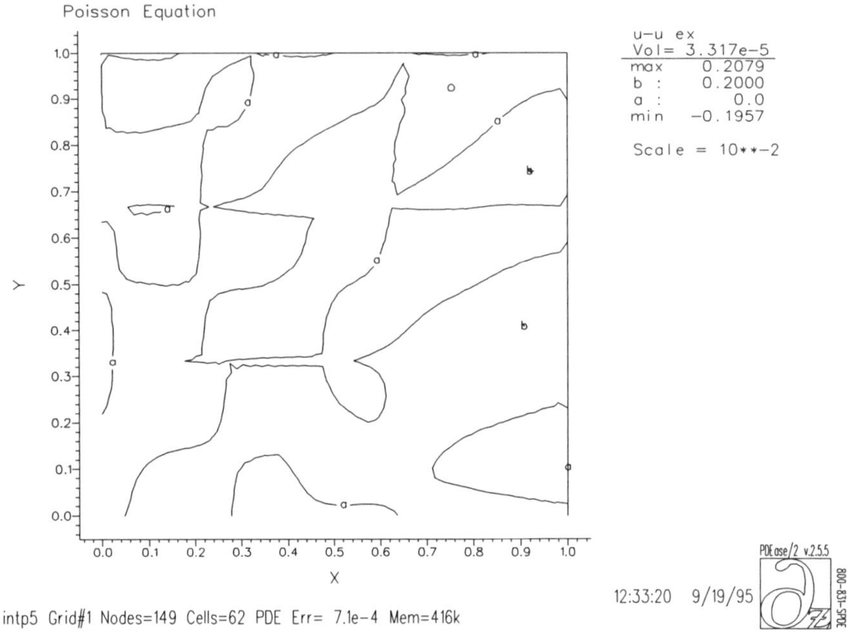

Solving for the Node Values

If all dependent variables are known at all the nodes of the grid, we may thus interpolate to obtain values at any point of the solution domain and also take suitable differences to approximate all derivatives of first and second order. In FEA, knowing the node values of the dependent variables is equivalent to having a solution to the problem.

It now remains to calculate the values pertaining to the nodes, and this is the most difficult task. A general, linear PDE of second order reads

$$\frac{\partial^2 u}{\partial x^2}+f_1\frac{\partial^2 u}{\partial x\partial y}+f_2\frac{\partial^2 u}{\partial y^2}+f_3\frac{\partial u}{\partial x}+f_4\frac{\partial u}{\partial y}+f_5 u+f_6=0$$

where f_k are functions of x and y. Using an extremely condensed notation we may even write

$E = 0$

if only we remember what E means. If $u(x,y)$ is an exact solution to the PDE then E remains zero over all of the solution domain D. Integrating over this domain we thus have

$$\iint_D E\,dxdy = 0$$

It is *not* true, however, that a function $u(x,y)$ which satisfies this relation must be a solution: E might take positive as well as negative values, which could cancel on integration. If, on the other hand, we take the square of the PDE, the condition

$$\iint_D E^2\,dxdy = 0$$

implies that $u(x,y)$ is a solution. In fact, it is possible use this relation for solving the equation numerically.

As we have already noted, any numerical solution may be presented by its node values u_i, which we may use to derive any other values of $u(x,y)$ and its derivatives. Hence, for a given PDE, the left side E of the equation may be regarded as a function of the node values and the coordinates x and y.

$$I(u_i) \equiv \iint_D E(u_i, x, y)^2\,dxdy = 0 \qquad \bullet$$

Of course, in view of the limited accuracy of numerical computation we could never hope to find a set of node values which makes the integral $I(u_i)$ exactly equal to zero, so we must be content with a set that minimizes the integral. This is equivalent to finding the minimum of a function of many variables, and there are standard methods available for solving this auxiliary problem. The effect of taking the square of E, however, is to make the integrand more complicated and necessary non-linear.

An alternative method, which only involves linear analysis as long as the PDE itself and it boundary values are linear, proceeds as follows. Instead of condensing the problem into a single integral expression as in the scheme above, we shall now introduce a whole set of equations of the same type (numbered $j = 1, 2, \ldots m$):

$$I_j(u_i) \equiv \iint_D W_j(x,y)E(u_i,x,y)\,dxdy = 0$$ •

The equation is hence no longer squared, and $W_j(x,y)$ is a weight function which may emphasize one subdomain inside the total solution domain D. We are free to choose the various weight functions as we please, but from the point of view of computations the simplest strategy would be to let the weight be either 0 or 1. In the first equation, for instance, we might use unit weight within one of the triangular cells and zero in all the others. The central idea is to force the trial function (eventually the solution) to satisfy the PDE over every subdomain.

One way of applying the above method would be to use unit weight in cell number j and zero elsewhere, which would give us m equations testing various parts of the solution domain.

The problem is that the number of cells (m) always is much smaller than the number of nodes (n). How many node values do we need to determine? Let us assume that there is only one dependent variable (u) and that the boundary conditions are specified by *value* all around. All the values on the boundary are hence known, and we only need to solve for values on the interior nodes. Even this number is generally larger than the number of cells, however, which means that we need some flexible means of increasing the number of equations.

Several methods of choosing weight functions are known. One set of rules which would give the right number of equations is the following. If node j is a midpoint, let $W_j = 1$ for the two cells sharing this point and choose $W_j = 0$ elsewhere. If node j is a corner, let $W_j = 1$ for the cells sharing this corner and let $W_j = 0$ elsewhere. This procedure provides the right number of equations, and all the sub-domains for integration become different and hence yield independent equations.

Naturally, we need more sophisticated weighting schemes in the case of several dependent variables. The method actually used by PDEase is not revealed by the software producer.

So far we have assumed that the *value* of the solution is given all along the boundary. If the problem involves *natural* conditions over part or all of the boundary, this information must be incorporated by special means. A well-known theorem which connects the integral of a PDE with the outward normal component of a field **F** is

$$\iiint \nabla^2 U dV = \iiint \nabla \cdot (\nabla U) dV = \iiint \nabla \cdot \mathbf{F} dV = \oiint \mathbf{F} d\mathbf{s}$$

Whenever we integrate the PDE over a cell at the boundary the volume integral of the Laplacian may be replaced by a surface integral involving the *natural* boundary condition. The other two surfaces of the cell yields integrals which may be matched against those of adjacent cells.

Whenever *natural* boundary conditions are involved it is important to remember that multiplying the PDE by a certain factor is bound to change the volume integral, and hence the surface integral over the boundary. Thus, if we multiply through by a factor x, say, the corresponding *natural* boundary condition also must be multiplied by the same factor. We have already seen cases where multiplication by a factor x is desirable in order to avoid very large local values.

If both the PDE and the boundary conditions are linear the result of the analysis is a system of linear equations. If the PDE or the boundary conditions are non-linear we obtain a system of non-linear algebraic equations. PDEase solves both types of systems by iterative methods, in order to avoid excessive demands on memory space.

Exercises

❑ Calculate and plot the function f= sin(x)*cos(y) and its second derivatives. Determine the errors (magnified if necessary) with respect to the exact derivatives.

❑ Investigate the effect of multiplying the equation of heat conduction (p.109) by the factor x*y.

Never Used the DOS Editor Before?

A text editor is a rudimentary word processor, with the same basic facilities as an ordinary typewriter. In addition, it permits you to erase, copy and move text. There are several text editors on the market, but an excellent product comes with the operating system of your PC, provided you have DOS 5 or 6.

Getting Started

The DOS editor normally is installed under the \DOS directory, which in turn should be one of the directories listed in the PATH statement of your AUTOEXEC (startup) file. To check that this is the case try typing EDIT (using lower case letters if you prefer). After a second or two a framed typing area should appear on the screen. If this is not the case you need to modify your startup file as follows.

Change directory by typing CD \DOS and then EDIT \AUTOEXEC.BAT (with a space before \). You should now see the editor frame and a few lines of text. (If not, you need practical help from some friend who is familiar with DOS.) Even if the basic functions of the editor are fairly obvious, you could obtain help by pushing the <F1> button of the keyboard. Thus you will also learn how to use the editor without a mouse, if you should need to know.

One of the lines on the screen begins with PATH. Using the arrows on the keyboard or the mouse, move your entering point to the end of the PATH line and add ;\DOS. Then close the file by pointing the mouse marker at *File* on the top ribbon, then pushing the left mouse button. Similarly, choose *Exit* and push the mouse button again. Answer the question which follows by *Yes*. The AUTOEXEC file is now changed, but to make it take effect you must also run it. To do this, simply type \AUTOEXEC, and then push <ENTER>. Now change directory again to \PDEASE or \PDE, if you have already created the latter directory for your descriptor files.

You should now be ready to edit descriptor files. Start by typing EDIT FUNC1.PDE. Enter the statements listed on p.4, exactly as you would do on a typewriter. Then point the mouse at *File* and click on the left button. Choose *Exit*, then answer *Yes*, and your file is saved on the hard disk and is ready for use.

Next time you edit a descriptor, you need not type the entire list from scratch. Much of it is always the same, and a few more lines may usually be taken over from a previous file. You may begin by typing COPY FUNC1.PDE FUNC2.PDE, and then you use the editor to modify the latter file as necessary.

When modifying a file you wish to replace a line, or a block of lines, by new statements. To do this you point the mouse at the beginning of the first line and then hold the mouse button pressed while sliding the mouse downwards. A block of lines then becomes marked by a different color and remains highlighted after you have released the button. Of course, you may also mark a suite of only a few words on a line by sliding sidewise.

There are several ways of erasing the block you have marked. If you start typing new text, the marked block automatically disappears. If you just want to get rid of it, you use the <Delete> button. Another manner is to click your mouse on *Edit* at the top of the frame. This gives you three options: *Cut*, *Copy* and *Clear*. Clicking on the first option, *Cut*, removes the text block but stores it for later pasting in some other position, while clicking on *Clear* is equivalent to pushing <Delete>. The middle option, Copy, just stores the block without removing it. Using these facilities you may thus erase blocks of text and move part of a descriptor to another location on the screen.

At this point a word of warning would be in order. When you have marked a block, it is susceptible to damage. Just touching a key makes the block disappear, and this could occur if a pen happens to fall on the keyboard. In this event, the best you can to is to choose *File*, *Exit* and answer *No* (i.e. do not save). Then at least you are back to where you were last time you saved that file. A wise policy hence is to save your file (*File*, *Save*) every time you have finished editing a line.

There are other ways to misbehave while editing. When you point the mouse at the beginning of a line, slide it downwards and then click on the button, you may inadvertently move the mouse slightly. Positioning the mouse while clicking at the same time is a little like handling Chinese chopsticks: your muscles had better work independently. This is supposed to be an acquired skill, but some of us never learn. The more professional you become, the less time you spend looking at the screen, and you may not notice that you have marked more than you intended until it is too late. When you find that a line has been lost, the only remedy is to exit the editor without saving and then resume the editing.

Another useful function of the DOS editor is to help you locate a given string of characters by clicking on *Search*. You may simply repeat this command to find the *n*:th occurrence of the same string. Choosing *Search, Change*, you may replace one string by another throughout the file, and with full control of every operation.

In case you are in favor of invigorating colors, *Option, Display* provides you with a palette. While you are at it, you may find it convenient to set *Tab Stops*=3 for indenting lines and spacing commands.

Copying From one File to Another

You can move the contents of an old file, or part of it, to the file you are creating. The basic tip is of course to copy the model file to a new name and to start from there.

There are cases, however, where you only want to insert part of a previous file. For this purpose there is a valuable feature of the DOS editor which seems to be largely unknown. If you are editing FILE2.PDE and you require fragments of FILE1.PDE, you type EDIT FILE1.PDE, mark part or all of that file and click on *File, Copy*. After that, you choose *File, Open* (no not exit!) and ask for FILE2.PDE. The block you copied is now available and may be placed anywhere within the new file by *File, Paste*.

If there are several parts you wish to transfer from a previous file, the most convenient device is to paste the entire model file at the end of the new file and enclose it by the comment symbols { }. This reference file is then ignored by PDEase and may be used as a source for *Copy, Paste* as long as it is needed.

Syntax of PDEase

Just inside the back cover of this book you will find a skeleton descriptor laying down the order of the various segments. It also contains examples of the most common commands, in a terse and suggestive form. It may be of value for refreshing your memory after a period of unrelated activity. It also refers you to pages where a given segment or command is demonstrated or explained. We shall now consider the various commands in detail, in the order in which they appear in a descriptor.

Title

This segment, which is optional, only contains the name of the problem within quotes. You may write, for instance,

```
title 'Atomic Diffusion in a Bar of Nuclear Fuel'
```

or

```
title
  'Atomic Diffusion in a Bar of Nuclear Fuel'
```

The format is as free as could be achieved without creating ambiguity: we are allowed to indent a line of text, or even distribute it over several rows. This flexibility is especially useful in the segments and commands that follow. The program also considers upper-case and lower-case characters to be equivalent.

Select

This segment may be omitted. We can use it, however, for setting certain parameters which control the detailed operation of PDEase. The statement

```
Errlim= 1e-4  (or Errlim= 1e-4 or errlim= 1e-4)
```

means that we wish the relative error of the solution, on the average, to be less than 1 part in 10^4. By this command we only request a limit for the average error, which means that the error locally may exceed this value. In order to extend our error control to all triangular cells of the domain, the program also takes the command

```
cellerrlim= 3
```

which assures that calculations continue until the relative errors in all cells have been reduced below 3 times the value of *errlim*.

There are two further commands that control the solution process, i.e.

```
converge= 1e-6
subconverge= 1e-6
```

The finite element analysis produces a system of algebraic equations, which PDEase solves by successive approximation. The above commands specify the accuracy which must be reached in this process before any refinement of the grid is allowed. These two conditions are essential in certain types of problems.

In difficult cases we may want to interrupt calculations before the above error conditions have been fulfilled. The program successively subdivides the cells where necessary, which creates ever more nodes, eventually flooding the primary memory of the PC. If we accept to be present during the run, we may always finish and obtain a solution by pushing the key <Esc> and then choose <F>

In case of your absence, the following commands may help to avoid overflow. The most useful one is

```
nodelimit= 1000
```

which limits the number of nodes at approximately 1000.

The statement

```
gridlimit= 3
```

assures that there will be only 3 successive refinements of the triangular grid.

Another statement limits the time spent on the calculations. For instance,

```
timelimit= 60
```

interrupts the solution process in an orderly manner after 60 minutes, if it should still go on. There is no guarantee, however, against memory overflow, which unfortunately would abort the solution.

There are also a few useful statements which control *plots*. Whenever cylindrical coordinates are used, we may avoid confusion by stating

```
alias(x)= 'Radius'            alias(y)= 'Z'
```

to provide correct labels for the axes.

We may easily customize our *contour plots* by the statement

```
contours= 20
```

which means that we request 20 contour curves over the problem domain used. We should not be astonished, however, to find that the actual number of curves produced turns out to be 18 or 22, since for convenience the value pertaining to each curve also should have a small number of decimals, a condition which PDEase takes into account.

Surface plots may take too long to draw using the default parameters. In particular, color painting produces a beautiful and informative result, but works very slowly. If we do not need this high quality we just specify, for instance,

```
paint= off
surfacegrid= 30
```

thereby inhibiting color coding and, by the second command, reducing the number of triangles on the surface to 30×30.

PDEase offers a similar command for *vector* plots. In some cases, arrows may be too close to be distinguishable. We control the number of arrows by typing, for example,

```
vectorgrid= 20
```

which yields 20×20 arrows in total over the domain.

The statement

```
stages= 10
```

makes PDEase run through the descriptor 10 times. It also provides an integer variable *stage*, which steps from 1 to 10 on executing these repeated runs.

The *stage* variable may be used in the *definitions* section to modify the numerical parameters in various ways. The most obvious use of this device is to solve for a range of parameter values. It is also possible to introduce a non-linear term gradually in the PDEs, so as to provide a good starting point for each iteration.

We may also supply explicit parameter values for each of the stages, such as,

```
errlim= staged( 1e-2, 1e-3, 1e-4)
```

which would avoid spending too much time on the first two iterations.

Versions 2.5 and later are able to calculate eigenvalues (*lambda*) and the corresponding eigenfunctions. In this case the *select* section must contain the command

```
modes= 10
```

if 10 eigenvalues are requested.

Coordinates

PDEase operates by default in Cartesian (*x,y*) space. Versions 2.5 and later permit us to specify cylindrical coordinates in a new *coordinates* section. The cylinder axis may point in either the *x* or the *y* direction, as specified by the commands

```
xcylinder( 'z', 'r')
```

or

```
ycylinder( 'r', 'z')
```

With cylindrical coordinates declared in this manner the program internally expands the differential operators grad, div and curl.

Variables

In this segment we simply list the names we have given to the dependent variables, i.e. the solutions to the PDE (or system of PDEs). A variable name must begin by a letter but could be long, if you so wish. Variable names often become more readable if separated by an underscore character, e.g. *carrier_concentration* or *c_conc*.

In the case of transient problems, which involve the time explicitly, it is often wise to declare a *range* of variation for the dependent variable, e.g.

```
temp(range= 300, 1000).
```

Definitions

Geometrical dimensions and materials properties are listed in this segment. For instance, we might type

```
electrical_conductivity= 1.3e8                         { Basic SI units }
```

where the curly brackets enclose the user's comments, ignored by the program.

If the descriptor contains more than one region, we wish to specify the materials properties for each one of these regions in the boundaries segment. This is allowed if we just declare the parameter in the definitions segment as follows

```
electrical_conductivity
```

In addition to these primary data we may define various functions of the parameters, of the independent variables x, y and t, as well as of the dependent variables declared in the previous section. These functional expressions may also involve derivatives, e.g.

```
jabs= cond*sqrt[ dx(U)**2+ dy(U)**2] ,
```

which defines the magnitude of the electric current density in a potential field U. It is expedient to define functions containing derivatives in this segment, rather than typing the entire expression in a plot command.

The arithmetic rules are close to those of FORTRAN, and the internal functions are also similar:

```
sqrt( ), exp( ), ln( ), log10( ),
sin( ), cos( ), tan( ), arcsin( ), arccos( ), arctan( ), sinh( ), cosh( ),
erf( ), erfc( ), abs( ), sign( ), ustep( ), upulse( , ), uramp( , ),
min( , ) and max( , ).
```

It is important to note that the two last functions, *min* and *max*, are different from the variables of the same name occurring in the plots. The function max(,) takes two arguments and just returns the largest one of these.

The discontinuous function *ustep*( ) may be used for conditional statements. A more direct means to this end is the *if-then-else* construct, e.g.

```
a= if r0>pi/2 then 0  else cos(r0)
```

PDEase not only handles elementary functions, but also functions defined by a table stored in an ASCII file. For instance, if the data are stored in a file named QUANT.P01, the following command makes it available as a function for use in the descriptor.

```
f= table( 'quant.p01' )
```

A table for this purpose may be created in a separate run by a contour plot, just by appending the command *print*.

Versions 2.5 and later allow us to create vectors in the *definitions* section, such as

```
E= -grad( U)
```

or

```
E= vector( Ex, Ey)
```

The symbol **E** may now be used with operators

```
div( E)
curl( E)
magnitude( E)
```

and if **D** has also been defined, PDEase accepts structures such as

```
s= dot( E, D)                                   { Scalar product }
vz= cross( E, D)                  { z component of vector product }
```

Versions 2.5 and later also let us use *integral* in the *definitions* section, not only under *constraints*. This means that we can access the variable "*Vol*" appearing in contour and surface plots. The command

```
intsq= integral( psi^2)
```

computes the integral

$$\iint \psi(x,y)\,dx\,dy$$

over the solution domain. If a region has been given a name (not only a number) under *boundaries*, an integral may also be taken over that sub-area. The string containing the name must then be entered as a second argument to *integral*.

Similarly, there is a command for computing the *line integral* along a curve. The statement

```
lint= bintegral( f, 'curve1')
```

thus integrates $f(x,y)$ along the line defined as *curve1*.

Initial Values

When a PDE is linear, i.e. the coefficients of the dependent variables and their derivatives are functions only of the independent variables, there is hardly any need for initial (trial or "guess") values for the dependent variables. For non-linear problems initial values are not always required, but the solution time may be shortened if you specify at least the order of magnitude of the solution variable. If a problem is strongly non-linear, it might even be necessary to supply a function which roughly approximates the solution. PDEase permits us to use *stages* to provide a good first approximation.

We may always watch the solution process by a *monitor* graph to judge if any additional clue is needed. Whenever the solution seems to converge well, there is no reason to supply an approximate solution. If no initial value is given, PDEase uses zero as the first approximation of the dependent variables.

For transient (time-dependent) problems a completely different kind of initial value is required. Here the initial value (or function) is part of the problem definition and must always be specified exactly.

Equations

A unique feature of this software is that we are free to enter a PDE, or a system of PDEs, using ordinary characters. PDEase then interprets these equations and adapts the algorithm to the problem.

A PDE consists of functions and derivatives, expressing some relation pertinent to the physical problem. For example, the following equation expresses mechanical equilibrium in the x direction of an elastic object.

```
dx[ c*( dx(u)+ mu*dy(v)) ]+  dy[ G*(dx(v)+ dy(u)) ] = 0
```

The first pair of square brackets contains normal stress and the second one shear stress, both being expressed in terms of the Cartesian components u and v of displacement (p.159). The coefficients c and G may generally vary in space; hence they should not be moved out of the brackets or parentheses where they initially appeared in the derivation. Even if there is no continuous variation of the elastic modulus, an object may consist of various materials with different elastic properties. The above equation should thus be left as it is, independently of the details of the problem.

The example above also shows how to use *square brackets*, together with parentheses, to collect terms into groups. This makes equations more readable and helps to explain their physical basis. The free syntax used by the program permits us to put in spaces, or groups of spaces, between terms for similar purposes. We may even distribute an equation over several lines, if it is long or if dividing it makes the structure clearer.

With versions 2.5 and later it is allowed to define expressions containing derivatives and use the corresponding symbols in the PDE. The above equation may thus also be written

```
dx[ sx ]+  dy[ sxy ] = 0
```

in terms of the stress components. In addition, they permit us to use the differential operators grad and div to simplify the equations.

There must be as many equations as there are dependent variables. If there is more than one, they must appear in a certain order, related to the nature of the problem and the list of dependent variables. For instance, if we have declared the dependent variable u (x displacement) before the variable v, then the PDE expressing equilibrium in the x direction must also appear first.

Constraints

This segment is optional and should only contain relations involving *integral*. It permits one to express the conservation of mass, energy or charge by integrals over the domain. Such conservation statements have many applications in chemistry, where we may specify

```
integral(a+b+2*c)= integral(a0+b0+2*c0)
```

to conserve the number of atoms in the solution domain.

Boundaries

This segment specifies the geometry of the object and the conditions on its boundary. In addition, it may contain expressions for materials properties. The declaration of the main boundary begins with

```
region 1
```

which has a different status than any following region of higher number. This first boundary defines the total extent of the object, and other regions (if any) are sub-domains of the first one.

Region 1 is thus unique in having contact with all of the space exterior to the object under study. This outside world imposes various conditions on the boundary of the object, such as a distribution of electric potential or current density. Thus, for region 1 we not only specify the geometry but also the boundary conditions of our problem. A typical clause for the main region would be

```
      region 1            cond= 1e8/temp              { Electrical conductivity }
start(0,0)
natural(U)= 0      line to (1,0)
value(U)= 3        line to (1,1)
natural(U)= 0      line to (0,1)
value(U)= 0        line to finish
```

The first line contains a value for the conductivity, which might equally well be supplied in the definitions segment. After giving the *start* coordinates (x,y) we specify *natural*(U)= 0, which means that along the first straight line segment to (1,0) the outward normal derivative of U is zero. Along the next segment the value of U is 3, and so on. The keyword *finish* is necessary to indicate that the polygon is finished, but the symbol *finish* is also equivalent to the coordinate pair of the *start* point. By specifying the coordinate points from start to finish we have thus described a closed polygon.

When specifying the boundary we move around the object in the positive sense (counter-clockwise). If we wish to cut out a hole in the object, on the other hand, we must circulate in the negative sense. A polygon may approximate virtually any continuous curve, and with enough patience we can use the same method to describe a quite complicated boundary. We can do much better, however, by using the command *arc*, once the start values have been given. Three point are enough to define a circular arc, and we could simply type

```
start(1,0)  arc to (0,1) to (-1,0)
```

but if we only specify two points on the curve we need to add the center coordinate:

```
start(1,0)  arc(center= 0,0) to (0,1)
```

There are three different statements for boundary conditions: *value*, *natural* and *point value*. The last variety is useful if normal derivatives only have been specified, in which case we may employ it to anchor the solution at a single point. That additional value must be supplied immediately after the coordinate pair defining the point, as in the following example.

```
start(0,1)    point value(u)= 0
```

For *region* 2 and higher, which are sub-domains of the main region, we use similar rules for drawing boundaries. The essential purpose of these secondary regions is to specify new materials properties. Such values usually already exist for the main region, but new properties specified for a sub-region always override the global ones. In this manner we can easily enter materials data for all the parts of the object under study.

Versions 2.5 and later contain a new command with interesting applications. Using *line* and *arc* statements we may define a curve within the solution domain, for instance

```
feature
start 'curve1' (0,-1) arc to (1,0) to (0,1)
```

Note that this construct must not be terminated by *finish* and that the curve should not be superimposed on a boundary. Boundaries may be named in a similar manner.

A curve created by *feature* may be used with an *elevation* plot or with *bintegral* to calculate a line integral. It also produces a finer grid over the region immediately surrounding it, and hence it is useful for enhancing node density in a specified part of the domain.

Time

Recent versions of PDEase accept a specified time range for transient problems, e.g.

```
time
  0 to 1 by 0.001
```

This statement is usually redundant, since the plot commands provide this information, but the program issues a warning if a time range has not been specified.

Monitors

This is an optional segment which may contain various plot commands, identical to those described in the next section. The difference is that these plots are not stored but appear on the screen only. The main purpose of the monitors is to let us observe how the calculations are proceeding, in particular how much change there is from one gridding to the next. Non-linear problems especially need watching.

The *elevation* plot takes the least time of all and may be sufficient in most cases, if well chosen. Sometimes the monitor plots may flicker too fast over the screen. To make a figure more readable, we may use an the additional command, e.g.

```
delay= 3
```

which stops the program for 3 seconds.

Plots

There are five different types of graphical display at our disposal. The statement

```
grid( x, y)
```

shows the latest grid. The two arguments need not simply be x and y, but can be any functions of the variables declared. A useful variant is

```
grid( x+100*u, y+100*v)
```

which displays the deformation of a strained object, exaggerated by a factor of 100.

Two types of plots permit us to visualize a three-dimensional surface:

```
contour( U)
surface( U) viewpoint( 1, 2, 45)
surface( U) interactive
```

A *contour* plot shows the curves of intersection between the surface $U(x,y)$ and a number of planes of constant U.

A *surface* plot displays the surface in perspective, from a *viewpoint* specified by three values: the x and y coordinates and the angle of elevation of the viewer's eye. If we omit the viewpoint command the program provides a reasonable default as a starting point for further improvement. The alternative command *interactive* lets us turn around the surface, by hitting four different keys, to explore which view best brings out the features of interest. Pushing the key <h> permits us to store the final view, or even several views.

The *contour* and *surface* plots also supply the integral of the surface, i.e. the volume between the surface and the *xy* plane. This value lets us obtain various *mean values* of the solution, or of any function of the solution.

The following type of plot takes two arguments, e.g.

```
vector( j_x,  j_y)
```

which displays a set of arrows indicating the direction of a field, the lengths being related to (but not proportional to) the magnitude.

After any of the above plots we may supply a *zoom* command to enlarge part of it, e.g.

```
contour( U)  zoom(3, 4,   2, 2)
```

positioning the lower-left corner of the viewing window by the first two coordinates and giving the window size by the width and the height.

Another useful appended command is *print*, which makes PDEase create a table containing the data for a plot. The line

```
contour( f) print
```

thus produces a readable file named ATOM.P01, if the descriptor requesting it is called ATOM.PDE and the plot is the first one spawned by the descriptor.

The *contour* and *surface* plots may take some time to generate. The *elevation* plot is much quicker but only gives us a curve along the surface, taken vertically above a specified, straight line in the *xy* plane. In other words, the *elevation* plot displays the curve of intersection between the surface and a plane perpendicular to the *xy* plane. An example of this type of plot is

```
elevation(temp) from (0,1) to (0,2)
```

where the second coordinate pair should have larger coordinate values. In addition to the curve we obtain the value of its integral (*Area*), i.e. the area between the curve and the line in the *xy* plane defined in the command.

If a vector field and a named *feature* have been defined, we may calculate the *line integral* of the normal or tangential component along the curve by an *elevation* plot, e.g.

```
elevation( normal( E)) on 'curve1'
elevation( tangential( E)) on 'curve1'
```

We may supply a title for any of the above plots by appending, for instance,

```
as 'Current density'
```

If we omit the title, the program will use the expression plotted as the default title.

In transient problems one usually wants to plot a field at given values of the time. We may achieve this by a flexible statement, for instance

```
for t= 1, 10, 100
      contour( temp)   surface( temp)
```

which yields three snapshots, each in the form of two plots.

Histories

As an alternative to plots at specific times, we may follow the evolution of a variable with time at given points in space. This gives rise to a number of curves, and a descriptor statement could read

```
history( temp) at (0,0) at (1,0) at (0,2)
```

Browsing and Hardcopy

Plots are not only displayed on the screen, but are stored for later viewing, in any order, by the separate program SEEPLOT (p.10). The plots are stored in readable character (ASCII) files and may be copied to a Hewlett-Packard plotter or a compatible device. We may also translate these plots (p.11) for presentations on PCL language printers, such as Hewlett-Packard LaserJet II/III or DeskJet. Finally, Word for Windows® provides a filter for transferring PDEase plots to documents.

References

As an example of the notation in this volume, the superscript[3p55] *means reference 3, page 55.*

[1] Kreyszig, E. *Advanced Engineering Mathematics, 3rd edn*, John Wiley and Sons, 1972.

[2] Synge, J. L. and Griffith, B. A. *Principles of Mechanics, 3rd edn*, McGraw-Hill, 1959.

[3] Cheng, D. K. *Field and Wave Electromagnetics, 2nd edn*, Addison-Wesley, 1989.

[4] Carslaw, H. S. and Jaeger, J. C. *Conduction of Heat in Solids, 2nd edn*, Clarendon, 1959.

[5] Crandall, S. E., Dahl, N. C. and Lardner, T. J. *An Introduction to the Mechanics of Solids, 2nd edn*, McGraw-Hill, 1978.

[6] Timoshenko, S. P. and Goodier, J. N. *Theory of Elasticity, 3rd edn*, McGraw-Hill, 1970 .

[7] Tritton, D. J. *Physical Fluid Dynamics, 2nd edn*, Oxford University Press, 1988.

[8] Gresho, P. M. and Sani, R. L. *International J. for Numerical Methods in Fluids*, Vol. 7, 1111-1145 (1987)

[9] Bickford, W. B. *A First Course in the Finite Element Method*, Richard D. Irwin, 1990.

[10] Pauling, L. and Wilson, E. B. *Introduction to Quantum Mechanics, McGraw-Hill*, 1935

Index

W

Z

Notice

PDEase® is continuously developing.
Different versions of the software may yield
slightly different results, within the limits of error.

Do not expect always to obtain results
identical
to those shown in the examples.

PDEase/2 v.2.5.5